Water Science and Technology Library

Volume 131

The aim of the *Water Science and Technology Library* is to provide a forum for dissemination of the state-of-the-art of topics of current interest in the area of water science and technology. This is accomplished through publication of reference books and monographs, authored or edited. Occasionally also proceedings volumes are accepted for publication in the series. *Water Science and Technology Library* encompasses a wide range of topics dealing with science as well as socio-economic aspects of water, environment, and ecology. Both the water quantity and quality issues are relevant and are embraced by *Water Science and Technology Library*. The emphasis may be on either the scientific content, or techniques of solution, or both. There is increasing emphasis these days on processes and *Water Science and Technology Library* is committed to promoting this emphasis by publishing books emphasizing scientific discussions of physical, chemical, and/or biological aspects of water resources. Likewise, current or emerging solution techniques receive high priority. Interdisciplinary coverage is encouraged. Case studies contributing to our knowledge of water science and technology are also embraced by the series. Innovative ideas and novel techniques are of particular interest.

Comments or suggestions for future volumes are welcomed.

Vijay P. Singh, Department of Biological and Agricultural Engineering & Zachry Department of Civil and Environment Engineering, Texas A&M University, USA
Email: vsingh@tamu.edu

All contributions to an edited volume should undergo standard peer review to ensure high scientific quality, while monographs should also be reviewed by at least two experts in the field.

Manuscripts that have undergone successful review should then be prepared according to the Publisher's guidelines manuscripts: https://www.springer.com/gp/authors-editors/book-authors-editors/book-manuscript-guidelines

Akhilesh Kumar Yadav · Saba Shirin
Editors

Emerging and Innovative Arsenic Removal Technologies for a Sustainable Future

Springer

Editors
Akhilesh Kumar Yadav
Department of Mining Engineering
Indian Institute of Technology (Banaras Hindu University)
Varanasi, India

Department of Environmental Engineering and Management
Chaoyang University of Technology
Taichung, Taiwan

Environmental Science and Engineering Department
Indian Institute of Technology Bombay
Mumbai, India

Department of Electronics and Communication Engineering
Manipal University Jaipur
Jaipur, India

Department of Civil Engineering
Manipal University Jaipur
Jaipur, India

Saba Shirin
Department of Environmental Science, School of Vocational Studies and Applied Sciences
Gautam Buddha University
Greater Noida, India

Department of Mining Engineering
Indian Institute of Technology (Banaras Hindu University)
Varanasi, India

ISSN 0921-092X ISSN 1872-4663 (electronic)
Water Science and Technology Library
ISBN 978-3-032-01175-6 ISBN 978-3-032-01176-3 (eBook)
https://doi.org/10.1007/978-3-032-01176-3

This Springer imprint is published by the registered company Springer Nature Switzerland AG
The registered company address is: Gewerbestrasse 11, 6330 Cham, Switzerland

Dedicated
to
Environmental Engineers and Scientists
around the World

Foreword

Arsenic contamination in water remains a major global environmental and public health concern. Its presence in drinking water can cause chronic illnesses, developmental issues, and serious health conditions. Tackling this challenge demands scientific innovation, sustainable technologies, and effective regulations to ensure access to safe water for all. Edited by Akhilesh Kumar Yadav and Saba Shirin, this book brings together leading experts to present a comprehensive study across 16 chapters. It covers arsenic occurrence, health impacts, and cutting-edge removal methods, with a strong focus on bioremediation using microorganisms and plant-based solutions. The book also explores low-cost adsorbents, ion exchange, and coagulation as alternative treatment options. It critically evaluates both traditional and emerging techniques, offering insights into their strengths, limitations, and potential for future application. Designed as a resource for scientists, policymakers, engineers, and environmentalists, the book bridges academic research with practical solutions. It aims to contribute meaningfully to public health improvements and environmental sustainability.

The editors and contributors are to be complimented for their dedication to this critical issue and for compiling such an insightful and forward-thinking volume. It is my hope that this book will inspire further research, technological innovations, and collaborative efforts to combat arsenic pollution and protect water quality worldwide.

Triloki Nath Singh, Ph.D.
Director
Indian Institute of Technology Patna
Patna, India
director@iitp.ac.in

Preface

In recent years, sustainable nanoparticles have shown promise in remediating arsenic-contaminated water. Arising from both natural and human-made sources, arsenic poses a global threat, particularly in developing countries where inadequate water treatment heightens health risks. Groundwater is especially vulnerable due to arsenic's increased mobility in low-oxygen conditions, often resulting from geothermal activity and the weathering of arsenic-bearing rocks. Chronic exposure to arsenic concentrations above 10 μg/L is linked to severe health issues, including cancers, cardiovascular diseases, neurological disorders, and skin lesions. Addressing this pressing challenge calls for innovative and sustainable water treatment technologies. This book explores arsenic contamination, its health implications, and cutting-edge removal strategies. It begins with an overview of arsenic sources and their adverse health impacts and then examines diverse remediation techniques such as biological and phytoremediation methods, nanotechnology-based solutions, and cost-effective adsorption materials. Conventional methods, i.e., ion exchange, coagulation, and electrodialysis are also reviewed for their effectiveness and limitations. Emphasis is placed on sustainable, integrated bioremediation technologies capable of large-scale application with minimal environmental impact. By presenting the latest research and perspectives from leading experts, the book aims to guide scientists, engineers, policymakers, and environmental professionals in developing practical and effective arsenic remediation strategies. The knowledge compiled here aspires to foster innovation and stimulate further research in the pursuit of safe, arsenic-free water.

We first like to thank the Almighty for His blessings throughout the completion of this project, enabling us to explore this challenging research domain with a technical perspective successfully. The authors are sincerely grateful to Springer for the opportunity to publish this work on such a contemporary topic of modelling and simulations. Special thanks goes to all the contributors who made this volume a valuable source of knowledge, reporting the latest findings in their respective fields of the 48 abstracts submitted, 33 was selected for full-chapter contributions, and 16 chapters were ultimately included based on the scope and quality of the submissions. The authors have shown tremendous effort throughout the process, from writing and revising based on comments to addressing Springer's evaluation reports and final proofreading. The review process adhered to a double-blind method, in which both authors and reviewers' identities were kept confidential to ensure unbiased decision making. We also thank the Springer publication team for their timely responses and guidance.

The editors welcome any feedback, comments, or suggestions to improve future editions.

Taichung, Taiwan — Akhilesh Kumar Yadav, Ph.D.
Greater Noida, India — Saba Shirin, Ph.D.

Contents

About the Editors

Akhilesh Kumar Yadav earned a Bachelor of Technology in Electronics and Communication Engineering from Chaudhary Charan Singh University in Meerut, India. He then obtained a Master of Technology in Environmental Engineering from Madan Mohan Malaviya Engineering College in Gorakhpur, India (affiliated with Dr. A. P. J. Abdul Kalam Technical University, Lucknow). His doctoral degree, with a research focus on air pollution, was awarded by the Indian Institute of Technology (Banaras Hindu University), Varanasi, India. Dr. Yadav has served at various institutions, including Indian Institute of Technology (Banaras Hindu University), Varanasi, India; Indian Institute of Technology Bombay, Mumbai, India; Bhabha Atomic Research Centre, Mumbai, India (as a research collaborator); Manipal University Jaipur, Jaipur, India and Chaoyang University of Technology in Taichung, Taiwan. He has authored/co-authored several research and review articles, book chapters, and edited/authored books, and holds a patent. Dr. Yadav received the Young Engineer Award from the Institution of Engineers (India) and the Young Scientist Award from VDGOOD Technology Factory, Kolkata. His research interests include air, water, and soil pollution; climate change; vulnerability; human health risk assessments; and GIS applications in environmental pollution and management. He is an active member of several reputed international professional bodies, including the IEI (Kolkata), ECI (New Delhi), AEACI (Mumbai), MSI (New Delhi), and IASTA (Mumbai).

Saba Shirin completed her doctoral degree at the Indian Institute of Technology (Banaras Hindu University) Varanasi, India, in 2019. Presently, she is working as an Assistant Professor (OCFD), at Gautam Buddha University, Greater Noida, India. Her research interests are water and wastewater treatment, air quality, risk assessment, environmental pollution, environmental monitoring, and solid waste management. She has published articles in peer reviewed journals and was a reviewer in reputed journals. She has attended many international and national conferences and workshops around the world. She is an Associate Member of the Association of Environmental Analytical Chemistry of India, Analytical Chemistry Division, Bhabha Atomic Research Centre, Mumbai, India.

Chapter 1
Arsenic in Water: Understanding the Silent Threat

Yusra Sharf and Rushda Sharf

Abstract In Minnesota, arsenic is found naturally in soil and rocks. Volcanoes, the weathering of minerals and ores that contain arsenic, and commercial or industrial processes all release arsenic into the atmosphere. Groundwater that could be used for drinking water may absorb trace amounts. Compared to organic and inorganic arsenic, inorganic arsenic is typically too toxic. In certain nations, inorganic arsenic can be found in private well water, copper-chromated arsenate-treated lumber, and industrial settings. For the vast majority, drinking water contaminated with naturally occurring inorganic arsenic is the source of contact. The biggest risk to public health is drinking water polluted with arsenic, which can increase the risk of cancer and other serious health consequences. The most common ways that arsenic exposure results in disease are through ingestion and inhalation. The symptoms of arsine gas poisoning differ significantly from those of other types of arsenic poisoning. Arsine rapidly linked itself to red blood cells after inhalation, causing irreversible destruction of the cell membrane. Neurological, respiratory, hematologic, cardiovascular, gastrointestinal, and other systems can all be adversely affected by arsenic. Arsenic causes cancer in several of organ systems. Skin, lung, prostate, bladder and liver angiosarcoma, are the most prevalent forms of cancer. The central nervous system, cell respiration, inhibition of the pyruvate dehydrogenase enzyme complex, and overt gastrointestinal disturbances may be caused by acute or subacute exposure to As (more than a few mg of inorganic arsenic/day). Chronic arsenic hyperpigmentation can affect mucous membranes in addition to the trunk, where it manifests itself as a thinly dappled "raindrop" manner. Arsenic is a teratogen and a reproductive toxin. The concentrations in the cord blood and maternal blood are comparable, and it is transferred across the placenta. The chapter focuses on causal inference, which is also known as hazard identification and clinical prevention in the terminology of risk assessment.

Y. Sharf (✉)
Department of Zoology, Aligarh Muslim University, Aligarh, India
e-mail: sharfyusra92@gmail.com

R. Sharf
Department of Environmental Science, Integral University, Lucknow, India

A. K. Yadav and S. Shirin (eds.), *Emerging and Innovative Arsenic Removal Technologies for a Sustainable Future*, Water Science and Technology Library 131,
https://doi.org/10.1007/978-3-032-01176-3_1

Keywords Arsenic exposure · Contaminated water · Carcinogenesis · Threats · Clinical prevention

1.1 Introduction

The concentration of arsenic (As) in the Earth's crust is 1.8 parts/million (ppm). It can be found as sulfides (As_2S_3) such as arsenopyrites (FeAsS) rather than in its elemental form. As is harmful in each form. Since it is a redox-sensitive element, pH and redox potential have an impact on its mobility in groundwater (Perkins, 2022). Worldwide, freshwater and marine environments contain arsenic, which poses a threat to aquatic life. This material occurs in both inorganic and organic forms. Variations in As toxicity have been present in monomet hylarsonate (MMA), arsenite, arsenate, and dimethyl arsenate (DMA). Deeper groundwater layers naturally occur at high As levels; this is referred to as "contamination." Globally, the Ganga–Brahmaputra region is the most polluted place on Earth and is a common element in groundwater. Arsenate (H_2AsO_4) and arsenite are the two predominant inorganic forms in water (H_2AsO_3). Arsenic is a critical environmental problem due to its bioaccumulation and movement through the aquatic food chain at different trophic levels. Arsenic bioaccumulates in aquatic organisms, such as fish, after prolonged exposure to low concentrations of the metal. When fish tainted with arsenic are consumed, humans may be exposed to the metal. The total amount of As that the gastrointestinal tract can easily absorb makes up N95% of the inorganic arsenic (iAs) in drinking water. However, up to 50–70% of the absorbed arsenic is removed by methylation in the kidneys, and this is subsequently eliminated through urine. When more As is consumed than eliminated, it tends to build up in tissues, including the hair, nails, and other body parts (Moulick et al., 2021). As is mostly present in the shallow groundwater zone, where clay and sulfite minerals combine to form in a reducing environment. During the summer session, groundwater levels rapidly decreased, leading to the oxidation of sulfides and their aeration into solution. The Worldwide problems associated with drinking contaminated water and groundwater contamination by As have been documented. Pigmentation, keratosis, and skin cancer are believed to be typical signs of chronic toxicity caused by long-term ingestion of tainted water. The most common type of As toxicity in humans is caused by interactions with various biological ligands that inactivate the enzyme system, and the symptoms of As exposure are keratosis in the soles and palms and hyperpigmentation, especially on the trunk. Some studies have reported that these skin lesions frequently appear 5–10 years after the first exposure (Adeloju et al., 2021). Arsenic is one of the most hazardous substances on Earth and is well known to cause cancer. Arsenicosis, is As-related health issues such as skin, bladder, kidney and lung cancer, probably diabetes, elevated blood pressure, and reproductive disorders, can be caused by excessive and prolonged (such as five to ten years) human ingestion of toxic food and drinking water (Fig. 1.1) As is currently thought to impact over 150 million people worldwide due to its progressively higher levels in drinking water (Shirin & Yadav, 2014; Stroud et al., 2011). Major river basins or large deltas around the world have been reported to be affected by arsenicosis. Examples of these areas include the Mekong Delta in Cambodia, the Duero Cenozoic Basin in Spain, the Paraiba do Sul Delta in Brazil (Mirlean et al., 2014), and the Hetao River Basin in Mongolia. Water is the primary pathway by which As is consumed in the human body, so it is important to comprehend how As gets contaminated in groundwater, the health risks that come with it, and how to mitigate the issue.

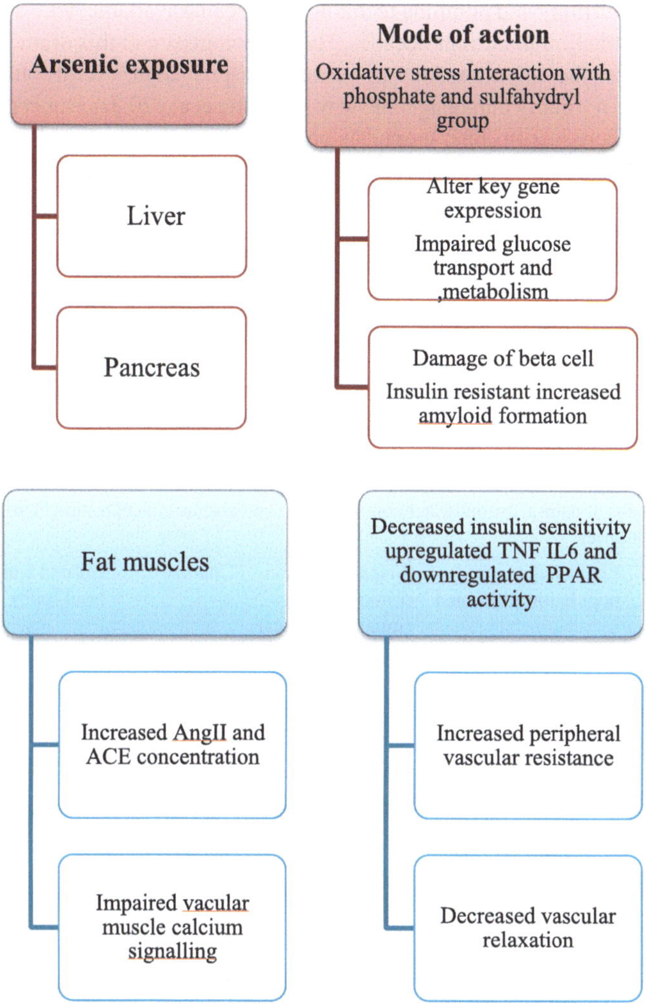

Fig. 1.1 Exposure to environmental arsenic and its involvement to human diseases

1.2 Arsenic Contamination in Groundwater Overview

The natural resource on the planet is groundwater. More than 2.5 billion people worldwide depend on groundwater for their drinking requirements; one of the most pressing problems facing human society today is the provision of clean drinking water. Although groundwater is thought to be safe, high levels of arsenic can be harmful to human health. About 230 million people around the world are faced with As poisoning. It has been extensively mined in many parts of the world since new, less expensive drilling and pumping technologies became available, with a noticeable increase in extraction in current decades. Hydrogeologists call this dramatic shift in groundwater use the "silent revolution" because it happened in several countries in an unplanned and normally unmanageable way (Bustaffa et al., 2020). But in the twenty-first century, there is a stronger emphasis on addressing global groundwater quality issues. The length of time that rocks and water come into contact, the type of aquifer, and the input from different natural and artificial sources all affect how chemically pure groundwater is As contamination of groundwater is also more common in wealthy nations like the US and Canada although concentrations are usually lower. A recent study found that asphyxia in groundwater affects 107 countries. Most As pollution hotspots are located close to present mountain ranges and deltaic regions in sedimentary basins. Groundwater accounts for 40% of Mexico's drinking water supply (North America), and considerable As concentrations of 0.10 ppb have been found several of the nation's groundwater basins (Table 1.1).

Table 1.1 Status of natural groundwater contamination and arsenic concentration in drinking water in different countries

Country	Estimated exposed population (millions)	Arsenic levels in drinking water (microgram/liter)	Country	Ground water arsenic level (ppb)
Argentina	2.0	< 1 to 7550	Cambodia	Up to 900
Bangladesh	35 to 77	< 10 to > 2500	Canada	1.5–738
Chile	0.4	600 to 800	Finland	17–980
China	0.5 to 2.0	< 50 to 4400	India	10–3200
Ghana	< 0.1	< 2 to 175	Japan	1–293
India	> 0.1	< 10 to > 800	Nepal	Up to 2620
Mexico	0.4	5 to 43	Pakistan	Up to 906
Taiwan	N/A	< 1 to > 3000	Tailand	1 > 5000
United State	> 3.0	< 1 to > 3000	United state	Up to 2600
Vietnam	> 3.0	< 0.1 to 810	Vietnam	< 1–3050

Sumant and Chaurasia (2024)

1.3 Mechanism of Arsenic Contamination

As is an inorganic compound that is mostly found in the terrestrial environment. Its trivalent As(III) and pentavalent As(V) forms have various toxicity levels. As(V) quickly changes into As(III) when it enters humans and animals. Then As(III) can be understood as follows: arsenite methyltransferase (AS3MT) facilitates the movement of the methyl donor using S-adenosylmethionine (SAM) in several phases in cells. It has been planned that non-enzymatic As(III) methylation is possible when glutathione and methylcobalamin are present. The conjugates of GSH As $(GS)_2$–OH and $As(GS)_3$ are produced when glutathione moieties replace their $As(OH)_3$ groups with their –OH groups (Singh et al., 2019).

1.3.1 Natural and Anthropogenic Sources of Arsenic Contamination

Geological formations such as volcanic rocks, soils, geothermal activity, and coal, are the chief natural sources. In surface water, and groundwater, they may contain inorganic arsenic. The normal reported range of arsenic levels in the earth is 1.5 to 5 mg/kg, although these values are known to vary. Arsenic is present in significant amounts in a number of mineral species that are found in sedimentary and magmatic rocks. Anthropogenic-related arsenic polluted groundwater has been studied in 54 countries. The primary sources of this contamination are coal mining, petroleum extraction, and mining (Fig. 1.2). Environmental risks related to the mining and smelting of arsenic leaks into surface and groundwater from slag pits, waste dumps, and mines. 74 countries around the world are affected by arsenic contamination related to mining, specifically coal mining. 17 nations around the world have been affected by arsenic linked to petroleum. The important source of As in India's groundwater are alluvial sediments, which are primarily derivable from Himalayan sediments as a result of erosion. Arsenic is mobilized when Fe^{3+}-oxyhydroxides dissolve reductively in a drop environment (Kumar et al., 2016).

1.4 Major Routes of Arsenic Exposure

Inhalation and ingestion are the two main ways that arsenic enters the body. The total exposure from all routes is known as the exposure dose.

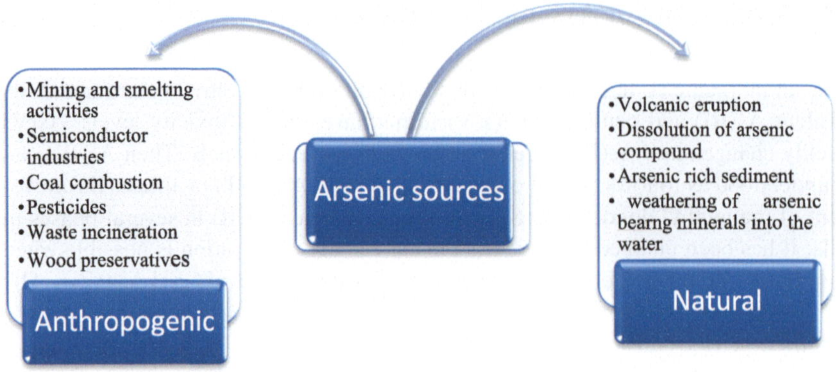

Fig. 1.2 Schematic diagram of natural and anthropogenic arsenic sources

1.4.1 Ingestion

The primary way that the American population is exposed to arsenic is through the consumption of arsenic-containing food. Consumption of drinking water, soil, and air is typically substantially lower. Adults in the United States are believed to consume 40 μg of arsenic per day on average through their diet. Poultry, fish, and meat make up 80% of the arsenic in food. Arsenic can also be found in fish, bivalve shellfish, and algae such as arsenobetaine and arsenocholine. Human toxicity from fish arsenic is minimal and it is quickly eliminated through urine. According to recent research, hijiki, a type of seaweed, has a high inorganic arsenic. Worldwide, the contamination of well water by bedrock has been connected to arsenic toxicity. Surface water will primarily contain arsenate (arsenic V), while deep wells will primarily contain arsenite (arsenic III). High levels of arsenic in the soil may pose a risk to ingestion, particularly in children who mouth while playing and have pica.

1.4.2 Inhalation

Arsenic trioxide is the mainly general form of airborne As in the workplace. In a US copper smelter, eight-hour averages of airborne arsenic were measured between 1943 and 1965, and the results showed a range of 6.9 to 20 mg/meter cubed. Because the recent acceptable exposure concentration is 0.01 mg/meter cubed, high-level occupational exposures are rare. Burning materials containing arsenic as a wood preservative can cause exposure that results in severe arsenic toxicity. The smaller particles are in the respiratory tract. Particles collected in the upper respiratory tract and swallowed after mucociliary clearance are absorbed by the gastrointestinal tract (Yip & Dart, 2001).

1.5 Risk of Overexposure to Arsenic

Furthermore, areas with arsenic mineral deposits that have been mined have noticeably higher arsenic in the soil and water. The Northwest and east, Southwest, and other regions close to geothermal activity have the highest naturally occurring groundwater arsenic concentrations in the United States (ATSDR, 2007). High arsenic can also be found in groundwater compared to surface water, resulting in arsenical pesticide runoff. The normal level of arsenic in natural groundwater is one part/ billion; however, in mining areas and areas with high soil arsenic in the soil these concentrations can reach 1000 ppb or higher. Wood that has been treated with copper chromate arsenate (CCA) to stop microorganism growth from causing rotting is frequently used in recreational structures for playgrounds, patio decks, and marine applications.

1.6 Arsenic Metabolisms

At the cellular level, red and white blood cells that convert arsenate (V) to arsenite (III) absorb arsenic into the bloodstream. Glutathione (GSH) is required for the process to occur. The liver methylates part of arsenite (As III) by transferring the methyl group from S-adenosylmethionine (SAM) to methyl arsenate (MMA V) and dimethyl arsenate (DMA V) enzymatically (Styblo et al., 2002). Although methylation has long been thought to be the primary method of detoxifying arsenic, a growing body of research has emerged in favour of alternative detoxification strategies in recent times. Certain animal species, for instance, excrete inorganic arsenic because they do not have arsenic methylation. This suggests that mammals may have additional and more significant systems for detoxifying arsenic. Furthermore, detoxification mechanisms, such as antioxidant defenses, resistance to apoptosis, or transport, have been proposed by other studies (Yoshida et al., 2004). Studies on the metabolism of arsenic have also revealed that trivalent methylated arsenic metabolites, namely monomethylarsonous acid (MMA III) and dimethylarsinous acid (DMA III), (Fig. 1.3) are "strangely able of interacting with cellular targets like proteins and DNA" and that methylation of inorganic arsenic can be a toxification process (Kitchin, 2001). High doses of arsenic appear to reduce human methylation efficiency. Arsenic is more likely to remain in soft tissues when an excessive amount of inorganic arsenic exceeds the liver's methylating capacity. About 50% of As eliminated in human urine is dimethylated, 25% is monomethylated, and the remaining 25% is inorganic. When urinary total arsenic levels increase, the methylated forms are elevated and at lower, the inorganic form is predominant (Caldwell et al., 2008). Humans are determined to have a biological half-life of less than 20 h for ingested fish arsenic and to completely remove the arsenic through urine in 48 h. Urine levels may be noticeably high, but blood levels may be normal because arsenic leaves the body quickly.

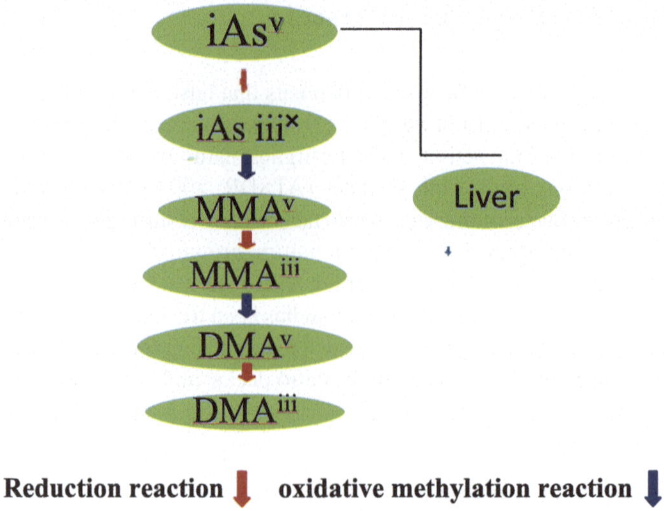

Fig. 1.3 The diagram shows inorganic arsenic metabolism via a route of oxidative methylation and reduction steps. Methylation occurs using the arsenic-3-methyltransferase enzyme, (AS3MT) and methyl donor S-adenosylmethionine (SAM). $^{\times}$ indicate highly toxic species

1.7 Arsenic Induces Toxicity

Arsenic poisoning has an epidemiological effect. It causes an overabundance of ROS to form, which harms organisms. In addition, cytotoxicity and genotoxicity have been linked to arsenic. Long-lasting arsenic effects cause skin lesions, cancer, peripheral vascular disease, and Blackfoot (Fig. 1.4). However, several investigations have documented arsenicosis as a result of high levels of As.

1.7.1 Cytotoxicity

Cytotoxicity arises when a cell exhibits abnormalities caused by hazardous substances or other contaminants. Arsenic produces ROS, which causes cytotoxicity. Cell's internal ROS levels rise sharply in response to elevated exposure to As. ROS overproduction leads to the oxidation of proteins and lipids, as well as the processes involved in mitochondrial damage. Shen et al. (2001) discovered that a mitochondria-dependent apoptotic pathway is responsible for ROS-induced oxidative stress. The activation of c-Jun N-terminal kinases (JNK), a significance of mitogen-activated protein kinases that inhibit a variety of cellular processes such as cell division, proliferation, and apoptosis, is how ROS elicit their cytotoxic effects, as has been extensively reported (Shen & Liu, 2006). Furthermore, ROS can modify signal transduction pathways, which in turn impacts a range of biological processes such as HIV

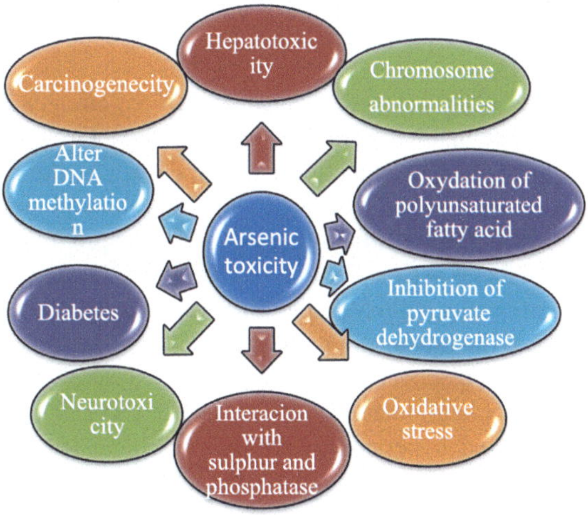

Fig. 1.4 Mechanism of action of arsenic toxicity that affects human health

activation, apoptosis, proliferation, and cell adhesion, tumor suppressor protein 53 is affected by arsenic, which causes cytotoxicity. Protein 53 is incredibly important for many different cellular processes, including transformation modulation, cell growth regulation, DNA synthesis, and repair and apoptosis (Ryan et al., 2001).

1.7.2 Genotoxicity

Genotoxicity, which is the result of damage to a cell's genetic material, causes mutations. When there is excess ROS within the cell, it reacts with other components of the cell to cause genotoxicity. Because ROS reacts with bases and deoxyribose in DNA to produce base lesions and strand breaks, genotoxicity occurs. Furthermore, ROS has a role in gene regulation mechanisms, DNA oxidation, altered DNA repair, and threats to gene stability. Interacts with zinc finger motif-containing DNA repair proteins, which are involved in transcription, and DNA–protein binding. As(III) bonds to PARP-1, its target molecule, affecting the zinc fingers and ultimately leading to oxidative DNA damage and single- and double-strand breaks in DNA (Zhou et al., 2011). In addition to causing DNA methylation loss, chronically high levels of arsenic exposure can also cause SAM depletion in cells. Next, genomic instability is affected by DNA hypomethylation (Zhao et al., 1997).

1.7.3 Carcinogenesis

Arsenic (As) induced mechanism of carcinogenesis is a well-known human carcinogen that has been thoroughly studied. Growing data have shown that arsenic disrupts signal transduction pathways, such as the Nrf2-mediated redox signaling pathway, the MAPK pathway, and the protein 53 signaling pathway, and inhibits a number of gene proliferation processes. The production of ROS caused by As is also a major factor in cancer initiation (Shi et al., 2004). Arsenic also contributes to several non cancerous multisystemic illnesses, like, hypertension, diabetes mellitus, dermatological disorders, and cardiovascular disease (Sharma et al., 2014). Furthermore, it has been discovered that the primary cause of diabetes is the inhibition of pyruvate and N-ketoglutarate dehydrogenases (Navas-Acien et al., 2006). Cardiovascular diseases and hypertension are mainly connected. Several mechanisms, including altered blood vessel vascular tone, endothelial dysfunction, and increased inflammatory activity, can result in arsenic-induced hypertension. Arsenic-induced ROS have been connected to cytokine production, apoptosis, and modifications in cell signaling. According to Eblin et al. (2006), these incidents lead to inflammation, which stimulates the synthesis of ROS and mutagenesis, ultimately advancing the pathophysiology of diseases caused by exposure to arsenic (Fig. 1.5).

1.8 Effect of Arsenic on Human Health

1.8.1 Acute Effects

Acute arsenic poisoning typically manifests as nausea, vomiting, and diarrhea, but can also cause cramping in the muscles, tingling and numbness in the extremities, and cruel cases, even death.

1.8.2 Long-Lasting Effects

Initial signs of long-term exposure to higher inorganic arsenic typically include skin lesions, lung cancer, hyperkeratosis, and pigmentation changes. According to the International Agency for Research on Cancer (IARC), humans can develop cancer from arsenic and its compounds. This conclusion applies to groundwater water containing arsenic. Prolonged exposure, to inorganic arsenic may also be harmful to, lung conditions, and development. Specifically, exposure to arsenic can cause myocardial infarction, which could be a significant cause of excess mortality.

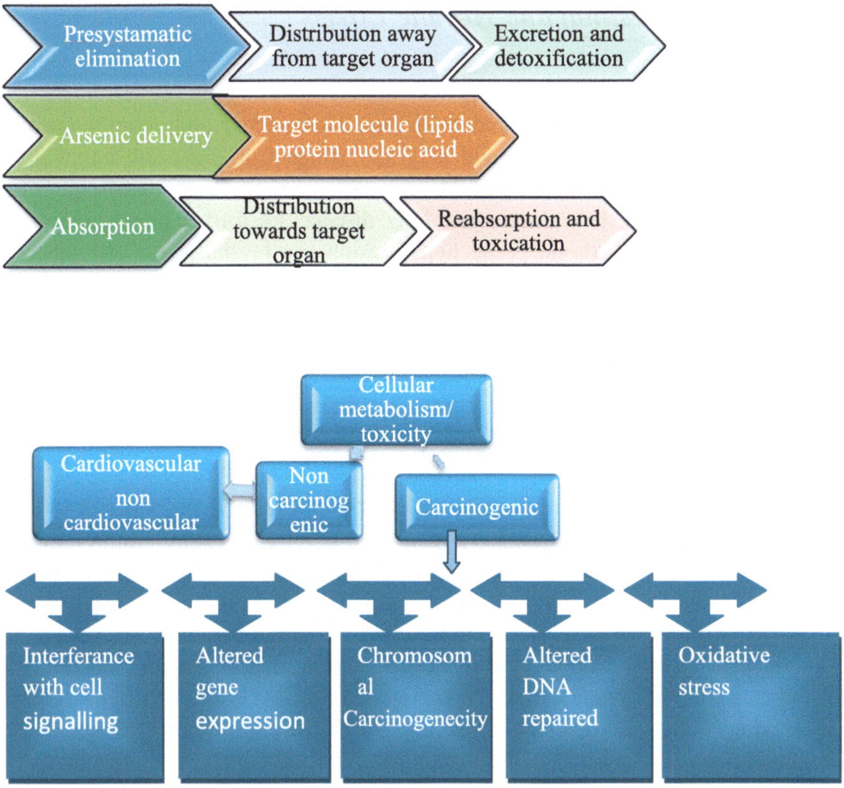

Fig. 1.5 Schematic diagram of the arsenic exposure through various routes following absorption into the bloodstream, arsenic is transported to the intended organs, where it causes both cancerous and non-cancerous side effects

1.8.3 Diabetogenic Effects

Arsenic interacts with sulfhydryl groups to cause enzymes disruption. As (III) is especially effective in this regard. The tricarboxylic acid cycle, the pyruvate oxidation pathway, gluconeogenesis, and oxidative phosphorylation are all inhibited by enzyme effects. Instead of competing with phosphate, arsenite prefers to bind to dithiol groups. High-energy bonds in molecules such as ATP are rapidly hydrolyzed when the As (V) anion takes the place of the phosphorus anion in phosphate.

By dissolving high-energy phosphate bonds, this procedure essentially "uncouples" the respiration of the mitochondria. As a result of arsenite's specific binding to hormone receptors with thiol groups, steroids cannot bind to these receptors (Kaltreider et al., 2001). The ability of arsenic to bind to and prevent the insulin receptor is believed to be involved in its diabetogenic effect.

1.8.4 Gastrointestinal and Hepatic Effects

The essential gastrointestinal lesion reveals to be a rise small blood vessels permeability, resulting in hypotension and fluid loss. The mucosa and submucosa may become extremely inflammatory and necrotic, which can cause the gut wall to perforate. Bloody diarrhoea is a possible symptom of hemorrhagic gastroenteritis. Elevated liver enzyme levels and toxic liver disease are further consequences of arsenic intoxication. Cirrhotic portal hypertension can be the result of long-term arsenic consumption. Chronic arsenic exposure and cirrhotic portal hypertension have a weak correlation (NRC, 2000).

1.8.5 Renal and Cardiovascular Effects

As part of its systemic toxicity, severe acute arsenic toxicity can result in tubular necrosis and renal failure. There have also been reports of chronic kidney failure related to cortical necrosis. Possible precipitating causes of renal injury include hemoglobinuria tubular injury, mayo globin uric tubular injury, hypotensive shock, and direct exposure to arsenic in tubule cells. Proteinuria may be the result of glomerular damage. The kidney is not the organ most susceptible to long-term arsenic poisoning, as another organ can be exaggerated at low levels of arsenic. There have been reports of cardiomyopathy outbreaks related to the consumption of tainted beer containing arsenic (Aposhian, 1989). According to 2004 study by Simeonova and Luster, arsenic exposure has been shown to contribute to atherosclerosis by inhibiting endothelial nitric oxide synthase, altering coagulation, and causing inflammation. Peripheral vascular disease and cardiovascular mortality have increased when people consume drinking water containing arsenic levels between 0.8 and 1.82 ppm, which is higher than the normal range (lower than 01 ppm), according to reports from Taiwan and Chile (Rossman, 2007).

1.8.6 Hematopoietic and Hematologic Effects

Arsenic poisoning can have an impact on the hematopoietic system both acutely and chronically. There may be a reversible bone marrow depression accompanied by pancytopenia. Chronic arsenic toxicity frequently results in anemia and leukopenia, which are often accompanied by mild eosinophilia and thrombocytopenia. Peripheral blood smears can show basophilic stippling in addition to normocytic or macrocytic anemia (Selzer & Ancel, 1983). Arsenic toxicity has been associated with a high frequency of spontaneous abortions and congenital malformations.

1.8.7 Reproductive Effects

Arsenic is a teratogen and a reproductive toxin. It readily crosses the placenta and cord blood that resembles mother blood. According to a 1979 study, there was a high frequency of spontaneous abortions and congenital malformations in women who worked in or lived close to a copper smelter where ambient arsenic was higher. Furthermore, it has been demonstrated that early childhood and prenatal exposure to arsenic increases the mortality risk in young adults from heart attacks, lung disease, and multiple cancers. Arsenic has an impact on children's health and is associated with poor pregnancy outcomes and infant mortality. Exposure to arsenic has been shown in numerous studies to have detrimental effects on memory, IQ, and cognitive development.

1.9 Mitigation of the Arsenic Problem

Therefore, to mitigate the problem of groundwater contamination, technological and socioeconomic factors must be taken into consideration. Switching to shallow tubewells from other tried mitigation strategies is the most preferred strategy (Ahmed et al., 2006). Large-diameter open wells, also known as dug wells (DWs), can be dug out of contaminated shallow aquifers to provide safe drinking water for free. Arsenic contaminated water is frequently the result of mining and industrial activity. Finding clean aquifers is crucial, but since many basins have similar geological makeup, it can be challenging to locate healthy aquifers at an affordable price, this is the situation in many parts of Bangladesh, Ethiopia, and India. However, in many cases, the elimination of pollutants is the only effective solution, necessitating the use of technology. In affected areas, ponds, lakes, and rivers can be reintroduced as a source of safe drinking water because they are typically low in As or free of As. The only things that have been shown to be effective in the fight against chronic arsenic toxicity are safe drinking water, healthy food, and regular exercise. The best ways to address the arsenic problem are to properly treat watersheds use water efficiently, raise public awareness, and keep costs low (Shakya & Ghosh, 2019). Mendoza-Chávez et al. (2020) state that electrocoagulation is believed to be a successful method for purifying arsenic-contaminated water and achieving drinking water quality requirements (Fig. 1.6). A probable technique to eliminate As from groundwater is to incorporate a coagulant and then watch for the formation of a floc. Positive-charged cationic coagulants reduce the negative charge of the colloids during particle aggregation, which leads to the formation of larger particles. To remove heavy metals from groundwater, another option is to employ a cement-based filter medium (Holmes et al., 2019; Shirin & Yadav, 2021). Other popular methods for treating arsenic-contaminated water include coagulation, chemical oxidation, ion exchange, and reverse osmosis. Arsenic oxidizes very slowly in the air, but under a variety of circumstances, chemicals such as permanganate and chlorine can quickly oxidize arsenite to arsenate

Fig. 1.6 Different technique used for the elimination of arsenic from water

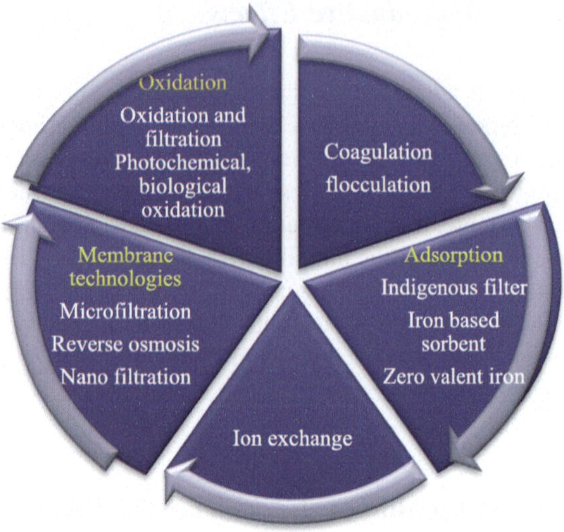

(Wegelin et al., 2000). However, to remove arsenite, AsIII, oxidizing agents must first preoxidize AsIII to AsV, which raises the cost of the process and can occasionally produce unhealthy byproducts. Biological oxidation: Research by Pallier et al. (2010) shows that certain microorganisms, such as Leptothrixochracea and Gallionellaferruginea, support and accelerate the biotic oxidation of iron, creating an environment that is favorable for arsenic adsorption. Bacteria are important players in the geochemical cycling of As, affecting its mobility and speciation through oxidation/reduction reactions. These bacteria, such as Bacillus arsenicoselenatis, Desulfutomaculum, and auripigmentum, are termed dissimilatory arsenate-reducing bacteria. According to Song et al. (2015), another method for removing arsenic from groundwater is a nano filtration membrane configuration.

1.10 Prevention and Control

Ensuring clean water for drinking, cooking, and agricultural irrigation is crucial to controlling further arsenic exposure in affected communities. While water with a high arsenic content is better suited for other tasks such as laundry and washing, water with a low arsenic content is best for drinking, cooking and gardening. This can be an inexpensive and effective method of rapidly reducing exposure to arsenic when coupled with effective education. Arsenic removal systems for small or household supplies are becoming more and more accessible and effective, but it is not yet clear how well these systems work over long periods. For interventions to be successful, community participation and education are essential. The risks of excessive exposure to arsenic and its sources, such as in crops (like rice) that absorb from irrigation water and food absorbs from cooking water, must be known to the community.

1.11 Summary

According to the currently published literature, contamination can effectively extend into the groundwater system during hydraulic fracturing because it mobilizes in aquifers and groundwater. Its contamination might therefore have a large effect on the entire world. Prolonged exposure to arsenic-rich water has been associated with serious health problems for people all over the world. Furthermore, long-term exposure to As may cause arsenicosis, which can lead to cancer, Blackfoot disease, and skin lesions. Consequences of toxicity in oxidative DNA damage, deletion mutations, chromosomal aberrations, cell cycle arrest, apoptosis, autophagy, and excessive cell proliferation.

1.12 Conclusion

The Polluted arsenic of groundwater has been considered a serious problem globally. Various countries in the world have seen the dissolution of naturally occurring groundwater due to because of biogeochemical mechanisms. In addition, we have looked at the problem of As poisoning in groundwater worldwide and provided a comprehensive summary of the disease and arsenic toxicity in humans and animals. Many remediation strategies based on contemporary and hybrid technologies for the elimination of arsenic in various regions of the world have been seriously reviewed to address the arsenic problem. Public awareness is believed to play an equal role in preserving and selecting mitigation strategies for As(V). Furthermore, a comprehensive study should focus on comprehending the arsenic origin, occurrence and distribution.

1.13 Prospect

Tap water contamination is a silent danger that can have detrimental effects on our health and well-being. The first line of defense against the different pollutants is knowledge of their possible effects on health. We can reduce the risks connected with contaminated tap water by purchasing water filtration systems, testing our water, frequently staying informed, and supporting clean water initiatives. Therefore, raising awareness and providing these individuals with the care they need continue to be extremely difficult tasks. To prevent harmful effects on human health, very few international policy measures have been implemented in the last ten years to reduce arsenic contamination of drinking water. We suggest that the World Health Organization and the United Nations Environment Program assess the worldwide arsenic poisoning crisis and initiate a global awareness campaign to educate the public, health workers, medical professionals, and administrators about this pressing issue.

Acknowledgements The authors are grateful to Dr. Rushda Sharf Integral University, Lucknow for her valuable advice. They also thank the Department Integral University Lucknow of Environmental Science India, for providing the necessary laboratory and library facilities.

References

Adeloju, S. B., Khan, S., & Patti, A. F. (2021). Arsenic contamination of groundwater and its implications for drinking water quality and human health in under-developed countries and remote communities—A review. *Applied Sciences, 11*(4), 1926.

Agency for Toxic Substances and Disease Registry (ATSDR). (2007). Toxicological profile for arsenic. Draft for Public Comment. Atlanta GA [updated 2007 August; accessed 2008 January 14].

Ahmed, M. F., Ahuja, S., Alauddin, M., et al. (2006). Ensuring safe drinking water in Bangladesh. *Science, 314*(5806), 1687–1688.

Aposhian, H. V. (1989). Biochemical toxicology of arsenic. *Review of Biochemical Toxicology, 10*, 265–299.

Bustaffa, E., Gorini, F., Bianchi, F., & Minichilli, F. (2020). Factors affecting arsenic methylation in contaminated Italian areas. *International Journal of Environmental Research and Public Health, 17*(14), 5226.

Caldwell, K. L., Jones, R. L., Verdon, C. P., Jeffery, M., Jarrett, J. M., Caudill, S. P., & Osterloh, J. D. (2008). Levels of urinary total and speciated arsenic in the US population: National health and nutrition examination survey 2003–2004. *Journal of Exposure Science and Environmental Epidemiology*, 1–10.

Eblin, K. E., Bowen, M. E., Cromey, D. W., et al. (2006). Arsenite and monomethylarsonous acid generate oxidative stress response in human bladder cell culture. *Toxicology and Applied Pharmacology, 217*(1), 7–14.

Holmes, R. R., Hart, M. L., & Kevern, J. T. (2019). Removal of arsenic from synthetic groundwater using sulfur-enhanced cement-based filtermedia. *Journal of Hazard. Toxic Radioactive Waste, 23*(3), 04019006.

Kaltreider, R. C., Davis, A. M., Lariviere, J. P., & Hamilton, J. W. (2001). Arsenic alters the function of the glucocorticoid receptor as a transcription factor. *Environment Health Perspectives, 109*, 245–251.

Kitchin, K. T. (2001). Recent advances in arsenic carcinogenesis: Modes of action, animal model systems, and methylated arsenic metabolites. *Toxicology and Applied Pharmacology, 172*, 49–261.

Kumar, M., Das, A., Das, N., Goswami, R., & Singh, U. K. (2016). Co-occurrence perspective of arsenic and fluoride in the groundwater of Diphu, Assam, Northeastern India. *Chemosphere, 150*, 227–238.

Mendoza-Chávez, C. E., Carabin, A., Dirany, A., Drogui, P., Buelna, G., & Meza-Montenegro, M. M., et al. (2020). Statistical optimization of arsenic removal from synthetic water by electrocoagulation system and its application with real arsenic-polluted groundwater. *Environment Technology*, 1–12.

Mirlean, N., Baisch, P., & Diniz, D. (2014). Arsenic in groundwater of the Paraiba do Sul delta, Brazil: An atmospheric source? *Science of the Total Environment, 482–483*, 148–156.

Moulick, D., Samanta, S., Sarkar, S., Mukherjee, A., Pattnaik, B., Saha, S., Awasthi, J. P., Bhowmick, S., Ghosh, D., Samal, A. C., et al. (2021). Arsenic contamination, impact and mitigation strategies in rice agro-environment: An inclusive insight. *Science of the Total Environment*, 149477.

National Research Council (NRC). (2000). *Arsenic in drinking water*. National Academy Press.

Navas-Acien, E. K., Silbergeld, R. A., Streeter, J. M., Clark, T. A., & Guallar, E. (2006). Arsenic exposure and type 2 diabetes: A systematic review of the experimental and epidemiologic evidence. *Environmental Health Perspectives, 114*(5), 641–648.

Pallier, V., Feuillade-Cathalifaud, G., Serpaud, B., & Bollinger, J. C. (2010). Effect of organicmatter on arsenic removal during coagulation/flocculation treatment. *Journal of Colloid Interface Science, 342*(1), 26–32.

Perkins, D. (2022). Sulfides and sulfosalts. Geosciences LibreTexts. https://geo.libretexts.org/Bookshelves/Geology/Mineralogy_(Perkins_et_al.)/09%3A_Ore_Deposits_and_Economic_Minerals/9.02%3A_New_Page/9.2.2%3A_Sulfides_and_Sulfosalts. Accessed October 11, 2022.

Rossman, T. (2007). Arsenic. In W. Rom & S. Markowitz (Eds.), *Environmental and occupational medicine* (4th ed., pp. 1006–1017). Lippincott Williams & Wilkins.

Ryan, K. M., Phillips, A. C., & Vousden, K. H. (2001). Regulation and function of the p53 tumor suppressor protein. *Current Opinion in Cell Biology, 13*(3), 332–337.

Selzer, P. M., & Ancel, M. A. (1983). Chronic arsenic poisoning masquerading as pernicious anaemia. *Western Journal of Medicine, 139*(2), 219–220.

Shakya, A. K., & Ghosh, P. K. (2019). Simultaneous biological removal of arsenic, iron, and nitrate from groundwater by a terminal electron accepting process. In *Wastewater reuse and watershed management: Engineering implications for agriculture, industry, and the environment* (p. 247).

Sharma, A. K., Tjell, J. C., Sloth, J. J., & Holm, P. E. (2014). Review of arsenic contamination , exposure through water and food and low cost mitigation options for rural areas. *Applied Geochemistry, 41*, 1–33.

Shen, H. M., & Liu, Z. G. (2006). JNK signaling pathway is a key modulator in cell death mediated by reactive oxygen and nitrogen species. *FreeRadical Biology and Medicine, 40*, 928–939.

Shen, H. M. C. F., Yang, W. X., Ding, J. L., & Ong, C. N. (2001). Superoxide radical-initiated apoptotic signalling pathway in selenite-treated HepG2 cells: Mitochondria serve as the main target. *Free Radical Biology & Medicine, 30*, 9–21.

Shi, H., Hudson, L. G., Ding, W., et al. (2004). Arsenite causes DNA damage in keratinocytes via generation of hydroxyl radicals. *Chemical Research in Toxicology, 17*, 871–878.

Shirin, S., & Yadav, A. K. (2014). Physico-chemical analysis of municipal wastewater discharge in Ganga river, Haridwar district of Uttarakhand, India. *Current World Environment, 9*(2), 536–543. https://doi.org/10.12944/CWE.9.2.39

Shirin, S., & Yadav, A. K. (2021). Assessment and suitability for the dumping of flyash in an abandoned mine. *Environmental Challenges, 4*, Article 100111. https://doi.org/10.1016/j.envc.2021.100111

Simeonova, P. P., & Luster, M. I. (2004). Arsenic and atherosclerosis. *Toxicology Applied Pharmacology, 198*, 444–449.

Singh, N., Nayak, J., Sahoo, S. K., & Kumar, R. (2019). Glutathione conjugated superparamagnetic Fe_3O_4-Au core shell nanoparticles for pH controlled release of DOX. *Materials Science and Engineering: C, 100*, 453–465.

Song, J., Jia, S. Y., Ren, H. T., Wu, S. H., & Han, X. (2015). Application of a high-surface-area schwertmannite in the removal of arsenate and arsenite. *International. Journal of Environment Science Technology, 12*(5), 1559–1568.

Stroud, J. L., Norton, G. J., Islam, M. R., et al. (2011). The dynamics of arsenic in four paddy fields in the Bengal delta. *Environmental Pollution., 159*, 947–953.

Styblo, M., Drobna, Z., Jaspers, I., Lin, S., & Thomas, D. J. (2002). The role of biomethylation in toxicity and carcinogenicity of arsenic: A research update. *Environment Health Perspectives, 110*, 767–771.

Sumant, K. V., & Chaurasia, S. R. (2024). Implicating the effects of consuming water with a high level of arsenic content: Highlighting the cause and consequences of arsenic contamination in drinking water. *Water Practice & Technology, 19*, 1071. https://doi.org/10.2166/wpt.2024.072

Wegelin, M., Gechter, D., Hug, S., Mahmud, A., & Motaleb, A. (2000). SORAS-a simple arsenic removal process. Water sanitation and hygiene: Challenges of the millennium. In *26th WEDC conference, Dhaka, Bangladesh*, pp. 255–258.

Yip, L., & Dart, R. (2001). Arsenic. In J. Sullivan & G. Kreiger (Eds.), *Clinical environmental health and toxic exposures* (2nd ed., pp. 858–865). Lippincott Williams & Wilkins.

Yoshida, T., Yamauchi, H., & Sun, G. F. (2004). Chronic health effects in people exposed to arsenic via the drinking water: Dose-response relationships in review. *Toxicology Applied Pharmacology, 198*, 243–252.

Zhao, C. Q., Young, M. R., Diwan, B. A., Coogan, T. P., & Waalkes, M. P. (1997). Association of arsenic-induced malignant transformation with DNA hypomethylation and aberrant gene expression. *Proceedings of the National Academy of Sciences of the United States of America, 94*, 10907–10912.

Zhou, X., Sun, X., Cooper, K. L., Wang, F., Liu, K. J., & Hudson, L. G. (2011). Arsenite interacts selectively with zinc finger proteins containing C3H1 or C4 motifs. *The Journal of Biological Chemistry, 286*(22855–22863), 2011.

Chapter 2
Challenge of Arsenic Contamination in Crop Lands and Food Chains in India and Need for Cost-Effective Solutions

Dhanya Punnoli and Kunnampally Jayarajan

Abstract Arsenic (As) an inorganic contaminant is critical to consider when determining if groundwater is suitable for cultivation and drinking. The WHO states that the issue of arsenic in crops and drinking water, including rice, is rising alarmingly worldwide. Contaminate water, food, and irrigation pose the largest risk to human health from arsenic. This research chapter examines the problem of arsenic contamination in Indian food chains and drinking water and the need for low-cost mitigation and management methods. In many Indian states, unconfined aquifers are widely used for agricultural and drinking water supplies, where arsenic contamination is rampant and causing harm. In India, West Bengal had the highest (9126) affected dwellings as on March 2019, followed by Assam (4301). However, after taking up JAL Jeevan Mission by central government, there is a substantial change in number of habitations affected by arsenic by 2021. Mitigation strategies to reduce hazards in high-risk locations are recommended in addition to control of iAs content in drinking water and food chains.

Keywords Arsenic · Water security · Health impacts · Sustainability · Global warming

2.1 Introduction

Arsenic pollution is a severe global issue with significant health impacts across the globe. Leaching, runoff, and rock and mineral weathering are widely recognized as the key causes of arsenic into the soil strata and groundwater resources. Anthropogenic sources can also release it into groundwater and soil. In situations of high temperature stress such as forest fires, burning vegetation, volcanism, and coal-fired power plants produce arsenic. In its inorganic form, arsenic is extremely poisonous.

D. Punnoli (✉) · K. Jayarajan
Department of Geography, Government College Chittur, Palakkad, Kerala, India
e-mail: dhanya.eptri@gmail.com

A. K. Yadav and S. Shirin (eds.), *Emerging and Innovative Arsenic Removal Technologies for a Sustainable Future*, Water Science and Technology Library 131,
https://doi.org/10.1007/978-3-032-01176-3_2

Consuming contaminated food, smoking tobacco, using polluted water in industrial processes, using filthy water for cooking and irrigation, and eating contaminated food can expose humans to significant doses of inorganic arsenic (WHO, 2018, 2023; Saravanabhavan et al., 2017). Both organic and inorganic forms of arsenic can be found in the environment. According to Gomez-Caminero et al. (2001), the most prevalent type of arsenic is inorganic (such as As^{3+} and As^{5+}), which is present in food, water, rock, minerals, and air.

Arsenobetaine, arsenocholine, arsenosugars, and arsenolipids are additional sources of organic forms of arsenic, although fish, shellfish, and algae are the primary sources. Observe significant health risks that coincide with various Sustainable Development Goals (SDGs). There are numerous underlying hindrances in translating global commitments into local actions.

The severity of exposure to the health impacts of arsenic is seen as acute or chronic problem. In a number of serious health conditions, including skin lesions, malignancies (particularly lung, bladder, and skin cancers), cardiovascular disorders, and developmental abnormalities in children, may arise from prolonged exposure to arsenic (Ravenscroft et al., 2009). Arsenic poisoning from acute exposure can cause symptoms like nausea, vomiting, and even death. When infected, arsenic can enter the brain and have harmful consequences. The research conducted by Tolins et al. (2014) has shown that prenatal and early postnatal exposure is associated with animal models with decreased brain weight, decreased numbers of glia and neurons, and altered neurotransmitter systems. According to some groundbreaking research, arsenic has been linked to diabetes and poor glucose tolerance. Diabetes and prolonged exposure to arsenic have been linked (Mazumdar et al., 2024).

2.2 How Arsenic Pollution Impacts Various SDGs Implementations

Rising levels of arsenic pose a significant challenge for achieving the SDGs by 2030. SDGs that are related to the arsenic issue are. SDG 3 that focuses on well-being and Good Health, SDG 6: Sanitation and Clean Water, SDG 1: No Poverty, SDG 13: Climate Action, SDG 2: Zero, etc. (Fig. 2.1). Integrating SDG goals into national and local policies ensures coherent and aligned efforts across different sectors. The majority of the population in developing and underdeveloped areas do not have access to health care services, hence lack monitoring or diagnosis. Unsafe drinking water is yet another challenge for many parts of the world. Without access to good quality drinking water, a life of dignity and good health is virtually impossible (UN, 2024). Unless poverty is eradicated, people affected by arsenic can have the capacity to afford not only health care costs but basic resources. Hence, it is a critical challenge for vulnerable low- and middle-income group people.

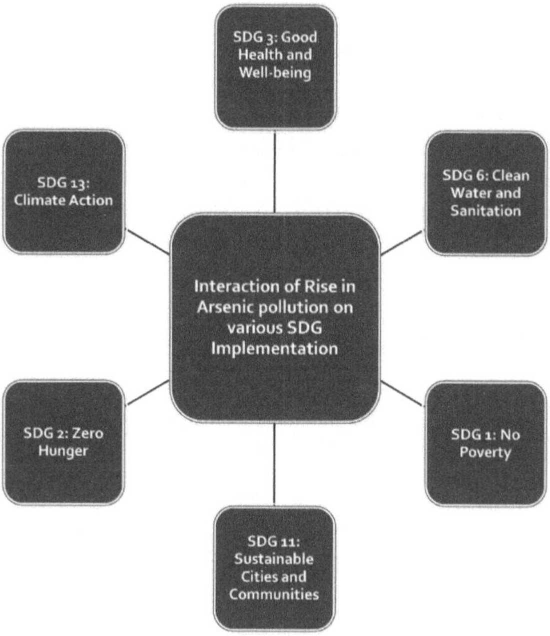

Fig. 2.1 Interactions of arsenic pollution to the implementation of various SGDs

When it comes to the food chain, it affects SDG2, reducing hunger and posing a risk to food safety and nutrition. It causes poor diet and malnutrition in susceptible communities. Unsafe irrigation and over-exploited groundwater resources are the causes of arsenic from affecting crop production and food quality. Getting clean water to both the urban and rural population is a serious growing concern. It is a policy prerogative globally and locally to ensure efficient waste management systems which can help reduce arsenic exposure; however, currently it has not been achieved in most of the developing world as part of SDG 6: Clean Water and Sanitation. Changes in water availability and agricultural practices caused by global warming may make arsenic pollution. Reducing the likelihood of arsenic-related health problems and mitigating these effects can be achieved by addressing climate change impacts such as droughts, dry spells, etc. There is an indirect interaction between SDG 13: Implementation of climate action and arsenic pollution.

2.3 Status and Management in India

Earlier, according to data published by NITI Aayog, the Water Management Index report, which was released in June 2018 states that 70% of India's water supplies are contaminated. Out of 122 countries, India ranks 120th in terms of water quality index. The research indicates that 600 million people are affected by the severe water issue. Nearly two lakh people die each year as a result of not having access to clean and safe to drink. West Bengal had the highest (9126) affected housing in March 2019, followed by Assam (4301) (Figs. 2.2 and 2.3). However, after taking up JAL Jeevan Mission by central government, there is a substantial change in number of habitations affected by arsenic. In the case of the state of Punjab, the government had announced, in accordance with the National Green Tribunal's directives, that short-term solutions were being implemented for all 563 habitations that were left out as of January 21, 2021. Despite the COVID-19 pandemics that continued during the 2021–2022, state administrations worked tirelessly to plan and grant all planned work to ensure water quality. However, the state had implemented temporary steps to provide safe drinking water in all communities affected by arsenic by June 30, 2021 (Fig. 2.4).

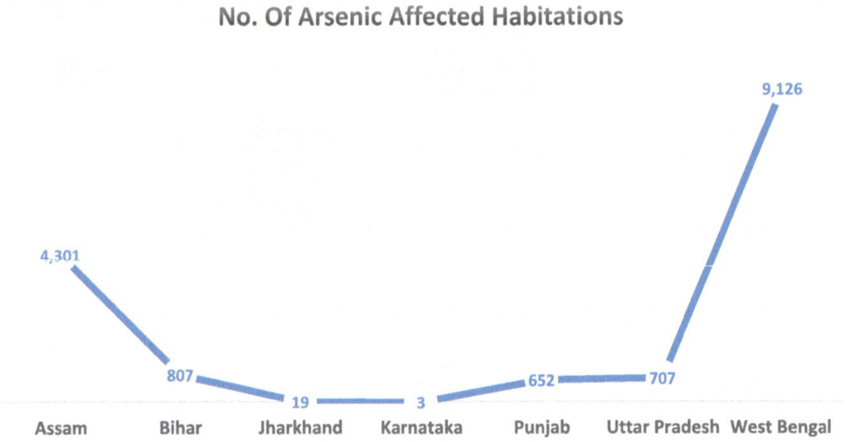

Fig. 2.2 States and number of arsenic-affected habitats in 2019

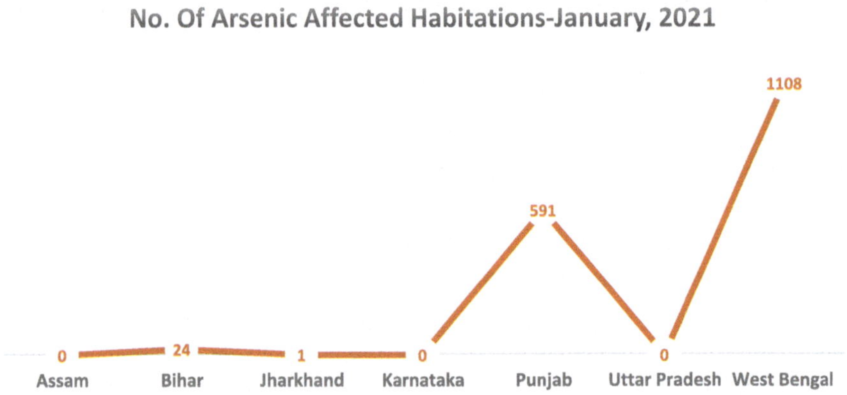

Fig. 2.3 States and number of habitats affected by arsenic in 2021

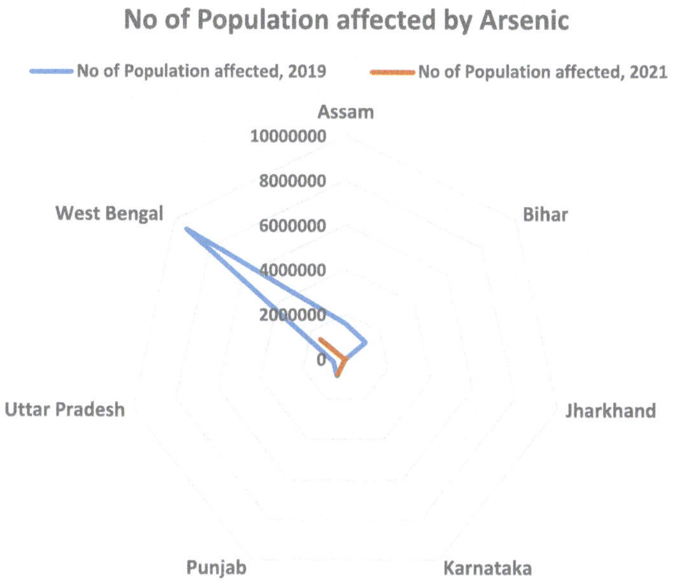

Fig. 2.4 Number of people affected by Arsenic in 2019 and 2021 in Indian states

It is evident from the data that states that have intensive agriculture activities, quarrying, and unconfined aquifers are heavily used for agricultural and drinking water supplies, where arsenic contamination is rampant and causing severe harm.

The crux of the problem is that low-income and marginalized populations are disproportionately affected by environmental contamination and often have less access to clean water and healthcare. Addressing these inequities is vital to improving health outcomes and ensuring that all communities have access to safe services. To reduce the amount of groundwater polluted with fluoride, the Department is educating people to drink surface water instead of groundwater and to implement appropriate groundwater management practices in such locations.

2.3.1 Global Warming and Arsenic Contamination

The relationship between arsenic and global warming is complex and involves several indirect connections, rather than direct causal links. Global warming can influence arsenic contamination and its impacts in several ways. Changes in hydrological cycles due to warming, there are likely fluctuations in both shallow and deep-water tables. Rising temperatures and changing precipitation patterns can worsen arsenic contamination in areas where groundwater is the key source of irrigation and drinking water if the water table interacts with arsenic-rich sediments. Arsenic from soils and sediments can be released into water supplies as a result of more frequent and severe flood events, also as a result of climate extremities. In impacted locations, this can increase the chance of spreading arsenic contamination. There is a literature warning that elevated temperatures have the ability to modify soil chemistry and microbial activity of the soil, which could lead to a greater release of arsenic from soil and sediments. For example, arsenic release into groundwater could be facilitated by alterations in pH or redox conditions in wet soils (Han et al., 2019; Shirin & Yadav, 2014). Another research group led by Podgorski and Berg (2020) had created a global prediction map of groundwater arsenic levels, exceeding 10 μg per liter, based on a random forest machine learning model based on multiple regional environmental factors and more than 50,000 aggregated data sets of observed groundwater arsenic content. Their research estimate that between 94 and 220 million individuals, primarily in Asia, may be exposed to high amounts of arsenic in groundwater by combining data on household groundwater consumption with universal arsenic forecast models. Because groundwater is being used more and more to sustain expanding populations and food production needs. At the same time, increased temperatures and altering water availability can affect how arsenic is taken up by crops, potentially leading to higher arsenic levels in food. Similarly, increasing contamination in food chains can have profound negative effects on health that can strain healthcare systems and undermine economic stability of people, state, countries, making attempts to combat pollution and climate change even more difficult.

2.3.2 Mitigation Measures to Minimize Arsenic Exposure

Mining operations, quarrying, and groundwater overexploitation and arsenic exposure are interconnected. There are various strategies to Address Arsenic-Related Health Issues. Periodic examination of surface and groundwater samples from overexploited, overexploited and critical groundwater zones must be carried out. If the water monitoring and necessary actions are not taken, the inhabitants must demand that their water quality monitored by the local government authorities. There are arsenic monitoring kits available. There are newer technologies to remove arsenic from the drinking water. Raising awareness among the public is critical in the Indian scenario. Educating the community living in the hot spots about risk factors would reduce the susceptible populations from exposure to arsenic. Public involvement should be used to educate local communities on ways to reduce their exposure to arsenic by disseminating pamphlets/health messages, and social and print media are essential. In addition to this, health care checks in vulnerable groups and areas are also critical (Fig. 2.1).

There are studies from Canada showing that the frequency of eating rice and seafood on the market (fish and shellfish) were substantially positively correlated with urinary inorganic-related As. Drinking lake water was specifically associated with uiA in children and young people. Consumption of berries and mushrooms from the area was specifically and positively correlated with As in adults (Cheung et al., 2020). Because anaerobic growth environments promote root uptake, the amount of As in rice varies significantly depending on the growing region (Mitra et al., 2017). Building green infrastructures such as rainwater storage and harvest tanks, treatment plants can improve climate resilience and foster adaptation. The provision of safe irrigation water for agriculture and aquaculture would also help to reduce the exacerbation of arsenic contamination. Another study published, states that eating some seafood may raise the levels of DMA in urine (Taylor et al., 2017). Consumption of seafood must therefore be taken into account as a potential confounding factor for risk assessments. Smith et al. (1998) and Yuan et al. (2010) have warned that in regions where arsenic is endemic, moderate to high levels of chronic exposure have been connected to an increase in skin, lung, bladder, and kidney malignancies (Fig. 2.5).

Just like arsenic, there are other geogenic contaminants, such as fluoride and nitrate, that are detrimental to irrigation as well as the health of people and animals.

Fig. 2.5 Measures to mitigate rising arsenic contamination (*Source* Author's Creation)

2.4 Conclusion

Rainwater harvesting systems are an excellent solution for avoiding excessive groundwater withdrawal to prevent the accumulation of pollutants in water-scarce locations. To solve the problem of arsenic contamination, governments, non-governmental organizations, international organizations, and local populations must collaborate. To solve arsenic contamination and its negative effects on well-being, partnerships can support information exchange, technological innovation, and resource mobilization.

Acknowledgements The first author would like to acknowledge the Higher Education Council of Kerala, Government of Kerala for providing project support in promoting climate change adaptation in rice growing belts.

References

Cheung, J. S. J., Hu, X. F., Parajuli, R. P., Rosol, R., Torng, A., Mohapatra, A., Lye, E., & Chan, H. M. (2020). Health risk assessment of arsenic exposure among the residents in Ndilǫ, Dettah, and Yellowknife, Northwest Territories, Canada. *International Journal of Hygiene and Environmental Health, 230*, 113623. https://doi.org/10.1016/j.ijheh.2020.113623

Gomez-Caminero, A., Howe, P., Hughes, M., Kenyon, E., Lewis, D. R., Moore, M., Ng, J., Aitio, A., & Becking, G. (2001). Environmental health criteria 224: Arsenic and arsenic compounds.

Han, Y.-S., Park, J.-H., Kim, S.-J., Jeong, H. Y., & Ahn, J. S. (2019). Redox transformation of soil minerals and arsenic in arsenic-contaminated soil under cycling redox conditions. *Journal of Hazardous Materials, 378*, Article 120745. https://doi.org/10.1016/j.jhazmat.2019.120745

Mazumdar, M., Wang, X., Biswas, S. K., Biswas, P. P., Farooque, A., Lee, M. S., North, C. M., Afroz, S., Husain, N., Islam, F., Mostofa, M. G., Mow, S., Liang, L., Hug, C., Ludwig, D. S., Quamruzzaman, Q., Fleisch, A. F., & Christiani, D. C. (2024). Arsenic exposure and measures of glucose tolerance in Bangladeshi adults: A cross-sectional study. *Environmental Epidemiology, 8*(5), Article e330. https://doi.org/10.1097/EE9.0000000000000330

Mitra, A., Chatterjee, S., Moogouei, R., & Gupta, D. K. (2017). Arsenic accumulation in rice and probable mitigation approaches: A review. *Agronomy.* https://doi.org/10.3390/agronomy7040067

Podgorski, J., & Berg, M. (2020). Global threat of arsenic in groundwater. *Science, 368*(6493), 845–850. https://doi.org/10.1126/science.aba1510. PMID: 32439786.

Ravenscroft, P., Brammer, H., & Richards, K. (2009). *Arsenic pollution: A global synthesis.* Wiley-Blackwell, p. 588. https://doi.org/10.1002/9781444308785.ch8

Saravanabhavan, G., Werry, K., Walker, M., Haines, D., Malowany, M., & Khoury, C. (2017). Human biomonitoring reference values for metals and trace elements in blood and urine derived from the Canadian Health Measures Survey 2007–2013. *International Journal of Hygiene and Environmental Health, 220*, 189–200. https://doi.org/10.1016/j.ijheh.2016.10.006

Shirin, S., & Yadav, A. K. (2014). Physico-chemical analysis of municipal wastewater discharge in Ganga river, Haridwar district of Uttarakhand, India. *Current World Environment, 9*(2), 536–543. https://doi.org/10.12944/CWE.9.2.39

Smith, A. H., Goycolea, M., Haque, R., & Biggs, M. L. (1998). Marked increase in bladder and lung cancer mortality in a region of northern Chile due to Arsenic in drinking water. *American Journal of Epidemiology, 147*, 660–669.

Taylor, V., Goodale, B., Raab, A., Schwerdtle, T., Reimer, K., Conklin, S., Karagas, M. R., & Francesconi, K. A. (2017). Human exposure to organic arsenic species from seafood. *Science of the Total Environment, 580*, 266–282. https://doi.org/10.1016/j.scitotenv.2016.12.113

Tolins, M., Ruchirawat, M., & Landrigan, P. (2014). The developmental neurotoxicity of arsenic: cognitive and behavioral consequences of early life exposure. *Annals of Global Health, 80*(4), 303–314. https://doi.org/10.1016/j.aogh.2014.09.005. Epub 2014 Nov 25. PMID: 25459332.

UN. (2024). https://www.unwater.org/publications/un-world-water-development-report-2024

WHO. (2018). Arsenic primer: Guidance on the investigation and mitigation of arsenic contamination. United Nations Children's Fund and the World Health Organization.

WHO. (2023). https://www.who.int/news-room/fact-sheets/detail/arsenic#cms. Accessed on July, 2024.

Yuan, Y., Marshall, G., Ferreccio, C., Steinmaus, C., Liaw, J., Bates, M., & Smith, A. H. (2010). Kidney cancer mortality. *Epidemiology, 21*, 103–108. https://doi.org/10.1097/EDE.0b013e3181c21e46.

Chapter 3
Smart Sensor Technologies for Arsenic Monitoring

Ramireddy Venkata Sai Reddy, Navaneeth K. Nampoothiri, and Sourav Maity

Abstract Arsenic poisoning of water supplies is a major worldwide health issue related to exposure to severe diseases, including cancer and neurological disorders. Concerning their complexity, expense, and nature, conventional identification methods could be useless. However, affordable, accurate, and real-time arsenic monitoring substitutes are provided by fresh advances in inventive sensor technology. These sensors improve the sensitivity and selectivity by means of materials including nanomaterials such as graphs, gold nanoparticles, and metal–organic frameworks (MOFs). These sensors have recently shown a rather remarkable capacity to detect arsenic in several water sources. Although electrochemical sensors based on images and gold nanoparticles show remarkable sensitivity and specificity, quartz crystal microbalance sensors provide a cheap and easy approach for monitoring surface water. Linking these sensors to cloud technology and the Internet of Things (IoT) allows real-time data processing and ongoing condition monitoring. This approach allows one to find and respond to arsenic pollution much faster. This is very important and is typically required in rural areas where traditional testing methods often fail. However, many challenges persisted in spite of these technical advances: preserving long-term sensor stability, lowering interference from coexisting contaminants, and managing legal approval processes. Current research is aimed at developing sensors more robust, compact, and energy efficient if it is to be utilized more extensively and regularly. Wearable sensors also let those living in high-risk environments track exposure in real time. We review the present situation of smart sensor technologies for arsenic detection, stressing its potential to greatly improve worldwide water safety monitoring. Emphasizing the importance of ongoing research and development to

R. V. S. Reddy
Centre for Biomedical Engineering, Indian Institute of Technology Delhi, Hauz Khas, New Delhi, Delhi, India

N. K. Nampoothiri
Department of Biochemical Engineering and Biotechnology, Indian Institute of Technology Delhi, Hauz Khas, New Delhi, Delhi, India

S. Maity (✉)
Centre for the Environment, Indian Institute of Technology Guwahati, Guwahati, India
e-mail: sourav.maity.92@gmail.com

A. K. Yadav and S. Shirin (eds.), *Emerging and Innovative Arsenic Removal Technologies for a Sustainable Future*, Water Science and Technology Library 131,
https://doi.org/10.1007/978-3-032-01176-3_3

solve current constraints and increase the implementation of these technologies in practical environments, it also addresses the difficulties and future directions in this subject.

Keywords Arsenic contamination · Smart sensors · Nanomaterials · Water quality monitoring · Environmental health

3.1 Introduction

Arsenic contamination, a ubiquitous worldwide problem, influences air, soil, and water in many different geographical areas. Natural processes include volcanic activity and human activity including mining, agricultural use of arsenic-containing herbicides, and industry operations release arsenic, a naturally occurring metalloid, into the atmosphere (Maity et al., 2020a). In areas such as South and Southeast Asia, where millions of people rely daily on groundwater for agriculture and drinking, arsenic in this resource is especially alarming (Dokania et al., 2024). Increased arsenic levels in drinking water create serious health hazards, so efficient monitoring of this contaminant becomes a major environmental and public health problem. According to Marghade et al. (2023), the preservation of public health and the ecosystem is contingent upon the ongoing development of consistent arsenic monitoring methods. These instruments help identify and quantify arsenic in different environmental contexts, allowing early intervention to lessen its detrimental effects. A basic tool for maintaining ecosystems and human health is helping to find pollution sources, assess the risk of exposure, and ensure compliance with safety guidelines. This is especially important in fields where arsenic is routinely released as a consequence of mining, smelting, and chemical manufacturing. Under these conditions, careful monitoring becomes absolutely essential not only to avoid direct exposure but also to lower inadvertent environmental emissions.

Still a serious threat around the world, arsenic pollution affects millions of people and disturbs ecosystems around them. For example, long-term arsenic-laced groundwater exposure has caused severe health problems, including skin lesions, malignancies, and cardiovascular diseases-posing a continuous public health emergency in areas such as Bangladesh and West Bengal. Similar trends have surfaced in Asia, Latin America, even industrialized nations where industrial activity has resulted in localized pollution (Maity et al., 2021b). Arsenic arrangement is so crucial in numerous spheres. Environmentally, it helps to preserve ecosystems and site control pollution on site. It helps to determine the degree of exposure and helps prevent chronic diseases connected to continuous arsenic intake (Dokania et al., 2023). The sector guarantees that the emissions and effluents meet legal criteria, therefore safeguarding local populations and workers from possible hazards. Arsenic monitoring is therefore a pillar of approaches meant to find sources of contamination, assess health and environmental issues, and guarantee compliance with the legal safety threshold. Chemically, arsenic exists in several forms, mostly inorganic and organic species, with inorganic arsenic (particularly arsenicite [As^{1+}] and arsenicate [As^{5+}]) being more poisonous and are found more often in the environment (Maity et al., 2021a). Environmental elements, including pH and redox potential, influence its behavior, including solubility and reactivity, thereby influencing its mobility and bioavailability. Arsenic pollution comes from several sources. In natural systems, it is sometimes linked to volcanic and geothermal activity, as well as the breakdown of arsenic-rich rocks. But in many affected areas, manmade inputs are more important, including mining, smelting, and arsenic-based agrochemicals (Maity et al., 2020b, 2024b; Patel et al., 2022). These activities highlight the need for regular monitoring and early intervention, as they significantly load arsenic into soil, water, and air. In the end, arsenic exposure causes major health hazards, mostly from contaminated drinking water. To address this problem calls for an approach that spans industrial control, public health, and environmental science is required. Different negative health effects, including skin lesions, malignancies (e.g., skin, bladder, and lung), cardiovascular problems, and neurological deficits, have been associated to continuous exposure to low amounts of arsenic. The toxicological mechanisms disturb oxidative stress and cellular processes, therefore causing DNA damage and cell death (Maity et al., 2022, 2023). These effects highlight how urgently good monitoring and mitigation techniques are needed to preserve public health. These elements define the speciation, modes of transportation, and ultimate destination of arsenic in the surroundings. To properly reduce arsenic pollution, effective monitoring and remedial plans require a thorough knowledge of these processes.

3.2 Traditional Methods for Arsenic Detection

Conventional analytical methods have essentially dominated arsenic detection and quantification; each has unique benefits and drawbacks depending on the sensitivity, specificity, and feasibility. The arrangement of arsenic in environmental and biological materials has made considerable use of conventional techniques such as atomic absorption spectroscopy (AAS), inductively coupled plasma mass spectrometry (ICP-MS), and colorimetric approaches (Fig. 3.1; Table 3.1). Detecting low quantities of metals, including arsenic (Bose et al., 2022a), is a commonly used method using atomic absorption spectroscopy (AAS). In multi-elemental analysis, including arsenic, coupled plasma mass spectrometry (ICP-MS) is known for its extraordinary sensitivity and precision. After the sample with an inductively coupled plasma source, ICP-MS analyzes the mass-to-charge ratio with a mass spectrometer. This approach offers ultra-low detection limits and the ability to concurrently identify numerous components, sometimes in the parts per trillion (ppt) range. Inductively coupled plasma mass spectrometry (ICP-MS) offers an unambiguous and strong match for high-throughput systems such as environmental surveillance or large-scale health studies. By contrast, colorimetric techniques are typically measured using a spectrophotometer by using arsenic reacting with target reagents to give a visible color shift. The Gutzeit method is one of the frequently used procedures in which arsine gas produced from the sample is combined with a mercuric bromide paper to produce a yellow-to-brown stain whose intensity corresponds to arsenic concentration (Lin et al., 2022). Although colorimetric techniques are less sensitive than AAS and ICP-MS, they are suitable for quick field screening appropriate and reasonably priced. AAS, ICP-MS, and high-performance liquid chromatography (HPLC) combined with many detection techniques are among the laboratory-based arsenic detection methods. The gold standard for arsenic study is these highly accurate and precise methods. Before detection, HPLC is sometimes employed with AAS or ICP-MS to separate several arsenic species. Known as HPLC-ICP-MS or HPLC-AAS, this hyphenated method enables the speciation of arsenic, hence separating between less harmful organic forms and more toxic inorganic arsenic species (arsenite and arsenic). Combining HPLC with AAS or ICP-MS improves the capacity to provide comprehensive information on arsenic speciation, which is crucial for risk assessment and regulatory compliance. Especially in remote or resource-limited environments, field-based colorimetric techniques provide a useful substitute for on-site arsenic testing. Usually, these techniques use portable tools to quickly check water samples for arsenic contamination. Portable test kits, such as the Hach EZ arsenic test kit, use colorimetric techniques to find arsenic in water. These kits measure using a handheld colorimeter or against a color chart by means of reagents that react with arsenic to generate a color change. These kits are convenient for quick evaluations and early decision-making during fieldwork due to their mobility and simplicity of use (Maity et al., 2025; Singh et al., 2022). That is why their sensitivity and accuracy fall short of what is possible with laboratory-based devices; occasionally, interference from other chemicals in the sample skews results.

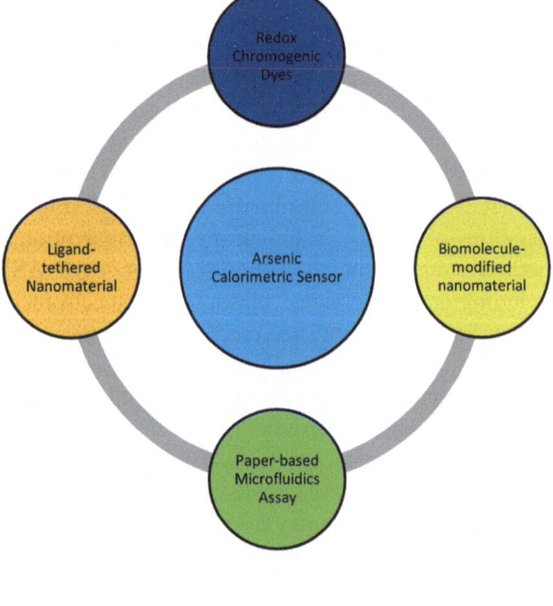

Fig. 3.1 An illustrative depiction of the essential optical colorimetric methods used to detect arsenic levels

Table 3.1 Biological methods for the detection of Arsenic (Pandey & Mishra, 2024)

Method	Advantages	Limitations	Working principle	Applications
Enzymes	High sensitivity and specificity, relatively inexpensive	Limited to specific arsenic species, may be susceptible to interference from other compounds	Enzymes catalyze a reaction that produces a measurable product	Water quality monitoring, food safety testing, and environmental monitoring
Aptamers	High sensitivity and specificity can be developed to detect different arsenic species, relatively stable under different conditions	It may be expensive to produce, and may be susceptible to interference from other compounds	Aptamers bind to arsenic specifically	Water quality monitoring, food safety testing, environmental monitoring, and biomedical research
Cells	High sensitivity and specificity can detect different arsenic species	It may be time-consuming and complex to implement, and may be susceptible to interference from other factors	Cells are monitored for their response to arsenic exposure	Environmental monitoring, biomedical research, and drug discovery

Although conventional techniques for arsenic detection have improved our knowledge of pollution, they have drawbacks. Laboratory-based methods, such as AAS, ICP-MS, and HPLC, call for advanced equipment, reputable laboratories, and knowledgeable staff to run and interpret the findings. This infrastructure may only be available in specific areas, especially in developing nations where arsenic pollution is generally the most severe. Widespread arsenic monitoring is hampered by the high cost of purchasing, maintaining, and running sophisticated analytical tools (Maity et al., 2025). Additionally, consumables and reagents are added to the total cost, which limits the availability of these approaches in environments with limited resources. Many traditional methods call for large-scale sample preparation, a time-consuming and contaminated-prone activity. Furthermore, influencing the accuracy and dependability of the results are interferences from other elements or chemicals in the sample matrix, which calls for strict quality control practices.

3.3 Smart Sensors for Arsenic Monitoring

A major worldwide problem, arsenic contamination particularly affects groundwater as the main supply of drinking water. Crucially, developing creative sensors to track arsenic levels and solve this problem is correct. Real-time detection made possible by these sensors is vital for public health protection (Maity et al., 2024a). Emphasizing their features, types, and components, this essay investigates the idea of smart and creative sensors, more especially their application in environmental monitoring for arsenic detection. A smart sensor gathers data, processes it, and tells other systems certain environmental conditions. Therefore, smart sensors are essential components of contemporary monitoring systems. The key traits distinguishing it as "smart" are its sensitivity, specificity, real-time monitoring ability, and flawless integration with data management systems. It guarantees the sensor's capacity to identify even minute levels of arsenic in water, so ensuring it stays safe for consumption. Sensitivity is therefore absolutely vital. Just as crucial is the specificity, which guarantees the sensor's reliable identification of arsenic free from the influence of other compounds. Real-time monitoring is another vital capacity since it allows one to continuously assess water quality and, should pollution levels rise, enable a quick response (Bose et al., 2022b). To enable their interaction with centralized monitoring systems, smart sensors can also have integrated data processing and transmission capabilities. This link helps the emerging field of big data analytics in environmental health management flourish. Environmental monitoring makes use of various clever sensors, all with particular advantages and uses. Among the most widely used instruments for arsenic detection are electrochemical sensors, whose high sensitivity and specificity help them to be used. Typically mixing arsenic with a sensing material, these sensors generate an electrical signal proportional to the arsenic concentration. On the other hand, optical sensors perform fast and noninvasive evaluations of arsenic-induced changes in light absorption or emission (Sardar et al., 2022). Especially helpful in differentiating between multiple arsenic species, biosensors, which employ biological molecules such as enzymes or antibodies as a detecting element—show amazing selectivity. Finally, ultrasensitive arsenic detection is accomplished with nanomaterial-based sensors that exploit unique features of nanomaterials, such as enhanced surface area and better reactivity (Gupta et al., 2019). Because they mirror the front edge of sensor technology, these sensors create new opportunities for downsizing and integration with portable electronics. Usually divided into three main components, the sensing device, the signal processing unit, and the data connection system, smart sensors meant for arsenic monitoring are arranged with the sensor's core is based on arsenic detection. Depending on the type of sensor, this element could be created from several materials: biological molecules in biosensors, optical fibers in optical sensors, or electrodes in electrochemical sensors (Maity et al., 2024a). Usually by amplifying and filtering the raw data from the sensing device, the signal processing unit turns them into useful signal, hence improving accuracy. This device might also incorporate necessary analysis tools and data storage, therefore, enabling the sensor to be more independent. Finally, the data transmission system helps the sensor interact with outside objects, such as computers or cellphones, usually by wireless technologies (Bose et al., 2022b). Smart sensors are absolutely essential in contemporary environmental monitoring systems, as remote monitoring and control is made possible by this connectivity. Including smart sensors with the Internet of Things (IoT) and big data analytics changes environmental monitoring. Large networks allow IoT-enabled sensors to be placed over polluted areas providing extensive coverage. These sensors constantly collect and send data to central servers from which real-time analysis is possible. After that, data analysis analyzes this data to find trends, project future developments, and offer useful information for controlling arsenic pollution. Smart sensor data, for instance, can be modelled to bring about the arsenic distribution in groundwater, forecast areas at risk, and help maximize intervention plans. Improving the capacities of individual sensors by merging IoT with big data, thus building a strong, scalable system for environmental monitoring.

3.4 Electrochemical Smart Sensors

The high sensitivity, selectivity, and fast reaction times of voltammetric and amperometric sensors have made them effective instruments for arsenic detection. These sensors record current (amperometry) by maintaining a constant potential or measuring the current response as a function of applied potential (voltammetry). As(III), the electrochemical behavior of arsenic enables these sensors to efficiently measure trace quantities of arsenic in different media. The development of modified electrodes with improved catalytic qualities has been an important focus of recent advances. For instance, the high surface area and significant conductivity of gold nanoparticles enhance electron transfer mechanisms, which is a major reason for the great sensitivity of gold nanoparticle-modified electrodes toward As(III). Including conducting polymers such as polyaniline has further improved sensor performance (Zayed et al., 2023) by introducing additional active sites for arsenic adsorption and reducing the overpotential needed for its oxidation. Especially those based on square wave voltammetry, amperometric sensors have shown promise in field applications due to their mobility and durability. The effective application of these sensors to monitor arsenic in groundwater offers a reliable and fairly inexpensive option for standard laboratory methods. Though problems with electrode fouling and interference from other metal ions still exist, more research on selective membrane coatings and signal processing techniques is hence essential to improve sensor performance. With continuous research aimed at overcoming current constraints to produce even more superior analytical performance, voltammetric and amperometric sensors constitute a major breakthrough in the electrochemical detection of arsenic overall. Because of their simplicity, low cost, and easy to downsize, potentiometric sensors—which track the potential difference between a working electrode and a reference electrode, have been intensively investigated for arsenic detection. These sensors produce a voltage proportional to the arsenic concentration by means of the selective interaction between arsenic ions and a sensing element, typically a chemically modified electrode. This discipline has been greatly improved by the introduction of ion-selective electrodes (ISEs) for arsenic detection. For example, excellent selectivity towards As(III) ions in the presence of competing ions such as phosphate and sulfate (Ruiz-Gonzalez, 2024) has been documented for electrodes changed with ionophore chemicals, such as dithiocarbamate derivatives. The sensitivity of potentiometric sensors was significantly enhanced by the incorporation of nanomaterials, such as graphene oxide and carbon nanotubes, into the sensing matrix. These materials increase the surface area available for arsenic binding and thereby strengthen potential responses by improving the electrical conductivity of the sensor. Recently, solid-contact ion-selective electrodes (ISEs) have improved this field by allowing the building of all-solid-state devices and thus removing the requirement for internal solutions. Strongiometric sensors can now be applied in remote or resource-limited environments where conventional methods would not be feasible because of these advances. Still, regular calibration needs, delayed response times, and electrode drift hint at additional improvement still needed.

When changes in electrical impedance upon analyte interaction are recorded, there are a sensitive and ever more attractive replacement impedimetric sensors that detect arsenic. Usually, these sensors detect changes in impedance under the influence of arsenic ions while running an alternating current (AC) to the electrode. Most of the fluctuations in impedance spectroscopy are due to two basic properties: charge transfer resistance and double layer capacitance. Increasing surface area and accelerating electron transfer kinetics (Khan et al., 2022) have helped nanostructured materials, including metal–organic frameworks and gold nanoparticles, have greatly enhanced sensor performance. For As(III), for instance, a detection limit as low as 0.1 ppb was obtained by means of an impedimetric sensor fitted with gold nanoparticles, which exceeded the WHO guideline of 10 ppb for arsenic in drinking water. Integration with microfluidic systems has further helped real-time, on-site arsenic monitoring to be more efficient, hence enhancing automation and throughput. Notwithstanding these benefits, there are still exist—particularly with respect to selectivity. The zeta potential measurements can be interfered with by organic waste and competing ions. To offset this, researchers are looking at selective recognition components such as molecularly imprinted polymers (MIPs) in concert with data processing methods motivated by machine learning. Together, impedimetric sensors indicate significant potential for sensitive and selective arsenic detection in environmental settings.

Nanomaterials have fundamentally transformed the discipline of electrochemical arsenic detection. Materials such as gold nanoparticles, graphene, carbon nanotubes, and MOFs are typically employed to improve electrode surfaces and thus their electrochemical performance. These nanostructures provide high surface areas for arsenic adsorption, enhanced electron transfer rates, and overall improved sensor conductivity. Ultratrace detection is made possible in particular, by significant affinity for arsenic compounds revealed by gold nanoparticles. Similarly, because of their large conductivity and surface area, carbon-based materials help to create very sensitive and selective sensors (Pashazadeh-Panahi et al., 2021). Since their changeable pore structures allow selective ion collection even in complex environmental matrices, MOFs have further broad applications. Researchers have also considered hybrid designs, that is, combining gold nanoparticles with graphene, to produce sensors with perfectly combined performance. Still unresolved, nevertheless, are issues such as batch-to-batch reliability of nanomaterial synthesis and the long-term stability of modified electrodes. Translation depends on addressing these issues by means of controlled fabrication and surface functionalization techniques (Fig. 3.2).

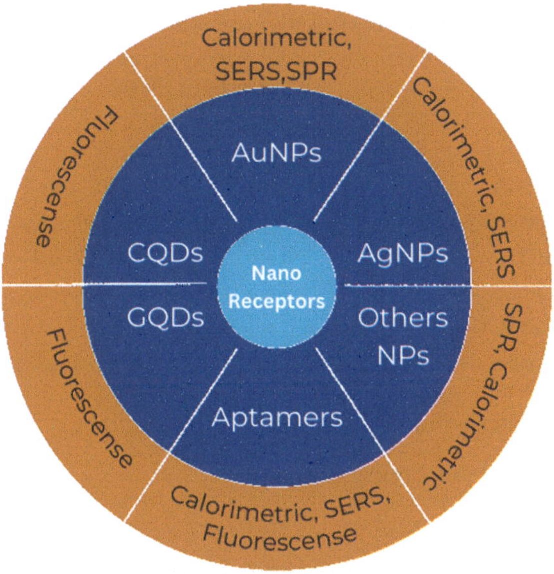

Fig. 3.2 A schematic diagram illustrating the classification of materials and phenomena for arsenic detection

3.5 Lab-Scale Advances into Commercially Viable Sensing Platforms

An innovative optical method for arsenic detection is the use of surface plasmon resonance (SPR) sensors. These sensors sense changes in the resonance angle when arsenic attaches to a functionalized layer on the sensor surface, which changes the local refractive index. SPR-based systems are suitable for environmental and public health monitoring because their great sensitivity allows them to measure arsenic at very low concentrations. Recent advances have focused on the addition of novel materials to increase sensitivity and selectivity (Table 3.2). Thiol-functionalized gold nanoparticles, for instance, improve arsenic binding affinity, thereby lowering detection limits and increasing accuracy of tests. Integrating SPR sensors with microfluidic devices has also allowed testing on-site with lower sample volumes and shorter turnaround times. However, elements such as high instrument prices and the difficulty of preparation of the sensor surface have hindered broad acceptance even with these developments. Using cost-effective platforms (Fig. 3.3) and investigating strong surface chemistries (Rai et al., 2022), a continuous effort seeks to solve these challenges.

Another exciting path for arsenic detection is provided by fluorescent sensors. These devices enable quantitative detection by means of fluorescent probes or quantum dots that release light upon excitation; arsenic ions either quench or boost this fluorescence. The design of probes with great selectivity for arsenic has recently attracted much attention using rhodamine-based dyes and thiolated quantum dots.

Table 3.2 Various materials for arsenic detection using the colorimetric principle (Sardar et al., 2022)

Material	Wavelength	Range	Detection limit	Absorption peak
NPs	533 nm	10–200 μg/l	8.5 μg/l	–
AgNPs	575 nm	0.5–30.0 mg/l	500 μg/l	–
AuNP (by glucose)	520 nm	1–14 μg/l	0.53 μg/l	2.5 a.u
MFE (Mangifera Indica)	280 nm	200–2000 μg/l	200 μg/l	1 a.u
Citrate-capped AuNPs	519 nm	4–100 μg/l	1.8 μg/l	0.65 a.u
AuNPs/GSH	532 nm	0–5 μg/l	0.11 μg/l	0.75 a.u
AuNPs + PDDA + Aptamer	Change 650/ 520 nm	$1.3*10^2$–$4*10^4$ nM	5.3 μg/l	1.1 A @ 2000 μg/l
AuNPs functionalized with GSH-DTT-CYs-PDCA	505 nm	2–20 μg/l	2.5 μg/l	0.753 A

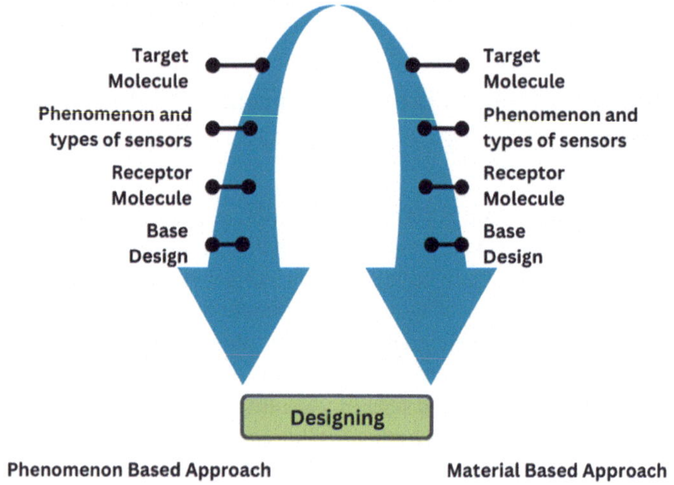

Fig. 3.3 A schematic diagram illustrating the approach for designing the sensor

A few of these technologies have attained detection limits falling in the parts-per-trillion level (Lai et al., 2023). The high surface area and strong optical characteristics of carbon dots and graphene quantum dots further enhanced sensitivity and stability. Fluorescence-based devices have a great advantage in field conditions: Fast, real-time detection is possible. Still, ongoing issues including photobleaching, matrix component interference, and probe synthesis's complexity need attention. More durable probe designs and multiplexed detection techniques are under investigation in current research to improve performance in challenging environmental samples. With possible uses in environmental analysis, public health protection as well as environmental monitoring, fluorescence-based sensors remain a viable weapon for arsenic monitoring. Simple and reasonably priced colorimetric smart sensors use visible color changes to detect arsenic. When these sensors interact with arsenic, the reagents utilized in them undergo a colorimetric reaction that results in a color change observable with the unassisted eye or with spectrophotography. The molybdenum blue method, which is typically employed as a reagent for arsenic detection, generates a blue-colored complex in the presence of arsenic (Gebremedhin et al., 2024). Recent developments in colorimetric sensors have focused on nanomaterial-based devices like gold nanoparticles. The change in color of the nanoparticles upon arsenic attachment to the surface is ascribed to the localized surface plasmon resonance effect. Low parts per billion (ppb) range detection constraints lead to increased sensitivity and selectivity of the improved sensors. Portable and user-friendly arsenic monitoring systems have been created by combining smartphones with colorimetric sensors, therefore allowing real-time data processing and sharing. Colorimetric sensors suffer from interference from other substances in the sample and the possibility of subjective interpretation of color changes, notwithstanding their simplicity and convenience of use (Bhat et al., 2024). Through the generation of more selective reagents and application of digital image processing techniques to standardize colorimetric readings, constant research helps to tackle these issues. Ultimately, smart colorimetric sensors offer a useful replacement for arsenic monitoring, particularly in environments with low resources where sophisticated analytical methods would not be accessible. Arsenic is detected, and molecular-level sample information is provided by Raman spectroscopy-based sensors through the inelastic scattering of light. These sensors search for the Raman shift, which is related to the vibrational modes of molecules or complexes, including arsenic. In this discipline, surface-enhanced Raman scattering (SERS) is a recent development whereby nanostructured materials including gold or silver nanoparticles augment the Raman signal by several orders of magnitude. SERS-based sensors, thus quite sensitive, may identify arsenic at parts per trillion (ppt) levels. Combining SERS enhancement with Raman spectroscopy is perfect for the detection of arsenic in complicated environmental materials. Field-deployable systems for on-site arsenic monitoring (Nurmamat et al., 2023) have been produced by means of a Raman sensor combined with a portable spectrometer. However, key obstacles are SERS substrate repeatability, interference from other chemicals, and data analysis difficulty. Constant research helps solve these challenges by producing more homogeneous and repeatable SERS substrates and improving data processing methods to increase arsenic detection accuracy and dependability. Raman spectroscopy-based sensors show tremendous promise for arsenic monitoring (Chen et al., 2024) with high sensitivity, specificity, and possibility for real-time analysis.

3.6 Biosensors and Biomimetic Sensors

Enzyme-based biosensors detect arsenic using biological enzymes (Fig. 3.4). These biosensors, which are usually electrochemical or optical, use the selectivity of enzymes toward certain compounds to transform arsenic-related biological events into quantifiable signals. One of the most frequently employed enzymes in arsenic detection is arsenicite oxidase, which catalyzes the oxidation of arsenicine (As(III)) to arsenicate (As(V)) with the consequent electron transfer recorded by the sensor. This exactly refers to the arsenic content and the produced signal (Puopolo et al., 2022). The sensitivity and stability of these biosensors have been greatly enhanced by developments in enzyme immobilization methods including nanomaterials such as graphene oxide and gold nanoparticles. By improving electron transfer rates and the general sensor's overall performance, these nanomaterials generate an environment fit for enzyme attachment. However, enzyme stability still presents difficulties for enzyme-based biosensors. Enzymes may not be used long-term under field conditions if they are susceptible to variations in temperature, pH, or other environmental elements. Furthermore, with time, enzyme degradation can cause sensor performance to drop, hence periodic enzyme layer replacement or calibration is needed. A contemporary approach for arsenic detection is aptamer-based sensors. Short, single-stranded DNA or RNA molecules called aptamers are used in these sensors Highly specific and selective, aptamers for arsenic are identified using SELEX (systematic evolution of ligands by exponential enrichment). Aptamers bind to arsenic and undergo a form change that can be seen by several means, including piezoelectric, optical, or electrochemical approaches. High selectivity, fast response times, and the capacity to operate in challenging environments, all of which let these sensors detect arsenic, have benefits for Recent developments in this sector have concentrated on enhancing the sensitivity and efficiency of the sensors by means of aptamers combined with nanomaterials such as graphene and gold nanoparticles For arsenic in the low parts per billion (ppb) range, aptamer-functionalized graphene oxide nanosheets, for example, have reached detection limits. These sensors are also reasonably affordable for ongoing monitoring since they are readily reusable. Still unresolved, though, are issues including nonspecific binding and the complexity of aptamer production. Research on building hybrid aptamer structures and applying signal amplification techniques to increase the selectivity and sensitivity of sensors is under continuous progress. Designed for use in environmental monitoring and point-of-care diagnostics, aptamer-based sensors show tremendous promise in precisely identifying arsenic. Usually genetically modified microorganisms, whole-cell biosensors use living cells, usually living cells—as the sensing element to detect arsenic. Inherent biological reactions of cells to arsenic exposure, that is, changes in gene expression or metabolic activity, can be monitored using a variety of transducers in these biosensors. Arsenic detection has been developed for bacterial cells altered to express a reporter gene, say luciferase or GFP, in response to arsenic binding to specific regulatory proteins. The correlation of the reporter gene expression level with arsenic concentration makes quantitative investigation possible. Whole cell biosensors have showed considerable promise, especially in complex environmental settings where chemical analysis may not be able to identify bioavailable arsenic (Maity et al., 2025). These systems are typically reasonably priced and can be reengineered to detect a variety of pollutants by modifying their fundamental genetic circuits. Having stated that, there are still main issues to solve mostly with cell viability, slower response times relative to other sensing platforms, and sensitivity to changing environmental variables that can influence signal stability. Recent studies have focused on creating more strong and stable microbial strains and combining whole cell biosensors with microfluidic platforms to increase sensitivity and lower reaction times, thus improving the performance of these biosensors. Combining the sensitivity of biological identification with the adaptability of genetic engineering, whole-cell biosensors present a special method for arsenic detection. Therefore, they have become a useful instrument for risk assessment and environmental monitoring. Synthetic materials that seek to mimic the recognition characteristics of natural biological receptors are molecularly imprinted polymers (MIP) (Yoo et al., 2020). They have an arsenic detection potential. Polymerization of functional molecules in the presence of a template molecule—such as arsenic or an arsenic-containing compound—generates MIPs. The template is removed during polymerization, leaving voids with dimensions, forms, and functional groups that resemble those of the target arsenic molecules. These imprinted cavities allow MIPs to selectively rebind arsenic to be detected using multiple methods, including electrochemical, optical and piezoelectric sensors (Table 3.3). MIPs are quite useful for detecting arsenic in complex environmental samples because of their remarkable selectivity and stability. Recent advances in MIP technology are designed to enhance sensor sensitivity by utilizing nanomaterials such as graphene and carbon nanotubes, thereby increasing the surface area. Especially sensitive MIP-based electrochemical sensors for arsenic have been developed with detection limits as low as parts per trillion (ppt). Moreover, MIPs can be used in harsh environments when traditional biosensors fail. However, to be suitable for general use, it is necessary to address issues such as the time-consuming and challenging synthesis of MIPs, as well as potential issues with template removal and reusability. Furthering their promise in arsenic detection, ongoing research seeks to maximize the synthesis process, investigate new template molecules, and improve the integration of MIPs with different sensor systems (Ovezova et al., 2024) (Fig. 3.5).

3.7 Conclusion

In this sense, the development of smart sensors for arsenic surveillance is a substantial advance in our capacity to safeguard the environment and public health from this ubiquitous contaminant. As we have seen, arsenic pollution affects ecosystems globally and millions of individuals, as well as other aspects. Never more urgent than the demand for dependable, reasonably priced, easily available monitoring solutions. Although useful, conventional techniques for arsenic detection have restrictions that

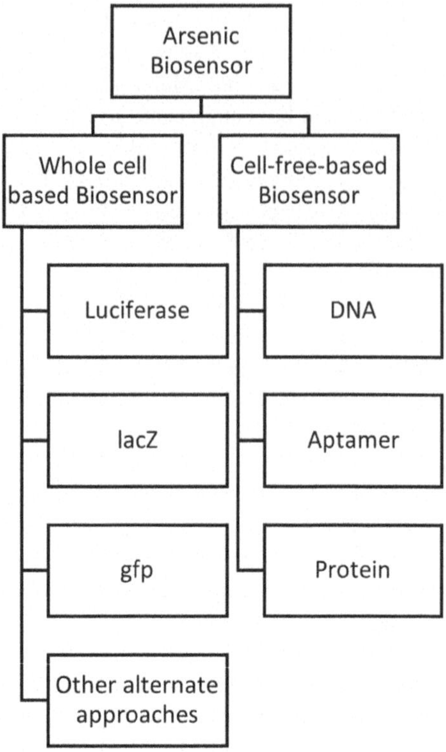

Fig. 3.4 A schematic representation of the classification of arsenic biosensors based on their biological components

Table 3.3 Different methods developed for the detection of Arsenic (Gebremedhin et al., 2024; Nurmamat et al., 2023; Pandey & Mishra, 2024)

Method	Deposition time	Electrode material	LOD/ppb
DPASV	600	Gold disk	0.02
DPASV	300	Gold disk	0.15
DPASV	240	Rotating gold disk	0.20
DPASV	240	Rotating gold disk	0.19
CCSV	240 (flow)	Gold ultramicroelectrode	0.20
ASV	200	Gold film modified on glassy carbon	0.56
SWV	300	Gold ultramicroelectrode array	0.013
Sonoelectrochemical SWV	120	Gold disk	0.75

prevent widespread use, particularly in resource-limited regions where arsenic pollution is generally most severe. Here, smart sensors are useful since they present a good substitute for real-time monitoring, sensitivity, and specificity. From electrochemical and optical sensors to biosensors and biomimetic sensors, the variety of smart

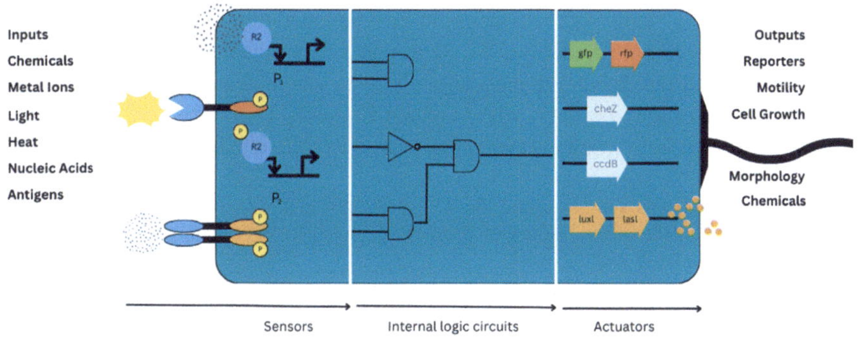

Fig. 3.5 A schematic illustration of a synthetic 'bioreporting' cell-based biosensor altered via genetic engineering

sensor technologies that we have covered each delivers special advantages. With their great sensitivity and the possibility of downsizing, electrochemical sensors present strong field use solutions. The sensitivity and ability to detect arsenic at low concentrations are particularly impressive in optical sensors that are based on surface plasmon resonance and fluorescence. Biomimetic sensors and biosensors offer an interesting frontier in arsenic detection. These sensors offer unparalleled selectivity and the possibility for continuous monitoring in demanding environmental matrix by means of biological identification via enzymes, aptamers, whole cells, or molecularly imprinted polymers. Using nanomaterials in several types of sensors has transformed general performance, selectivity, and sensitivity. From graphene and carbon nanotubes to gold nanoparticles, these materials are pushing the possibilities in arsenic detection. However, it is important to acknowledge that there are obstacles. Active areas of research cover issues including sensor stability, interference from other drugs, and the requirement of even greater downsizing and cost reduction. The pathway from laboratory development to wide field use is only sometimes straightcut; hence, continuous efforts are required to close this difference. Looking forward, these smart sensors have quite a lot of influence. Imagine a situation where real-time arsenic monitoring is as a regular as weather check-up. Environmental authorities can track and react to pollution events in this planet with hitherto unheard-of speed and accuracy, and communities have instant access to data about the safety of their water source. Including smart sensors linked to the Internet of Things (IoT) and big data analytics opens up further more opportunities. The day when large sensor networks will be able to offer complete, real-time data on arsenic levels over wide areas is not far away. Using modern algorithms, this data could expose hitherto undetectable trends and patterns, thus guiding more successful tactics for the reduction and avoidance. However, all problems cannot be resolved by technology. Good use of smart sensors for arsenic monitoring requires a multidisciplinary approach including legislators, engineers, public health authorities, and scientists. Moreover, very important will be community involvement and education to make sure that the advantages of modern technologies find those most in need. The development for arsenic monitoring finally reflects technological advancement, improved quality of life, and environmental protection. We approach a time when arsenic poisoning is no longer a secret threat but a manageable chore as we hone and use these sensors. Although the road may be long, with constant research, ingenuity, and dedication, we can hope for a time when everyone has access to safe, clean water.

Acknowledgements The authors would like to express our gratitude to the Indian Institute of Technology Delhi and the Indian Institute of Technology Guwahati for providing the infrastructural support needed to proceed with the work.

References

Bose, S., Maity, S., & Sarkar, A. (2022a). Nano-materials as biosensor for heavy metal detection. In *Food, medical, and environmental applications of nanomaterials* (pp. 493–526). Elsevier. https://doi.org/10.1016/B978-0-12-822858-6.00018-2

Bose, S., Maity, S., & Sarkar, A. (2022b). Review of microbial biosensor for the detection of mercury in water. *Environmental Quality Management, 31*(4), 29–40. https://doi.org/10.1002/tqem.21742

Chen, R., Ren, S., Li, S., Han, D., Qin, K., Jia, X., Zhou, H., & Gao, Z. (2024). Micro–/nanostructures for surface-enhanced Raman spectroscopy: Recent advances and perspectives. *Advances in Colloid and Interface Science*, 103235. https://doi.org/10.1016/j.cis.2024.103235

Dokania, P., Maity, S., Nayak, T., Jayantu, S., & Sarkar, A. (2023). Microbial and biosorbent-mediated hexavalent chromium removal from the tailing waters released from chromite mines. In *Asian mining congress* (pp. 337–344). Springer Nature Switzerland. https://doi.org/10.1007/978-3-031-46966-4_27

Dokania, P., Maity, S., Patil, P. B., & Sarkar, A. (2024). Isothermal and kinetics modeling approach for the bioremediation of potentially toxic trace metal ions using a novel Biosorbent Acalypha wilkesiana (Copperleaf) Leaves. *Applied Biochemistry and Biotechnology, 196*(5), 2487–2517. https://doi.org/10.1007/s12010-023-04678-5

Gebremedhin, K. H., Kahsay, M. H., Wegahita, N. K., Teklu, T., Berhe, B. A., Gebru, A. G., Tesfay, A. H., & Asgedom, A. G. (2024). Nanomaterial-based optical colorimetric sensors for rapid monitoring of inorganic arsenic species: A review. *Discover Nano, 19*(1), 38. https://doi.org/10.1186/s11671-024-03981-2

Gupta, N., Renugopalakrishnan, V., Liepmann, D., Paulmurugan, R., & Malhotra, B. D. (2019). Cell-based biosensors: Recent trends, challenges and future perspectives. *Biosensors and Bioelectronics, 141*, Article 111435. https://doi.org/10.1016/j.bios.2019.111435

Khan, Z. G., Patil, M. R., Nangare, S. N., Patil, A. G., Boddu, S. H., Tade, R. S., & Patil, P. O. (2022). Surface nanoarchitectured metal–organic frameworks-based sensor for reduced glutathione sensing: A review. *Journal of Nanostructure in Chemistry, 12*(6), 1053–1074. https://doi.org/10.1007/s40097-022-00480-7

Lai, L., Yan, F., Chen, G., Huang, Y., Huang, L., & Li, D. (2023). Recent progress on fluorescent probes in heavy metal determinations for food safety: A review. *Molecules, 28*(15), 5689.

Lin, H., Dai, H., & He, L. (2022). Toward a greener approach to detect inorganic arsenic using the Gutzeit method and X-ray fluorescence spectroscopy. *Analytical Science Advances, 3*(9–10), 262–268. https://doi.org/10.1002/ansa.202200014

Maity, S., Biswas, R., & Sarkar, A. (2020a). Comparative valuation of groundwater quality parameters in Bhojpur. *Bihar for Arsenic Risk Assessment. Chemosphere, 259*, Article 127398. https://doi.org/10.1016/j.chemosphere.2020.127398

Maity, S., Biswas, R., Verma, S. K., & Sarkar, A. (2021a). Natural polysaccharides as potential biosorbents for heavy metal removal. In *Food, medical, and environmental applications of polysaccharides* (pp. 627–665). Elsevier. https://doi.org/10.1016/B978-0-12-819239-9.00012-9

Maity, S., Bose, S., Dokania, P., Lohar, S., & Sarkar, A. (2024a). A comprehensive review of arsenic contamination in India with an emphasis on its detection through biosensors and bioremediation from the aqueous system. *Environmental Quality Management, 33*(4), 427–457. https://doi.org/10.1002/tqem.22100

Maity, S., Dokania, P., Goenka, M., Patil, P. B., & Sarkar, A. (2025). Health-risk assessment of groundwater arsenic levels in Bhagalpur, India, and development of a cost-effective paper-based arsenic testing-kit. *CLEAN–Soil, Air, Water, 53*(1), e202300291. https://doi.org/10.1002/clen.202300291

Maity, S., Dokania, P., Goenka, M., Rahul, S., Are, R. P., & Sarkar, A. (2024b). Techno-economic feasibility and life cycle assessment analysis for a developed novel biosorbent-based arsenic bio-filter system. *Environmental Geochemistry and Health, 46*(3), 79. https://doi.org/10.1007/s10653-023-01839-7

Maity, S., Nanda, S., & Sarkar, A. (2021b). Colocasia esculenta stem as novel biosorbent for potentially toxic metals removal from aqueous system. *Environmental Science and Pollution Research, 28*(42), 58885–58901. https://doi.org/10.1007/s11356-021-13026-1

Maity, S., Patil, P. B., SenSharma, S., & Sarkar, A. (2022). Bioremediation of heavy metals from the aqueous environment using Artocarpus heterophyllus (jackfruit) seed as a novel biosorbent. *Chemosphere, 307*, Article 136115. https://doi.org/10.1016/j.chemosphere.2022.136115

Maity, S., Sarkar, D., Poddar, K., Patil, P., & Sarkar, A. (2023). Biofilm-mediated heavy metal removal from aqueous system by multi-metal-resistant bacterial strain Bacillus sp. GH-s29. *Applied Biochemistry and Biotechnology, 195*(8), 4832–4850. https://doi.org/10.1007/s12010-022-04288-7

Maity, S., Sinha, D., & Sarkar, A. (2020b). Wastewater and industrial effluent treatment by using nanotechnology. In *Nanomaterials and environmental biotechnology* (pp. 299–313). https://doi.org/10.1007/978-3-030-34544-0_16

Marghade, D., Mehta, G., Shelare, S., Jadhav, G., & Nikam, K. C. (2023). Arsenic contamination in Indian groundwater: From origin to mitigation approaches for a sustainable future. *Water, 15*(23), 4125.

Nurmamat, X., Zhao, Z., Ablat, H., Ma, X., Xie, Q., Zhang, Z., Tian, J., Jia, H., & Wang, F. (2023). Application of surface-enhanced Raman scattering to qualitative and quantitative analysis of arsenic species. *Analytical Methods, 15*(37), 4798–4810. https://doi.org/10.1039/D3AY00736G

Ovezova, M., Yılmaz, F., Göktürk, I., Güler, K. Ç., & Denizli, A. (2024). Molecularly imprinted polymers for pharmaceutical and biomedical analysis. *Journal of Pharmaceutical and Biomedical Analysis Open*, 100038. https://doi.org/10.1016/j.jpbao.2024.100038

Pandey, S., & Mishra, S. (2024). A review of sensing technologies for arsenic detection in drinking water. *International Journal of Environmental Science and Technology*, 1–24. https://doi.org/10.1007/s13762-024-05912-1

Pashazadeh-Panahi, P., Belali, S., Sohrabi, H., Oroojalian, F., Hashemzaei, M., Mokhtarzadeh, A., & de la Guardia, M. (2021). Metal-organic frameworks conjugated with biomolecules as efficient platforms for development of biosensors. *TrAC Trends in Analytical Chemistry, 141*, Article 116285. https://doi.org/10.1016/j.trac.2021.116285

Patel, D., Pamidimukkala, P., Chakraborty, D., & Yadav, A. (2022). Bharuch district, Gujarat, India: Factor analysis and geographical distribution of water quality characteristics. *Environmental Nanotechnology, Monitoring & Management, 18*, Article 100732. https://doi.org/10.1016/j.enmm.2022.100732

Puopolo, R., Gallo, G., Limauro, D., Contursi, P., & Fiorentino, G. (2022). A new strategy for As(V) biosensing based on the inhibition of the phosphatase activity of the arsenate reductase from Thermus thermophilus. *International Journal of Molecular Sciences, 23*(6), 2942.

Rai, P. K., Islam, M., & Gupta, A. (2022). Microfluidic devices for the detection of contamination in water samples: A review. *Sensors and Actuators a: Physical, 347*, Article 113926. https://doi.org/10.1016/j.sna.2022.113926

Ruiz-Gonzalez, A. (2024). Ion-selective electrodes in the food industry: Development trends in the potentiometric determination of ionic pollutants. *Electrochem, 5*(2), 178–212.

Sardar, T., Pandey, S., Mishra, S., & Bhore, S. (2022). Sensing material and design of an optical sensor for detection of arsenic-A review. *IEEE Sensors Journal, 22*(8), 7442–7450. https://doi.org/10.1109/JSEN.2022.3151661

Singh, V. P., Maurya, S. P., Singh, R., & Yadav, A. K. (2022). An interdisciplinary modeling approach for dynamic adaptive policy pathways. In S. P. Maurya, A. K. Yadav, & R. Singh (Eds.), *Modeling and simulation of environmental systems: A computation approach* (pp. 343–353). CRC Press. https://doi.org/10.1201/9781003203445-22

Yoo, H., Jo, H., & Oh, S. S. (2020). Detection and beyond: Challenges and advances in aptamer-based biosensors. *Materials Advances, 1*(8), 2663–2687. https://doi.org/10.1039/D0MA00639D

Zayed, M. A., Hussein, M. A., El-Shishtawy, R. M., Albukhari, S. M., El-Said, W. A., & Elshehy, E. A. (2023). Molybdenum oxide grafted-polyaniline nanocomposite modified ITO electrode for electrochemical sensing of arsenic oxyanion. *Journal of Materials Research and Technology, 24*, 503–513. https://doi.org/10.1016/j.jmrt.2023.02.195

Chapter 4
Emerging Innovations in Arsenic Detection and Removal

Rishagni Chetia, Kevizali Neikha, and Amrit Puzari

Abstract Heavy metal contamination has a detrimental impact on human health and surrounding, making it a major global environmental issue. Metal and other industries are improperly disposed of in water bodies and soil which eventually causes heavy metal pollution in the atmosphere. These toxic metal pollutants affect biological functions and growth as well as bioaccumulation in different organs leading to serious life-threatening diseases such as cancer. Arsenic is one of those toxic heavy metal pollutants that can be responsible for a number of health issues including skin diseases and cancers, bladder cancer, lung cancer and cardiovascular diseases. Arsenic toxicity is mainly caused by the consumption of contaminated water containing more than the allowed amount of 'Arsenic'. According to the WHO, 220 million people worldwide are in a crisis of tainted drinking water having > 10 μg/L of arsenic approximately. Therefore, the utilization of economically viable, highly efficient, innovative techniques for the detection and elimination of arsenic is an important exploration objective, as well as extensive research activities are being conducted worldwide. In recent times, advanced functional materials have achieved significant attention as potential materials for the design of cost-effective techniques for the detection and removal of arsenic. The unique functional characteristics of such materials can be an added advantage for the development of such techniques. The proposed book chapter will highlight different research activities conducted so far for the development of such materials and provide an overview of the novel methods designed to identify and eliminate arsenic with the aid of new materials.

Keywords Heavy metals · Arsenic toxicity · Public health · Water contaminant · Advanced functional materials

R. Chetia · K. Neikha · A. Puzari (✉)
Department of Chemistry, National Institute of Technology Nagaland, Chumukedima, Nagaland, India
e-mail: amrit09us@yahoo.com

A. K. Yadav and S. Shirin (eds.), *Emerging and Innovative Arsenic Removal Technologies for a Sustainable Future*, Water Science and Technology Library 131,
https://doi.org/10.1007/978-3-032-01176-3_4

4.1 Introduction

Although 71% of the Earth's surface is covered by water, practically less than 1% water is available for human use. The rest 97.3% is present as saltwater in sea, 1.75–2.0% in the glaciers, and 0.0001% in present is the air as vapor or clouds. Even the small amount of water available is not always safe for intake. Firstly, due to the existence of multiple contaminants, water obtained from natural sources is not suitable for direct consumption. Globally, contaminated water becomes a major threat to human existence. The majority of the human population is suffering from the consequences of contaminated water. Both anthropogenic and natural activities are responsible for water contamination. The water becomes contaminated by the industrial waste that continuously pours toxic heavy metals into waterbodies along with other contaminants, and the situation is further aggravated by coal mining, agricultural waste, sewage sludge, etc. Thus, aquatic and human life is continuously affected. More than permissible limit [established by World Health Organization (WHO)] of heavy metal contaminants in drinking water can result in acute and chronic diseases causes which can be nonfatal such as physical and muscle weakness or they can be fatal to brain, neurological disorder and even cancer.

Toxic metals are referred to those elements which are toxic or harmful to living organisms. Most of them have higher atomic weight than other elements, so they are also known as heavy metals. Radioactive elements, metals like lead, iron, zinc, copper, chromium, and metalloids such as mercury and arsenic are considered heavy metals. Contamination of drinking water by these heavy metals because of the high concentration of heavy metals and metalloids in drinking water, the health of millions of people has been negatively impacted by contaminated water. Metal toxicity varies by many factors, such as quantity, exposure time and route. There are some metals like iron, which is essential for humans and non-toxic up to a certain amount. According to the Bureau of Indian Standards (BIS) and the WHO guideline, water contains 0.3 mg/L of iron, which is safe to drink. Another harmful heavy metal is Lead (Pb^{+2}), which causes acute diseases such as allergies, kidney failure, cancer, and mental disorder. Zinc is essential for many biological processes but is very dangerous for plants and aquatic animal in a trace amount. Mercury is another type of metalloid, which is toxic to human as to the well as environment. Arsenic (As^{+3} and As^{+5}) is another harmful heavy metal that causes various respiratory disorders, chronic liver disease, and serious diseases.

Arsenic is considered one of the most noxious elements on the planet. Approximately 3 billion metric tonnes of arsenic are produced by the 27 member states of the European Union, among which approximately 12,000 tones are released annually into the environment. Arsenic in groundwater is caused by a variety of processes, including human activity, biological activity and geochemical activity. Soil and drinking water may become contaminated by man-made discharge of arsenic from different industrial and agricultural processes. In water, arsenic can be found in the form of dissolved or particles. Most of the arsenic-based insecticides, mineral deposition in groundwater and improper disposal of arsenic-containing chemicals contaminate water and provide a harmful effect on environment. More than 105 countries globally have water contamination with arsenic. Approximately 220 million people worldwide are in a crisis of consuming drinking water that has approximately > 10 μg/L of arsenic. Since 2000, Asian countries including Bangladesh, Afghanistan, Pakistan, Korea, Cambodia, Nepal, India, and China have a notable level of groundwater contaminated with arsenic.

The intake of arsenic more than its permissible limit causes multiple negative impacts to humans and animals. Arsenic poising causes its toxicity on multiple organ systems including the cardiovascular, hematologic, nervous, and respiratory systems. Additionally, it has been associated with various cancers like bladder, liver, lung, and skin cancers. Numerous industrial and domestic activities contaminate water and affect human health and the environment due to its carcinogenic consequences. Since ancient time there are several traditional as well as scientific ways, such as boiling, filtration, and chemical disinfection were adopted to treat polluted water. However, the design of cutting-edge and novel technologies is essential for the detection and removal of toxic pollutants such as arsenic.

Over the past few decades, many conventional methods have been explored for the removal of arsenic. Some commonly used techniques include adsorption, membrane-based filtration, coagulation, flocculation, electrochemical methods, and ion exchange method. However, these methods can detect or adsorb heavy metals singularly; hence, they perform poorly in wastewater treatment, which limits their large-scale application. Each of the methods has certain limitation along with their advantages. Therefore, it is very necessary to develop such material, which can detect and remove toxic materials simultaneously and also improves their efficiency. To get past these obstacles, functional materials such as nanomaterials, polymers, and porous materials are excellent alternatives of detection and removal for toxic pollutants. In this chapter, we highlight different research activities conducted so far to develop such materials and will provide an overview of the novel methods developed for the detection and removal of arsenic using different functional materials.

4.2 Background

Arsenic is known as a human carcinogen. It is a silver-like crystalline mineral having atomic weight of 74.9 amu and a density 5.72 g/cm^3. About 1.5 ppm of arsenic are present in the earth's crust, 3 ppm in air, and 10 ppm in fresh water. For humans, inorganic arsenic has been the main cause of the majority of hazardous exposures. Inorganic As^{+3} is approximately two to ten times more hazardous than As^{+5} in terms of toxicity. Arsenic is the only metalloid that can mobilise very easily under both oxidizing and reducing conditions at a pH range between 6.5 and 8.5. Exposure to arsenic compounds has different health effects depending on their amount, path, lifestyle and heredity. Globally, approximately 140 million people are affected by arsenic poisoning. Children and the foetus are especially vulnerable to arsenic poisoning. Minimizing the interaction with arsenic is essential to minimize the risk of health issues. Adsorption is most common method for extracting compounds because of its high efficiency, ease of handling and cost-effective nature. The characteristic property of adsorbate has significant impact on its efficiency. Traditional methods involving the use of sand, clay, or other natural materials are known, and these methods employ low-cost adsorbate for the removal of arsenic along with commercially available materials like activated carbon, zeolite, etc.

Biochar, the solid material produced through biomass pyrolysis under a limited supply of oxygen, is an environmentally friendly, cost-effective adsorbent that can extract toxic materials from contaminated water. The membrane-based method has also shown promising results in separating 'As' through filtration or electrical repugnance by the membrane. Another conventional technique to remove arsenic is the ion exchange method, which can be removed through ion exchange in physicochemical processes. This method is mainly applicable for the elimination of hazardous chemicals such as arsenate and selenite from the contaminated water. Electrochemical methods are commonly used to remove toxic material like arsenic. It attracts charged particles towards oppositely charged electrode surface under influence of an electric field. Another popular technique for removal of arsenic from soil and groundwater is coagulation-flocculation. Currently, the development of novel functional materials for the intervention of heavy metals such as arsenic with a low detection limit, high sensitivity, and more efficient has acquired a number of research attention. In practical implementation, functional materials consist of unique advantages and capabilities to significantly reduce the issue of arsenic contamination.

4.3 Causes of Arsenic Contamination

Arsenic is a highly toxic metalloid commonly found in soils and water bodies. It is widely distributed both as a free element and in blended form in sulfide ores and it is ranked at 56th most abundant mineral in the earth's mantle. Generally, it has four valencies including As^0, As^{+3}, As^{-3}, and As^{+5}. Numerous inorganic and organic forms of arsenic, such as arsenite (As^{+3}), arsenate (As^{+5}) dimethylarsinic acid, and trimethylarsine oxide, are found in the nature. Most hazardous human hazardous exposures have been mainly caused by inorganic arsenic. The most common form is arsenopyrite; when it oxidizes releases arsenic which is toxic to human health and the environment. It has adverse effects on human and environmental health, as well as a number of harmful toxicity mechanism (Rehman et al., 2021; Yadav et al., 2014). Arsenic contamination is caused by various natural and anthropogenic factors. When it comes to anthropogenic sources, industrial activities lead to arsenic contamination in water, soil, and air due to fossil fuel combustion, metal smelting, and mining. Volcanic activity and extreme temperature operations such as coal-fired power plants are the primary sources of arsenic in the atmosphere.

Arsenic is ingested by millions of people around the world through natural resources found in vegetables, poultry, shellfish, and food treated with arsenic-containing water. Inorganic arsenic is found in ecosystems naturally, and the most

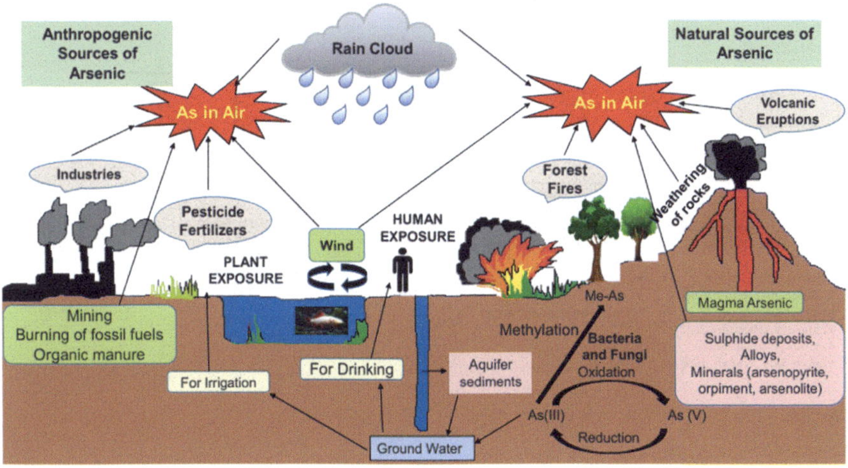

Fig. 4.1 How the food chain is contaminated with arsenic (reprinted from Rehman et al. 2021 with permission from Journal of Hazardous Materials. Copyright 2021, Elsevier)

significant trophic level in the food chain is where arsenic enters plants through their root systems from soil and groundwater. Compared to other cereal crops, rice can absorb arsenic into its grains at much faster rates. The land used to grow rice can be contaminated by arsenic-based fertilizer and farm water. Arsenic concentration in paddy fields is increasing, which is becoming a matter of concern due to the gradual entry into biodiversity and human nutrition chain. Animals get arsenic from affected water, grasslands, vegetables, and other types of foliage. Figure 4.1 illustrates how arsenic affects the food chain. Since humans are at the top of the food chain, there could be several reasons they are exposed to arsenic. Humans can be exposed to arsenic in several ways, including through a diet that contains high arsenic, contaminated water, or by inhalation of dust containing dust. People are most frequently exposed to arsenic through food and water generated from crops grown in contaminated soil.

4.4 Impact of Arsenic Contamination

According to WHO, arsenic has been ranked as most toxic carcinogenic pollutant in the environment. Aquatic species are particularly vulnerable to arsenic poisoning, since surface water is one of the primary environmental reservoirs of arsenic. The primary reason of bioaccumulation in the essential organs of aquatic species is their continuous contact with arsenic environment. Aquatic organisms can absorb arsenic through their skin, mucosal membranes, or by food. Inhalation of arsenic causes significant physiological and biochemical changes that include apoptosis, oxidative stress, immune system abnormalities, reduced reproduction, and destruction of cells

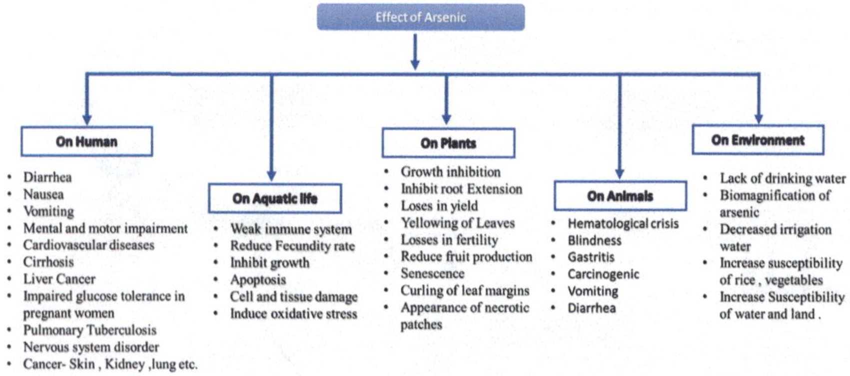

Fig. 4.2 Effects of arsenic contamination on our ecosystem

and tissues. It is also toxic to fish organs including skin, liver, kidney, lung, muscle, and gastrointestinal tract. Exposure to arsenic can affect a number of organ systems, including the psychological, gastrointestinal tract, respiratory, cardiovascular, and hematologic systems, in both acute and subacute manner. When plants are exposed to contaminated soil or water, arsenic develops in various plant parts and can have a harmful effect on the plant. When exposed to arsenic it experiences oxidative stress due to the emergence of responsive oxygen species, which damage the DNA as well as their cell in plants. The increase of the arsenic level in land and water can limit the regular growth of plants. It will also hamper the photosynthesis process, which will decrease the growth and yield of agricultural products. Arsenic contamination also has effects on its roots because roots are the initial point of contact with arsenic. Arsenic toxicity also disturbs morphology and physio-biochemical mechanism of plant, resulting fading of leaves, yellowing of leaves and drop in chlorophyll content (Fig. 4.2).

Furthermore, in animals, arsenic toxicity affects the gastrointestinal tract and cardiovascular system, leading to plasma transudation, blood loss, diarrhea, dehydration, depression, and cardiovascular failure. Arsenic can have multiple harmful consequences on human body organs including cardiovascular system, respiratory, gastrointestinal, endocrine and neurological. Muscle soreness and weakness are the first signs of acute arsenic poisoning, which is followed by severe stomach pain, nausea, vomiting diarrhea, and hallucinations. Acute poisoning also includes anaemia, leukopenia, and bone marrow. Inhalation of arsenic gas causes coughing, bronchitis, and shortness of breath. Prolonging arsenic exposure causes cancer in various organs that induces skin, lung, kidney, and liver. Several investigations have a shown that arsenic have harmful effect on the hormones and reproductive system. It mostly affects the male reproductive system including the decrease in testosterone and gonadotropin levels. Acute exposure to arsenic also causes significant brain damage in babies, with peripheral sensory fibers more affected than other fibers.

Therefore, from experiment and epidemiologic evidence, we can mention that arsenic has many adverse effects on human health as well as on the environment.

4.5 Functional Materials for Arsenic Detection and Removal

To overcome the obstacles of conventional methods, it is essential to generate innovative materials and methods with exceptional performance. In aid of detecting and eliminating arsenic from contaminated water, advanced functional materials with added advantages are introduced (Li et al., 2023). Discussion in this chapter also focuses on recent advancement of functional materials such as nanomaterials, polymers, and porous materials, for detection and removal of arsenic. The design of more durable materials encourages the development of more effective techniques for the remediation of contaminated water.

4.5.1 Nanomaterials

Materials whose sizes or dimensions range from 1 to 100 nm are called nanomaterials. The use of nanomaterials has garnered significant attention because of their high efficiency and adaptability. Metal nanoparticles, carbon-based materials and nanocomposites are particularly effective because of their significant area, high reactivity, and modular properties. Techniques like adsorption using metal oxides (e.g., Iron oxide NPs) and hybrid nanomaterials (e.g., nanocomposites) have shown promise in arsenic remediation (Zheng et al., 2024a, 2024b). These materials possess not only good adsorption capacities but also facilitate easy recovery and reuse, enhancing their practical applications. Some recent research article highlight the modification of these nanomaterials by doping and atom replacement (Yu et al., 2024).

Nanomaterials are generally categorized into three types on their dimensionality. These can be zero-dimensional (e.g., nanoparticles and quantum dots), one-dimensional (e.g., nanowire and nano rods), and two-dimensional (e.g., nanofilm). Metal oxide nanoparticles, carbon-based nanomaterials, and biobased nanomaterials are some common nanomaterials used in the removal of arsenic from water. These materials remove arsenic by adsorption (due to ion exchange, surface complex formation, and electrostatic attraction), precipitation (formation of insoluble compounds) and redox reaction (conversion of As^{+3} to As^{+5}).

4.5.1.1 Metal-Oxide Nanomaterials

Several metal-oxide nanomaterials such as iron oxide nanoparticles (IONPs), titanium dioxide nanoparticles, and zinc oxide nanoparticles have emerged as highly effective agents for arsenic removal in water mainly because of their high surface area, strong adsorption, and catalytic properties. These nanomaterials facilitate multiple mechanisms including adsorption, oxidation reduction, and photocatalysis, to effectively remove arsenic species.

IONP is widely studied in this field because it is selective to both As^{+3} and As^{+5} (Sahu et al., 2024). The successful outcome in IONP have inspired researchers to enhance arsenic adsorption capabilities through various methods, such as creating composites and combining it with other metal oxides. The ZrO_2–Fe_3O_4 nanoparticle converts As^{+3} to As^{+5} by photocatalytic oxidation directed by hydroxyl and superoxide radicals. The As^{+5} later coordinates with the precipitate formed by the nanoparticle (Sun et al., 2017). The interaction of arsenic with iron oxide was analyzed by FT-IR and APS studies of As-O and As-O-Fe bonds replacing metal-hydroxyl bonds by arsenic (Cheng et al., 2016; Shokri et al., 2016; Ge et al. 2017). In another work, Fe_3O_4 was used in the removal which was explained by XANES and EXAFS techniques. Adsorption of As^{+3} caused vacant and defect sites at (111) on the Fe_3O_4 surface (Liu et al., 2015). These interactions were also confirmed in another work where the bonding between arsenic and IONPs was confirmed under different pH and arsenic concentrations. It was confirmed that the As^{+3} interact by chemisorption process towards iron oxide nanomaterials.

Titanium oxide (TiO_2) is another well-known nanomaterial that is used in the catalysis of organic compounds and organic reactions. In presence of UV light, TiO_2 acts as photocatalyst and in dark as adsorbent (Guan et al., 2012). It was also observed that under UV light, the TiO_2 is more dispersed and the oxidative photocatalysis rate is higher at $pH = 8$. The authors also observed that the oxidation of As^{+3} was completed within 2 h of irradiation (Lee & Choi, 2002). In another work, it was shown that the photocatalytic oxidation rate is enhanced by the deposition of platinum (Pt) on the surface of TiO_2. In this case, Pt doping in the TiO_2 nanoparticle facilitates fascharge-pair recombination utilizing the valence electrons present in the conduction band (Ryu & Choi, 2004). Mechanism of photocatalytic oxidation involving O^{-2} and $OH^{\cdot}$ radicals were studied considering the degradation reaction of ROX, using isopropyl alcohol and superoxide dismutase. It was observed that addition of isopropyl alcohol and superoxide dismutase inhibits the degradation of ROX. Superoxide dismutase scavenges the superoxide anion radical by converting it into H_2O_2, while isopropyl alcohol can trap the hydroxyl radical, which otherwise establishes the dominant role of O_2 and $OH^{\cdot}$ in photocatalytic degradation attacking the As–C bond (Yuan et al., 2022).

Zinc oxide nanoparticles are another material utilized in arsenic removal. The primary mechanism involved in arsenic removal using ZnO nanomaterials includes adsorption, where arsenic is attracted and held onto the surface of the nanoparticles by electrostatic interactions. Photocatalysis by oxygen radical species can oxidize As^{+3}. ZnO is not only used in elimination of heavy metals in water but also in removal of other pollutants, photodegradation and antibacterial applications (Pérez Adassus et al., 2024). Similarly to the mechanism suggested with TiO_2 nanoparticles, ZnO nanomaterial also shows the formation of an As–O bond replacing Zn–O linkage. This was observed using FT-IR analysis that confirmed the bond of As–O (853 cm^{-1}) and As–OH (784 cm^{-1}) (Samad et al., 2016). ZnO also acts as a suitable adsorbent for As^{+3} where the highest sorption capacity was 52.64 mg/g in neutral solution containing 0.4 g/L of arsenic content. Further analysis by TGA confirms the loss of OH^- and $CO_3{}^{2-}$ groups at two events occurring spontaneously (Yuvaraja et al., 2018).

4.5.1.2 Carbon-Based Nanomaterials

The common carbon-based nanomaterials, such as graphene oxide (GO), Carbon Nanotubes (CNTs), activated carbon, and carbon quantum dots (CQDs), were widely used for the removal of arsenic from water due to their high surface area, strong adsorption, and functionalization potential. In this section, we will discuss more composites prepared and carbon-based nanomaterials for arsenic removal.

GO has abundant oxygen-containing groups that effectively adsorb arsenic through electrostatic interactions and hydrogen bonding. The presence of oxygen groups not only behaves as a platform for adsorbing arsenic but also enhances composite formation with other nanomaterials such as iron oxide. This results in an increase in the surface area from 129 to 341 m^2g^{-1} with a concomitant increase in arsenic adsorption. This composite absorbs approximately 147 mg/g of As^{+3} and 113 mg/g of As^{+5} with removal efficiency from approximately 110 ppb to less than 0.02 ppb. The mechanism involved in the composite involves both physical and chemical adsorption due to active site on the surface and ion exchange respectively (Su et al., 2017). Another GO compound involving GO/$CuFe_2O_4$ material reported an adsorption of 51.64 mg/g of As^{+3} and 124.69 mg/g of As^{+5}. This compound showed good adsorption in wide range of pH levels and about 96% of arsenic removed from water with good regeneration and reusability (Wu et al., 2018).

Similarly, CNTs also offer high adsorption sites and can be modified to enhance arsenic removal efficiency. The increase in the amount of CNTs increases the arsenic adsorption when the arsenic concentration was maintained at 10 mg/L. With the increase in active sites of the CNTs, the adsorption also gradually increases (Naghizadeh et al., 2012). CNT doped with iron oxide showed the formation of As–O, As_2O_3, and As_2O_5 bonds as confirmed from XPS with respective binding energies of 43.6, 44.5, and 45.4 eV respectively. The presence of FeO greatly improved arsenic adsorption in contaminated water (Vadahanambi et al., 2013). A recent work similar to the former was studied by preparing Fe-doped CNT with magnetically enhanced property. The objective of this work was to remove arsenic under alkaline conditions using Fe^{+5} doped CNT. However, a decrease in As^{+3} removal efficiency from 97.76% to 91.18% was observed due to the presence of several ions such as Cl, HCO_3 and $SO_4{}^{2-}$ in water (Zheng et al., 2024a).

Another derivative of carbon-based nanomaterial, such as CQDs which offers unique electronic properties and surface functionalities, showed good arsenic removal potential. Apart from arsenic, it is also used for fluorescence detection of other heavy metals in Arabica coffee such as Fe^{+3}, Pb^{+2} and Cr^{+3} (Nazar et al., 2024; Sgreccia et al., 2024). CQD is also used in detection of As^{+3} and As^{+5} at very low concentration about 11×10^{-9} M and 11.2×10^{-9} M respectively with high adsorption of 159 mg/g. This study also showed that the adsorption is fast and exhibits good interaction between the arsenic and CQD. Under basic conditions, the adsorption and desorption undergo several cycles showing good recyclability and reusability of the material (Yadav et al., 2023). Another work involving the water filtration using sodium doped CQDs was reported, which showed higher arsenic removal efficiency. Na-CQD was prepared as a thin-film nanocomposite (TFN). About 0.05% sodium-loaded TFN exhibits removal efficiency of 97.5%, 98.2%, and 99.5% towards SeO_3^{2-}, SeO_4^{2-}, and $HAsO_4^{2-}$ respectively. The presence of Na^+ ion on the surface increases the hydrophilicity property, improving the efficacy of the water filtration process (He et al., 2018). CQDs are also used in the detection of arsenic in water. A composite consisting of thiol-functionalized fluorescent CQD was also prepared for the sensitive and selective detection of As^{+3} in water. The fluorescence turn-on mechanism provides a reliable means of detecting low concentrations of arsenite, with high selectivity and sensitivity, making it as a promising tool for environmental monitoring (Pooja et al., 2017).

4.5.1.3 Bio-based Nanomaterials

Biobased nanomaterials are a type of nanocomposites that are made up of inorganic nanomaterials combined with biomolecules such as chitosan, cellulose, etc. Alginate-based nanocomposites are increasingly utilized for the removal of arsenic from contaminated water since they are eco-friendliness, abundance, and high adsorption capacity. These materials are often incorporated with nanoparticles, such as iron oxide, which enhance their affinity for arsenic. The removal mechanism typically involves a combination of adsorption and ion exchange processes. Arsenic ions (As^{+3} or As^{+5}) are attracted to functional groups (e.g., hydroxyl, amine, carboxyl) on the surface of the nanomaterial, where they undergo complexation or redox reactions. As mentioned in the former section of metal oxide nanomaterials, Fe_3O_4 nanoparticles can oxidize As^{+3} to As^{+5} which is more easily adsorbed. The high surface area and reactive sites of the biobased nanomaterials facilitate these interactions, leading to effective arsenic sequestration and improved water quality. The mechanism involved an interaction between the functional groups of the biomolecules and the arsenic present. For example, the presence of carboxyl groups (−COOH and –COOR) in alginate can form complexes with arsenic ions making them suitable adsorbents. This has shown As^{+5} adsorption capacity of 81.3 mg/g when the Fe-biomass ratio was 10%wt. (Meng et al., 2017). The interaction site of these materials is crucial because the sorption potential of As^{+3} should be preferred during material development. The sorption capacity of chitosan can be enhanced by chemical modification through the integration of selective functional groups such as $-NH_2$, –SH, –OH, and –COOH. These functional groups interact with the As^{+3} ion by electron donation, cation exchange, surface complexation, and acid–base interaction (Cathell et al., 2008; Ezeh & Harrop, 2012).

4.5.2 Polymer Materials

Organic polymers are the potential choice for chemical sensors because of their various reactive groups. The most promising features of these materials are their extraordinary sensitivity and selectivity. Moreover, they consist of a hydrophilic polymeric matrix and fluorescent probes. Polymer probes offer notable recoverability, low-cost effectiveness, and advantages in detection. Along with this, these polymer materials have some other benefits, including more active regions and more binding sites, good mechanical properties, reusability, and hydrophilic in nature. Fluorescent materials can be loaded and encapsulated in polymers to increase their stability and make them suitable for a range of environmental uses. Biodegradable polymers, because of their ability to degrade to harmless materials, are always preferred.

Chitosan is one of those biosorbents that effectively remove heavy metal ions from the environment. Chitosan is non-toxic, biodegradable, bioactive, and biocompatible. For example, utilising an in-situ doping technique, Zeng et al. (2020) reported. report on a naturally occurring polysaccharide chitosan-functionalized iron nanosheet. This material is designed to attract As^{+3} from aqua medium, with an optimum adsorption of 108.6 mg/g. Furthermore, Sierra-Trejo et al. (2020) studied the removal of As^{+5} using a modified chitosan gel that was synthesized using tungstate coagulation and molybdate techniques. Pereira et al. (2020) Investigated a bio adsorbent (CEDA), which is developed by tosylation of microcrystalline cellulose and then substituting the tosyl group with ethylenediamine in a nucleophilic approach. The optimum adsorption capacity shown by the polymer material is 1.62 mmol/g. Wei et al. (2019) reported Fe based chitosan adsorbents for As^{+3} purification from water. Fe-Chitosan adsorbate works effectively in a wide pH range of 3 to 9, with a maximum adsorption ability of 52.7 mg/g. Ayub et al. (2020) reported a Chitosan based magnetic biosorbent using chitosan as matrix and magnetic (Fe_3O_4) nanoparticle for As^{+3} and As^{+5} removal. The maximum adsorption capacity of the adsorbent for As^{+3} and As^{+5} species is 18.87 and 26.13 mg/g, respectively. As^{+5} and As^{+3} was removed with 99.5% and 99% efficiency at pH 6.7, respectively, at optimal biosorbent concentrations of 1.5 and 2 g/L. Wang et al. (2014), which have an adsorption capacity of 35.7 mg/g and 35.3 mg/g of As^{+5} and As^{+3} at 6.8 respectively.

4.5.3 Porous Materials

Porous materials are a class of organic–inorganic hybrid polymers, containing a large surface area with a tuneable pore size. Based on their pore size, these materials can be categorized into 3 types, microporous materials (pore size less than 2 nm), mesoporous materials (approx. 2–50 nm), and macroporous materials (more than 50 nm). These porous materials exhibit good thermal stability along with high adsorption capacity. Because of their functional advantages, MOFs and COFs are renowned porous materials that are suitable for detection and removal of arsenic. The main advantage of these materials as luminous materials in the detection and removal process is that they offer the same heavy metal ion interaction site, which increases the efficiency of their interaction with heavy metal ions.

4.5.3.1 Metal–Organic Framework

Because of the many structural advantages, such as surface area, large pore size, topology, and functionality, metal–organic frameworks (MOFs) have gained much emphases for the identification and elimination of heavy metal like arsenic (Haldar et al., 2020). Adsorption is a frequently used process that has been used for several decades to remove toxic contaminants from the water. MOF partially compensated the need for effective adsorbent to remove pollutant from water. These porous materials have exceptional porosity and large surface area, and due to their versatile characteristic, they are finding applications in energy storage, sensors, drug delivery, gas storage and separation, catalysis, environmental remediation, and luminescence (Yusuf et al., 2022). MOFs are also used for detection and elimination of toxic pollutants such as arsenic because of their distinctive property. In contract to traditional adsorbents, such as active carbon, MOFs have higher surface area. This method allows for the quick detection and elimination of heavy metals in water because of their internal surface area, chemical controllability, and tendency for the adsorption of significant molecule (Fig. 4.3).

Among the various transition metal-based MOFs, Fe-based MOFs show good performance for detection as well as removal of arsenic (Malhotra et al., 2024)

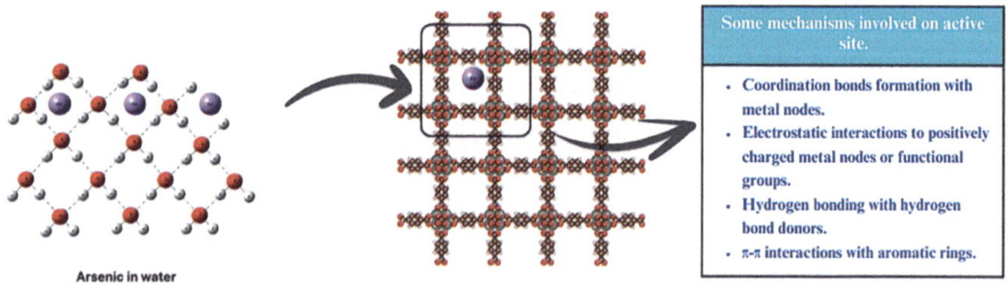

Fig. 4.3 Schematic representation of adsorption of arsenic in MOF on active sites

due to their internal pore environment. Zhu et al. showed an investigation on the elimination of As^{+5} using a Fe-BTC MOF for the first time (Zhu et al., 2012). This Fe-BTC MOF was generated by a solvothermal process under autogenous pressure, with iron serving as the metal core and benzene tricarboxylic acid as the organic linker. To find the ideal pH for adsorption, the absorption of As^{+5} as a pH index was studied in the range of 2–10. An adsorption ability of 12.3 mg/g at pH 4 is the optimal condition. The MOF shows 6.5 times better adsorption capacity than Fe_2O_3 nanoparticle towards removal of As^{+5}. Later Li et al. (2014) investigated a Zn-based MOF ZIF-8, showing an adsorption capacity of 76.5 mg/g at low equilibrium concentration of 9.8 μg/L, which is much higher than traditional adsorbent such as active carbon (3 mg/g) and zeolite (13.92 mg/g) whereas ZIF-8 functionalized ZIF-8 shows capacity of 72 to 83.5 mg per gram of adsorbent (Massoudinejad et al., 2018). At pH 7, it shows optimum results. In addition to ZIFs, the MIL family is a well-liked class of MOFs that has been investigated for arsenic removal. Li et al. (2014) studied an MIL-53 (Al) MOF with an adsorption capacity of 106 mg/g at pH 8 in 11 h. This MOF contains aluminium as the metal site and dicarboxylic acid as the organic node. Hou et al., (2018) reported an environment-friendly route for synthesis MIL-88B, which is an Fe-based MOF containing terephthalic acid as linker. The MOF was synthesized at an ambient temperature without any heat source and without using any organic solvents or without any acid–base. It has shown an excellent adsorption capability of 156 mg/g for As^{+5}. Xie et al. (2017) synthesized binary functionalized NH_2–Fe-MOFs for detection and elimination of As^{+5} from contaminated water. The Fe-MOF nano octahedra are suitable for the quantitative analysis of arsenate due to its remarkable selectivity towards arsenate, range (0.1–50 μM), high sensitivity (detection limit as low as 4.2 ppb) and quick reaction time (< 1 min). The material shows an adsorption capability of 125 mg/g within 60 min. Using two Zr-based MOF He et al. (2019) studied elimination of As^{+3} and As^{+5}. At high pH the material's As^{+3} and As^{+5} removal ability were 205 mg/g and 68 mg/g, respectively. Yu et al. (2019) showed that Zn-MOF-74 having an adsorption capability of 325 mg/g and 211 mg/g, respectively, for the removal of As^{+3} and As^{+5}. Because of their distinct physiological properties and greater arsenic removal effectiveness compared to other traditional materials, MOFs have attracted a great deal of interest for further research from both academic and industrial sectors. The potential for MOF-based materials to be offered for commercial use in the near future is due to their wide range of potential uses in water treatment.

4.5.3.2 Covalent-Organic Framework (COF)

Similarly to MOF, COF is a type of porous material whose structure is formed by covalent bonds with small elements such as H, B, C, N, and O. In addition, they include more surface area, porosity, crystallinity, adjustable pore size, and tailored functionality. Owing to these properties, these materials show potential applications in energy storage, catalysis, sensor, drug delivery, and adsorption etc. Various synthetic techniques are employed, such as hydro/solvothermal synthesis, microwave

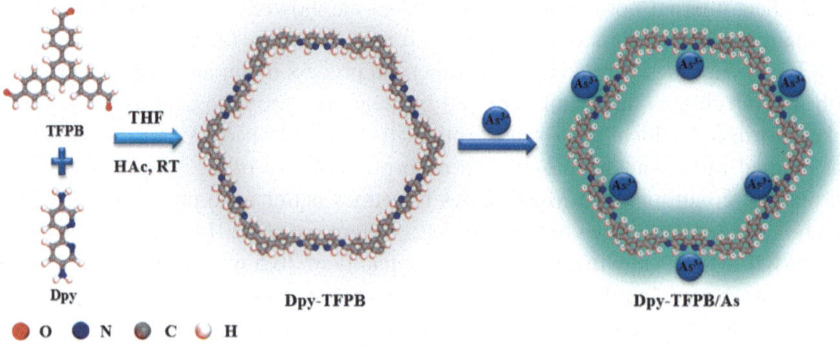

Fig. 4.4 Schematic representation of COF as arsenic adsorbent (reproduced from Yin and Liu (2021) with permission from Journal of photochemistry and photobiology A. Copyright 2021, Elsevier)

synthesis, mechanochemical synthesis, etc. COFs have high chemical and thermal stability. They are ideal functionalized materials for the identification and elimination of Arsenic-like heavy metals. The adjustable pores of COF materials can easily accommodate guest molecules such as arsenic. In a recent study, Yang et al. (2020) reported an EB-COF@Br which has adsorption capability of 53.1 mg/g and can adsorb phosphate and arsenate from water (Fig. 4.4).

Yin and Liu (2021) developed a Bipyridine based COF (Dpy-TFPB) for the detection of As^{+3}. The materials show an extreme low detection limit of 8.86 nmol/L. The Dpy-TFPB shows good performance, showed quick visualization, favorable sensitivity and rapid response. Roxarsone is a source in arsenic contamination of poultry and other animals. It is commonly used as an antibacterial and growth promoting additive for animal. Chen et al. (2022) studied for the first time the use of COF as for fluorescence detection and adsorbents for the identification and elimination of organic arsenic, that is, roxarsone (ROX) from water. Two isorecticular COFs functionalized with amidoxime were synthesized. After PSM the fluorescence quenching efficiency of both COFs was enhanced to 98% and the limit of detection (LOD) was estimated at 6.5 nM and 12.3 nM. The adsorption capacities of both materials shown up to 732 mg/g and 787 mg/g. Furthermore, a 1-D COF (EP-COF), which acts as a fluorescent probe for ROX detection, was studied by Liu, C. et al. (C. Liu et al., 2024). For ROX detection, it shows a good detection limit of 4.5 mmol/L. Recently, a sulphonated COF (TFPOTDB-SO_3H) was developed for elimination of As^{+3} from water by Khosravani et al. (2024), which was synthesized by combining 2,5-diaminobenzene sulfonic acid and triazine based linker in alkaline medium at room temperature. The material adsorbed up to 97.1% within 10 min and after four successive adsorption–desorption cycles, it retains an adsorption efficiency of 344.8 mg/g, which accounts for approximately 88.86% of Arsenic adsorption.

4.6 Conclusion

Recent advances in development of these type of advanced materials for the detection and eradication of arsenic have been documented in this chapter. To make a systematic description, functional materials that can efficiently detect and remove certain heavy metals are classified on their structural characteristics and properties. It has been observed that, since a variety of heavy metals are commonly present in wastewater, the competing adsorption between these ions can impact the efficiency of such materials toward removal of Arsenic ion. Arsenic contamination in water can pose a potential human health concern and hazard, therefore, the development of highly efficient materials for the simultaneous removal as well as detection of heavy metal ions is always considered as a priority area of research. With extensive study and research, these functional materials may be effectively promoted for industrial use in the near future for concurrently identification and eliminate toxic heavy metals.

Acknowledgements The authors sincerely acknowledge the Director and Department of the Science and Humanities of National Institute of Technology Nagaland, India.

References

Ayub, A., Raza, Z. A., Majeed, M. I., Tariq, M. R., & Irfan, A. (2020). Development of sustainable magnetic chitosan biosorbent beads for kinetic remediation of arsenic contaminated water. *International Journal of Biological Macromolecules, 163*, 603–617. https://doi.org/10.1016/j.ijbiomac.2020.06.287

Cathell, M. D., Szewczyk, J. C., Bui, F. A., Weber, C. A., Wolever, J. D., Kang, J., & Schauer, C. L. (2008). Structurally colored thiol chitosan thin films as a platform for aqueous heavy metal ion detection. *Biomacromolecules, 9*(1), 289–295. https://doi.org/10.1021/bm700845z

Chen, H., Liu, W., Cheng, L., Meledina, M., Meledin, A., Van Deun, R., Leus, K., & Van Der Voort, P. (2022). Amidoxime-functionalized covalent organic framework as simultaneous luminescent sensor and adsorbent for organic arsenic from water. *Chemical Engineering Journal, 429*. https://doi.org/10.1016/j.cej.2021.132162

Cheng, W., Zhang, W., Hu, L., Ding, W., Wu, F., & Li, J. (2016). Etching synthesis of iron oxide nanoparticles for adsorption of arsenic from water. *RSC Advances, 6*(19), 15900–15910. https://doi.org/10.1039/c5ra26143k

Ezeh, V. C., & Harrop, T. C. (2012). A sensitive and selective fluorescence sensor for the detection of arsenic(III) in Organic Media. *Inorganic Chemistry, 51*(3), 1213–1215. https://doi.org/10.1021/ic2023715

Ge, X., Ma, Y., Song, X., Wang, G., Zhang, H., Zhang, Y., & Zhao, H. (2017). β-FeOOH nanorods/carbon foam-based hierarchically porous monolith for highly effective arsenic removal. *ACS Applied Materials and Interfaces, 9*(15), 13480–13490.

Guan, X., Du, J., Meng, X., Sun, Y., Sun, B., & Hu, Q. (2012). Application of titanium dioxide in arsenic removal from water: A review. *Journal of Hazardous Materials, 215–216*, 1–16. https://doi.org/10.1016/j.jhazmat.2012.02.069

Haldar, D., Duarah, P., & Purkait, M. K. (2020). MOFs for the treatment of arsenic, fluoride and iron contaminated drinking water: A review. *Chemosphere, 251*, Article 126388. https://doi.org/10.1016/j.chemosphere.2020.126388

He, X., Deng, F., Shen, T., Yang, L., Chen, D., Luo, J., Luo, X., Min, X., & Wang, F. (2019). Exceptional adsorption of arsenic by zirconium metal-organic frameworks: Engineering exploration and mechanism insight. *Journal of Colloid and Interface Science, 539*, 223–234. https://doi.org/10.1016/j.jcis.2018.12.065

He, Y., Zhao, D. L., & Chung, T. S. (2018). Na+ functionalized carbon quantum dot incorporated thin-film nanocomposite membranes for selenium and arsenic removal. *Journal of Membrane Science, 564*(June), 483–491. https://doi.org/10.1016/j.memsci.2018.07.031

Hou, S., Wu, Y. N., Feng, L., Chen, W., Wang, Y., Morlay, C., & Li, F. (2018). Green synthesis and evaluation of an iron-based metal-organic framework MIL-88B for efficient decontamination of arsenate from water. *Dalton Transactions, 47*(7), 2222–2231. https://doi.org/10.1039/c7dt03775a

Khosravani, M., Dehghani Ghanatghestani, M., Moeinpour, F., & Parvaresh, H. (2024). New sulfonated covalent organic framework for highly effective As(III) removal from water. *Heliyon, 10*(3), Article e25423. https://doi.org/10.1016/j.heliyon.2024.e25423

Lee, H., & Choi, W. (2002). Photocatalytic oxidation of arsenite in TiO2 suspension: Kinetics and mechanisms. *Environmental Science and Technology, 36*(17), 3872–3878. https://doi.org/10.1021/es0158197

Li, J., Wu, Y. N., Li, Z., Zhang, B., Zhu, M., Hu, X., Zhang, Y., & Li, F. (2014). Zeolitic imidazolate framework-8 with high efficiency in trace arsenate adsorption and removal from water. *Journal of Physical Chemistry C, 118*(47), 27382–27387. https://doi.org/10.1021/jp508381m

Li, M., Shi, Q., Song, N., Xiao, Y., Wang, L., Chen, Z., & James, T. D. (2023). Current trends in the detection and removal of heavy metal ions using functional materials. *Chemical Society Reviews, 52*(17), 5827–5860. https://doi.org/10.1039/d2cs00683a

Liu, C. H., Chuang, Y. H., Chen, T. Y., Tian, Y., Li, H., Wang, M. K., & Zhang, W. (2015). Mechanism of arsenic adsorption on magnetite nanoparticles from water: Thermodynamic and spectroscopic studies. *Environmental Science and Technology, 49*(13), 7726–7734. https://doi.org/10.1021/acs.est.5b00381

Liu, C., Jia, C., Gan, S.-X., Qi, Q.-Y., Jiang, G.-F., & Zhao, X. (2024). A luminescent one-dimensional covalent organic framework for organic arsenic sensing in water. *Chinese Chemical Letters,* 109750. https://doi.org/10.1016/j.cclet.2024.109750

Malhotra, M., Kaur, B., Soni, V., Patial, S., Sharma, K., Kumar, R., Singh, P., Thakur, S., Pham, P. V., Ahamad, T., Le, Q. V., Nguyen, V. H., & Raizada, P. (2024). Fe-based MOFs as promising adsorbents and photocatalysts for re-use water contained arsenic: Strategies and challenges. *Chemosphere, 357*, Article 141786. https://doi.org/10.1016/j.chemosphere.2024.141786

Massoudinejad, M., Ghaderpoori, M., Shahsavani, A., Jafari, A., Kamarehie, B., Ghaderpoury, A., & Amini, M. M. (2018). Ethylenediamine-functionalized cubic ZIF-8 for arsenic adsorption from aqueous solution: Modeling, isotherms, kinetics and thermodynamics. *Journal of Molecular Liquids, 255*, 263–268. https://doi.org/10.1016/j.molliq.2018.01.163

Meng, F., Yang, B., Wang, B., Duan, S., Chen, Z., & Ma, W. (2017). Novel dendrimerlike magnetic biosorbent based on modified orange peel waste: Adsorption-reduction behavior of arsenic. *ACS Sustainable Chemistry and Engineering, 5*(11), 9692–9700. https://doi.org/10.1021/acssuschemeng.7b01273

Naghizadeh, A., Yari, A. R., Tashauoei, H. R., Mahdavi, M., Derakhshani, E., Rahimi, R., Bahmani, P., Daraei, H., & Ghahremani, E. (2012). Carbon nanotubes technology for removal of arsenic from water. *Archives of Hygiene Sciences, 1*(1), 6–11.

Nazar, M., Hasan, M., Wirjosentono, B., Gani, B. A., & Nada, C. E. (2024). Microwave synthesis of carbon quantum dots from Arabica coffee ground for fluorescence detection of Fe^{3+}, Pb^{2+}, and Cr^{3+}. *ACS Omega, 9*(18), 20571–20581. https://doi.org/10.1021/acsomega.4c02254

Pereira, A. R., Soares, L. C., Teodoro, F. S., Elias, M. M. C., Ferreira, G. M. D., Savedra, R. M. L., Siqueira, M. F., Martineau-Corcos, C., da Silva, L. H. M., Prim, D., & Gurgel, L. V. A. (2020). Aminated cellulose as a versatile adsorbent for batch removal of As(V) and Cu(II) from mono- and multicomponent aqueous solutions. *Journal of Colloid and Interface Science, 576*, 158–175. https://doi.org/10.1016/j.jcis.2020.04.129

Pérez Adassus, M. B., Spetter, C. V., Mansilla, A. Y., Salcedo, M. F., & Lassalle, V. (2024). Zinc oxide/iron oxide nanocomposites for photodegradation and adsorption of pollutants in water. *ACS Applied Nano Materials*. https://doi.org/10.1021/acsanm.4c00983

Pooja, D., Saini, S., Thakur, A., Kumar, B., Tyagi, S., & Nayak, M. K. (2017). A "Turn-On" thiol functionalized fluorescent carbon quantum dot based chemosensory system for arsenite detection. *Journal of Hazardous Materials, 328*, 117–126. https://doi.org/10.1016/j.jhazmat.2017.01.015

Rehman, M. U., Khan, R., Khan, A., Qamar, W., Arafah, A., Ahmad, A., Ahmad, A., Akhter, R., Rinklebe, J., & Ahmad, P. (2021). Fate of arsenic in living systems: Implications for sustainable and safe food chains. *Journal of Hazardous Materials, 417*(May). https://doi.org/10.1016/j.jhazmat.2021.126050

Ryu, J., & Choi, W. (2004). Effects of TiO_2 surface modifications on photocatalytic oxidation of arsenite: The role of superoxides. *Environmental Science and Technology, 38*(10), 2928–2933. https://doi.org/10.1021/es034725p

Sahu, U. K., Mandal, S., & Pu, S. (2024). Overlooked adsorptive route and challenges in arsenic decontamination using iron oxide nanomaterials. *Chemistry Africa, 7*(4), 1707–1727. https://doi.org/10.1007/s42250-023-00866-2

Samad, A., Furukawa, M., Katsumata, H., Suzuki, T., & Kaneco, S. (2016). Photocatalytic oxidation and simultaneous removal of arsenite with CuO/ZnO photocatalyst. *Journal of Photochemistry and Photobiology a: Chemistry, 325*, 97–103. https://doi.org/10.1016/j.jphotochem.2016.03.035

Sgreccia, E., Gallardo Gonzalez, F. S., Prosposito, P., Burratti, L., Sisani, M., Bastianini, M., Knauth, P., & Di Vona, M. L. (2024). Heavy metal detection and removal by composite carbon quantum dots/ionomer membranes. *Membranes, 14*(6), 134. https://doi.org/10.3390/membranes14060134

Shokri, E., Yegani, R., Pourabbas, B., & Kazemian, N. (2016). Preparation and characterization of polysulfone/organoclay adsorptive nanocomposite membrane for arsenic removal from contaminated water. *Applied Clay Science, 132–133*, 611–620. https://doi.org/10.1016/j.clay.2016.08.011

Sierra-Trejo, P. V., Guibal, E., & Louvier-Hernández, J. F. (2020). Arsenic sorption on chitosan-based sorbents: Comparison of the effect of molybdate and tungstate loading on As(V) sorption properties. *Journal of Polymers and the Environment, 28*(3), 934–947. https://doi.org/10.1007/s10924-020-01654-6

Su, H., Ye, Z., & Hmidi, N. (2017). High-performance iron oxide–graphene oxide nanocomposite adsorbents for arsenic removal. *Colloids and Surfaces a: Physicochemical and Engineering Aspects, 522*, 161–172. https://doi.org/10.1016/j.colsurfa.2017.02.065

Sun, T., Zhao, Z., Liang, Z., Liu, J., Shi, W., & Cui, F. (2017). Efficient removal of arsenite through photocatalytic oxidation and adsorption by ZrO_2–Fe_3O_4 magnetic nanoparticles. *Applied Surface Science, 416*, 656–665. https://doi.org/10.1016/j.apsusc.2017.04.137

Vadahanambi, S., Lee, S. H., Kim, W. J., & Oh, I. K. (2013). Arsenic removal from contaminated water using three-dimensional graphene-carbon nanotube-iron oxide nanostructures. *Environmental Science and Technology, 47*(18), 10510–10517. https://doi.org/10.1021/es401389g

Wang, J., Xu, W., Chen, L., Huang, X., & Liu, J. (2014). Preparation and evaluation of magnetic nanoparticles impregnated chitosan beads for arsenic removal from water. *Chemical Engineering Journal, 251*, 25–34. https://doi.org/10.1016/j.cej.2014.04.061

Wei, Y., Yu, X., Liu, C., Ma, J., Wei, S., Chen, T., Yin, K., Liu, H., & Luo, S. (2019). Enhanced arsenite removal from water by radially porous Fe-chitosan beads: Adsorption and H_2O_2 catalytic oxidation. *Journal of Hazardous Materials, 373*, 97–105. https://doi.org/10.1016/j.jhazmat.2019.03.070

Wu, L. K., Wu, H., Zhang, H. Bin, Cao, H. Z., Hou, G. Y., Tang, Y. P., & Zheng, G. Q. (2018). Graphene oxide/$CuFe_2O_4$ foam as an efficient absorbent for arsenic removal from water.

Chemical Engineering Journal, 334(Nov 2017), 1808–1819. https://doi.org/10.1016/j.cej.2017.11.096

Xie, D., Ma, Y., Gu, Y., Zhou, H., Zhang, H., Wang, G., Zhang, Y., & Zhao, H. (2017). Bifunctional NH2-MIL-88(Fe) metal-organic framework nanooctahedra for highly sensitive detection and efficient removal of arsenate in aqueous media. *Journal of Materials Chemistry A, 5*(45), 23794–23804. https://doi.org/10.1039/c7ta07934f

Yadav, A. K., Sahoo, S. K., Mahapatra, S., Kumar, A. V., Pandey, G., Lenka, P., & Tripathi, R. M. (2014). Concentrations of uranium in drinking water and cumulative, age-dependent radiation doses in four districts of Uttar Pradesh, India. *Toxicological & Environmental Chemistry, 96*(2), 192–200. https://doi.org/10.1080/02772248.2014.934247

Yadav, S., Choudhary, N., & Vasavdutta Sonpal, A. R. P. (2023). Carbon dots-embedded silica tubes: An excitation-independent yellow-emitting turn-on probe for simultaneous detection and removal of inorganic arsenic with in vivo tracking in living organisms. *Small (Weinheim an der Bergstrasse, Germany), 20*(11), 2307491.

Yang, C. H., Chang, J. S., & Lee, D. J. (2020). Covalent organic framework EB-COF:Br as adsorbent for phosphorus (V) or arsenic (V) removal from nearly neutral waters. *Chemosphere, 253*(V), 126736. https://doi.org/10.1016/j.chemosphere.2020.126736

Yin, Y., & Liu, G. (2021). A covalent organic framework containing bipyridine groups as a fluorescent chemical probe for the ultrasensitive detection of arsenic (III). *Journal of Photochemistry and Photobiology a: Chemistry, 421*(17), Article 113528. https://doi.org/10.1016/j.jphotochem.2021.113528

Yu, P., Xing, J., Tang, J., Wang, Z., Zhang, C., Wang, Q., Xiao, X., & Huang, W. (2024). Polyethyleneimine-modified iron-doped birnessite as a highly stable adsorbent for efficient arsenic removal. *Journal of Colloid and Interface Science, 661*(January), 164–174. https://doi.org/10.1016/j.jcis.2024.01.163

Yu, W., Luo, M., Yang, Y., Wu, H., Huang, W., Zeng, K., & Luo, F. (2019). Metal-organic framework (MOF) showing both ultrahigh As(V) and As(III) removal from aqueous solution. *Journal of Solid State Chemistry, 269*(V), 264–270. https://doi.org/10.1016/j.jssc.2018.09.042

Yuan, Y., Yuan, X., Luo, L., & Xie, X. (2022). Photolysis of roxarsone in aqueous solutions containing goethite under simulated sunlight irradiation: Kinetics, mechanism, and degradation pathways. *ACS ES T Water, 2*(5), 830–840. https://doi.org/10.1021/ACSESTWATER.2C00030

Yusuf, V. F., Malek, N. I., & Kailasa, S. K. (2022). Review on metal-organic framework classification, synthetic approaches, and influencing factors: Applications in energy, drug delivery, and wastewater treatment. *ACS Omega, 7*(49), 44507–44531. https://doi.org/10.1021/acsomega.2c05310

Yuvaraja, G., Prasad, C., Vijaya, Y., & Subbaiah, M. V. (2018). Application of ZnO nanorods as an adsorbent material for the removal of As(III) from aqueous solution: Kinetics, isotherms and thermodynamic studies. *International Journal of Industrial Chemistry, 9*(1), 17–25. https://doi.org/10.1007/s40090-018-0136-5

Zeng, J., Qi, P., Shi, J., Pichler, T., Wang, F., Wang, Y., & Sui, K. (2020). Chitosan functionalized iron nanosheet for enhanced removal of As(III) and Sb(III): Synergistic effect and mechanism. *Chemical Engineering Journal, 382*(July), Article 122999. https://doi.org/10.1016/j.cej.2019.122999

Zheng, R., Xu, Z., Qiu, Q., Sun, S., Li, J., & Qiu, L. (2024). Iron-doped carbon nanotubes with magnetic enhanced Fe(VI) degradation of arsanilic acid and inorganic arsenic: Role of intermediate iron species and electron transfer. *Environmental Research, 244*(Nov 2023). https://doi.org/10.1016/j.envres.2023.117849

Zheng, Y. P., Li, W., Liu, T. H., Li, X., Tang, Y. F., Zhi, D., Peng, B., & Zhou, Y. Y. (2024b). Current progress in the development of biosensors based on nanomaterials for the detection of inorganic arsenic. *International Journal of Environmental Science and Technology*. https://doi.org/10.1007/s13762-024-05881-5

Zhu, B. J., Yu, X. Y., Jia, Y., Peng, F. M., Sun, B., Zhang, M. Y., Luo, T., Liu, J. H., & Huang, X. J. (2012). Iron and 1,3,5-benzenetricarboxylic metal-organic coordination polymers prepared by solvothermal method and their application in efficient As(V) removal from aqueous solutions. *Journal of Physical Chemistry C, 116*(15), 8601–8607. https://doi.org/10.1021/jp212514a

Chapter 5
Removal of Arsenic from Water Using Adsorption and Phytoremediation Techniques

Archana Kushwaha, Zeenat Arif, and Bineeta Singh

Abstract Arsenic is a hazardous metal created by humans as well as occurring naturally in the environment. At even low levels, arsenic is a human carcinogen in water with a detrimental impact on human health and atmosphere as the well as ecosystem. Due to this effect, the World Health Organization (WHO) and the US Environmental Protection Agency (USEPA) established a drinking water arsenic limit of 0.10 parts per million (ppm) to safeguard consumers of public water systems. Many disorders, including black foot disease and a rare peripheral vascular disease, can be caused by an excess of arsenic ions. Arsenic contaminants can be eliminated from the ecosystem using a variety of techniques such as membrane filtration processes, electrokinetic processes, electrocoagulation, precipitation, reverse osmosis, phytoremediation, and adsorption. The adsorption method of removing arsenic has shown to be the most effective of all these techniques because of its ease of use, simplicity, cheap cost, and straightforward operating process, as well as the introduction of creative and useful materials, for example carbon nanotubes (CNTs), graphite oxides (GO), metal–organic frameworks, and some bio waste adsorbents. In this chapter, many aspects influencing the effectiveness of adsorption are highlighted along with the adsorptive removal of arsenic utilizing adsorption methods.

Keywords Arsenic · Electrocoagulation · Adsorption processes · Optimization · Wastewater

5.1 Introduction

Arsenic (As) is a metalloid that can occur in both organic and inorganic forms and is carcinogenic to humans (Weerasundara et al., 2021). Human health will be in danger if humans consume water contaminated with arsenic for a prolonged period

A. Kushwaha · Z. Arif (✉) · B. Singh
Department of Chemical Engineering, HBTU Kanpur, Kanpur, India
e-mail: darifIt@hbtu.ac.in

A. K. Yadav and S. Shirin (eds.), *Emerging and Innovative Arsenic Removal Technologies for a Sustainable Future*, Water Science and Technology Library 131,
https://doi.org/10.1007/978-3-032-01176-3_5

of time (Liu et al., 2020). Arsenic has the following oxidation states as arsenic (V), arsenic (III), arsenic (0), and arsenic (-III). Although arsenic (III) is more hazardous to humans and more reactive than As(V), nothing is known about how it adsorbs from water (Navarathna et al., 2019). Inorganic arsenic compounds are more harmful than organic ones. Globally, drinking water contaminated with arsenic poses a risk to more than 150 million people. Arsenic is found in water (naturally) owing to both natural geologic processes and human activity, including the mining industry, metallurgical, electronic, agricultural, pharmaceutical and chemical industry etc. (Yao & Yang, 2020). According to earlier research, people's skin, lungs, digestive systems, photosynthetic and respiratory systems, and neurological systems would all suffer negative consequences from prolonged exposure to water tainted with arsenic. According to WHO, the maximum limit of arsenic in drinking water is 0.01 mg/L due to its toxicity (Lata & Samadder, 2016).

The most common techniques for eliminating As from drinking water include membrane, coagulation & flocculation, electrocoagulation, reverse osmosis, and adsorption technologies. The adsorption technique's many benefits, including financial rewards, convenience of use, and a wide variety of substances that have been shown to operate as adsorbents and used arsenic removal from wastewater (Alka et al., 2021; Shirin & Yadav, 2014). Additional methods include phytobiological remediation, electrochemical treatments, ozone oxidation, and bioremediation. Nanofiltration membrane separation and Reverse osmosis technique both have exceptional percentages of arsenic removal, in particular, reverse osmosis can remove over 95% of arsenic (Rathi & Kumar, 2021). The current investigation examines the origins and causes of arsenic, as well as methods for determining its concentration and frequency. The main goals are to provide several arsenic removal techniques, in addition to techniques to determining the amount of arsenic present and tracking its current advancements in arsenic removal. The benefits and losses of several arsenic removal methods are as follows in Table 5.1 (Darban et al., 2022, Hansdah, 2016).

5.2 Removal of Arsenic from Wastewater Using Adsorption Technique

Adsorption has been shown to be a successful method for the removal of a variety of solutes from wastewater. Adsorption is a surface process that leads to the transfer of molecules, atoms, or ions from a fluid bulk to a solid surface. Here, a solid, liquid, or gaseous surface can exist (Verma et al., 2014).

- This method is dependent on the adsorption fundamental, which states that surface functional groups, chemical bonding, or electrostatic forces all work together to draw arsenic pollutants to the surface of the adsorbent.
- Specific adsorbents, such as activated carbon, iron oxides, alumina, or other modified materials made to possess a strong affinity for As species, are frequently utilized in practical applications of adsorption methods for As removal.

Table 5.1 Arsenic using different techniques: benefits and drawbacks

Methods	Advantages	Disadvantages
Coagulation and flocculation	• Low investment • Easy and Inexpensive	• Large amount of sludge produced • Difficulty of handling and disposal
Precipitation	• Simple and Inexpensive • Low cost	• Higher amount of sludge produced • Inadequate for treating low levels of heavy metal concentrations
Electrocoagulation	• No other chemical additions • Less sludge production	• High electricity consumption • Labor-intensive work
Reverse osmosis	• The energy requirement is very low • Water softening	• High Installation cost • Leads to the entire system being clogged
Adsorption	• Easy operation • Eco-friendly	• Inadequate selectivity • Waste product

- Overall, the adsorption technology for arsenic removal is a promising solution for both large-scale water treatment facilities and smaller, decentralized systems as it provides an efficient and successful way to treat arsenic-contaminated water sources (Rajendran et al., 2022).

5.2.1 *Types of Adsorptions*

(a) Physical adsorption: Molecules attach to the surface of a solid material by a process called physical adsorption, or physisorption, which is caused by intermolecular interactions like van der Waals forces. This type of adsorption occurs in the absence of a chemical link beneath the surface material, the adsorbent and the adsorbate (the substance being adsorbed).

(b) Chemical adsorption: Molecules attach to the surface of a solid substance by chemical bonding in a process called chemical adsorption, or chemisorption. Strong covalent or ionic bonds are formed between the adsorbate, or substance being adsorbed, and the adsorbent, or surface material, in the chemisorption process, in contrast to physical adsorption, which depends on Van der Waals forces (Khan & Ganai, 2020).

The chemical and physical adsorption process is shown in Fig. 5.1.

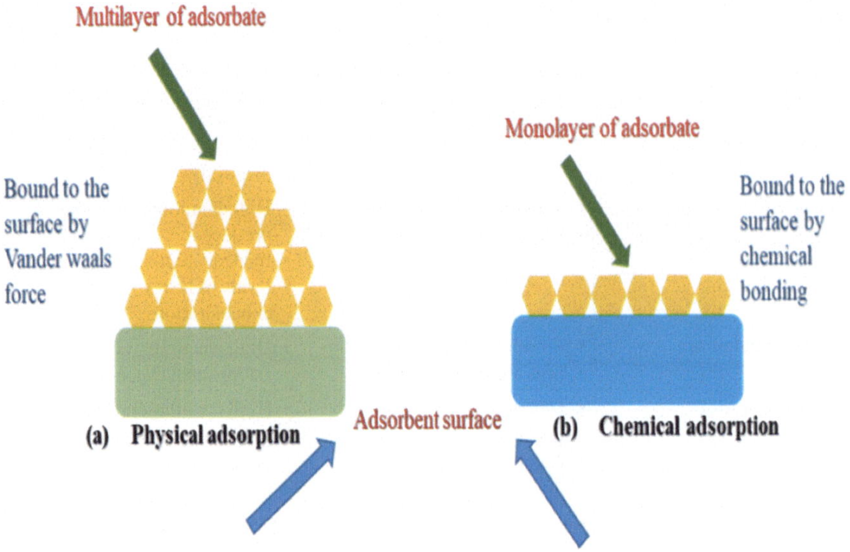

Fig. 5.1 Adsorption types **a** physical adsorption **b** chemical adsorption

5.2.2 Adsorbent

An "adsorbent" refers to a material or substance that adsorbs another substance onto its surface. There are two types of adsorbents, as shown in below.

1. Natural absorbents are substances derived from natural sources. For example, natural fibers, volcanic rocks, soils, activated carbon from Coconut Shells, Clay Minerals, Zeolites, Wood and plant-based adsorbents, chitosan and Tannins, agricultural and industrial wastes, etc. (Yeo et al., 2021).
2. Synthetic adsorbents are artificially produced materials designed specifically for adsorption applications. For example, Activated Carbon, Ion exchange resins, silica gel, polymeric absorbents, Molecular Sieves, Synthetic Zeolites etc. (Mohan & Pittman, 2007).

5.2.3 Arsenic Removal Using Low-Cost Adsorbents

Globally, arsenic-related issues are becoming more and more severe. (Yin et al., 2017). Arsenic is found mostly in groundwater, varying in concentration. Most ecological arsenic contamination originates from anthropogenic and volcanic activities, biological activity, geochemical interactions, natural weathering rocks and minerals and various human activities and it is widely known to have detrimental effects on both people and wildlife (da Silva et al., 2019). Arsenic is hazardous and does not serve useful metabolic purpose, it can lead to cancer, as well as diseases

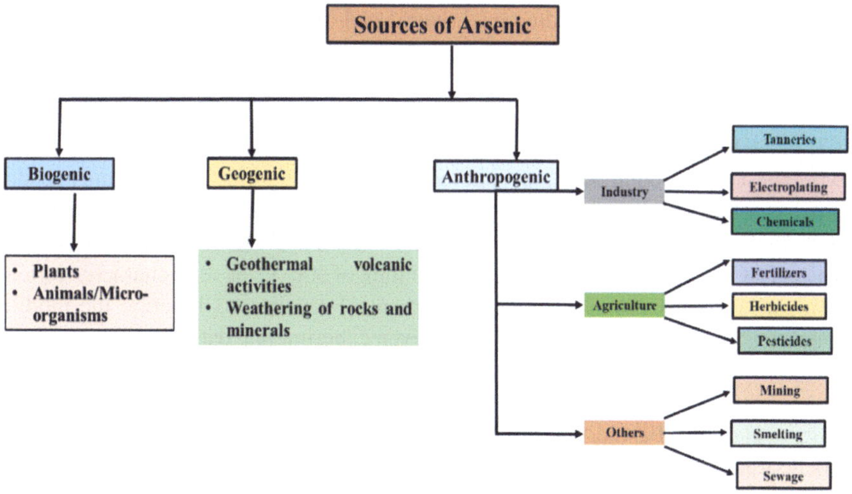

Fig. 5.2 Sources of arsenic contamination of both natural and anthropogenic

of the skin, circulatory system, and neurological system (Podgorski & Berg, 2020). Approximately 40 million people around the world are affected by arsenic exposure from drinking water that exceeds the maximum permitted limit of contamination is 0.01 mg L^{-1} (Sarkar & Paul, 2016). The sources of arsenic contamination of both natural and anthropogenic is shown in Fig. 5.2 (Altowayti et al., 2021). Thus, it became quite evident to treat wastewater contaminated with As before final discharge.

Two types of adsorbents used for As removal i.e. Conventional and Natural adsorbents.

Conventional adsorbents: used for As removal, such as commercial and synthetic activated carbons, zeolites, biochar and iron oxide-based sorbents.

Natural adsorbent: Natural adsorbents are materials derived from natural sources that have the ability to adsorb (or trap) substances from a fluid or gas phase onto their surfaces. These materials are often used in various applications where environmental sustainability and biodegradability are priorities. Here some examples of natural adsorbents, i.e., activated carbon from biomass, clay minerals, peat and Humic Substances, Chitosan, Natural Fibers, Fruit Peels and Agricultural Waste etc. (Asere et al., 2019).

Different authors in the literature reported the removal of As using different natural adsorbent material. Saif et al., (2021) reported an easy and environmentally beneficial method of producing copper oxide nanoparticles using *Phyllanthus emblica* leaf extract as fuel used for As adsorption. Nanoparticles were characterized by using several characterization techniques. The hydrodynamic size and zeta potential of the copper oxide nanoparticles were found to be 80 nm and 28.6 mV, respectively. According to Langmuir equation, the max. The sorption capacity of copper oxide for As(V) is 1.17 mg/g. In another study by Sahu et al. (2019), *Aegle marmelos* tree leaves, often known as the Indian bael, were used to create a new magnetic bioadsorbent that was employed to removal of As(V) from fluid solution. AMP@Fe_2O nanocomposite was produced by pyrolysis method and used in a batch adsorption procedure to remove As(V). Several experimental techniques, including XRD, FTIR, BET, FESEM, TEM, HRTEM etc. techniques were used to investigate the produced AMP@Fe_2O_3 nanocomposite and concluded that the Langmuir isotherm and pseudo I order kinetic models is more suitable for this system. Verma et al., 2019 reported that the remaining biomass of *Citrus limetta* was used to create two magnetic biosorbents of PAC-500 and PPAC-500, at a temperature of 500 °C. Several characterization techniques were used to describe the biosorbents. Furthermore, kinetics, thermodynamics, and isotherms were applied to the given experimental data. The capacities of As(III) sorption adsorbent of two magnetic adsorbents were 714.28 and 526.31 μg/g.

Gupta et al., (2019) reported an inexpensive and environmentally activated bioadsorbent of Jatropha seed cake (AC-JSC) prepared from Jatropha seed cake to remove As from contaminated water without any chemical modification. The performance efficiency was assessed by examining a range of experimental factors, such as dose of adsorbent material, beginning arsenic concentration, contact time, and medium pH. The arsenic content was successfully reduced to the acceptable level (WHO: 10 μg/L) by AC-JSC. Adsorption kinetics were described by the adsorption isotherm models, which were estimated as follows: Langmuir (b = 146 for As^V, and b = 96.1 for As^{III}) and Freundlich (Kf = 0.671, n = 0.350) for As^V and (Kf = 0.964, n = 0.250) for As^{III}. According to thermodynamic experiments carried out in the current system, As(V) and As(III) exhibit spontaneous adsorption (G: − 8.58 and − 8.60 Kilo Joule per mol). Several other natural adsorbents and their adsorption capacity are listed in Table 5.2.

5.3 Commercial and Synthetic Activated Carbons

Allen and Whitten in 1998 conducted an evaluation of the synthesis and properties of activated carbon from various carbonaceous sources. A lot of commercial activated carbon has been used to remove arsenic (III) and (V) from wastewater (Navarro & Alguacil, 2002). With this commercial carbon, an enormous arsenic adsorption capacity (2860 mg/g) was achieved. Arsenic treatment also made use of some activated carbons loaded with copper and metallic silver (Rajaković, 1992).

Synthetic activated carbons

Using carbonization, which involves slowly heating a substrate below 600 °C without the presence of air, activated carbons are created. Volatiles are eliminated in this way. Activation occurs either chemically or physically. The activation is finished by treatment with chemical activators, including KCNS, $ZnCl_2$, H_2PO_4, etc. or oxidizing agents, for example, steam, carbon dioxide, or oxygen at a high temperature (Mondal & Garg, 2017).

Table 5.2 List of different natural adsorbents and their adsorption capacity

S. No.	Natural adsorbent	Adsorption capacity (mg/g and %)	References
1	*Pongamia pinnata* microporous biochar	13.87 mg/g	Indurkar et al. (2023)
2	Pomelo peel	15.28 mg/g	Nguyen et al. (2022)
3	Eucalyptus barks	0.944 mg/g	De Almeida Ohana et al. (2022)
4	*Citrus limmeta* (peel and pulp)	2000 μg/g	Verma et al. (2019)
5	Magnetic Nanoparticles/onion peel (MNp-OP)	1.86 mg/g	Nikić et al. (2019)
6	Water melon rind	98%	Shakoor et al. (2018)
7	*Zizyphus jujuba* seeds	80–90%	Ziarati et al. (2019)
8	Perilla leaf biochar	97–100%	Niazi et al. (2018)
9	Tecomella undulate leaves	70.8–98.5%	Brahman et al. (2016)
10	Neem leaves/$MnFe_2O_4$ composite biosorbent	85.217 and 88.154%	Podder and Majumder (2016)

Zeolites

Alumino-silicates make up the bulk of the minerals that make up zeolite. The extraction of As(III) and (V) from fluid solutions has been accomplished using both natural and artificial zeolites. When the initial pH is in the alkaline zone, the addition of MnO_2 considerably improves the amphoteric characteristics, which reduces the equilibrium ph. Therefore, the removal effectiveness of As(V) of zeolite improves when treated with MnO_2 throughout a broad pH range (4.0–9.0) (Yu et al., 2018).

Iron-Based Adsorbents

Iron-based adsorbents are compounds that are specially made of iron and are used to adsorb or take out, particular chemicals from a solution or gas phase. When it is essential to remove pollutants or target compounds, these adsorbents can be used in a range of industrial and environmental applications (Cantu et al., 2016). Due to their abundance in Earth, great efficacy in arsenic remediation, and kindness towards the environment, iron-based adsorbents have garnered attention. The adsorption efficiencies of iron substances, for the removal of arsenic, are critically examined. Additionally, in an external magnetic field, saturated materials may be readily distinguished from water owing to the magnetic properties of particular iron-based adsorbents, for example Iron oxy-hydroxides, Iron oxy-hydroxides doped composite adsorbents, Nano-scale zero-valent iron (nZVI) adsorbents etc. (Podder & Majumder, 2016). Several types of iron-based nanoparticles for the removal of arsenic are listed in Table 5.3.

Table 5.3 List of several types of iron-based nanoparticle for Removal of arsenic

S. No.	Adsorbent	Synthesis technique	Adsorption capacity (mg/g and %)	References
1	Bimetallic Fe/Ni nanoparticles	Calcination	87.3%	Lin et al. (2022)
2	Cerium-iron nanoparticles	Hydrothermal	80%	Vences-Alvarez et al. (2022)
3	γ-Fe_2O_3@starch (starch based)	Co-precipitation	8.60 mg/g	Siddiqui et al. (2020)
4	Fe/Cu Nanoparticles	Chemical reduction	19.68 and 21.32 mg/g	Babaee et al. (2018)
5	FeOOH	Impregnation and coagulation	8.67 mg/g	Lin et al. (2018)
6	Fe_3O_4	Precipitation	8.25 mg/g	Chiavola et al. (2016)
7	Synthetic Fe_7S8 nanoparticles	Solvothermal	14.3 mg/g for As(III) and 31.3 mg/g for As(VI)	Cantu et al. (2016)
8	$Fe_3$04 nanoparticles	Hydrothermal	16.56 mg/g for arsenic (V), and 46.06 mg/g for arsenic (III)	Feng et al. (2012)

5.4 Parameter Effecting Performance of Arsenic Adsorption

pH: pH affects the adsorption method under several circumstances. It could significantly influence the adsorption process because it could change the charge of the surface of the adsorbent and As(V). The range of pH 2–11 was investigated to determine the optimum adsorption pH. According to the results, the adsorption capacity of As(V) stabilized at pH 2–4, decreased rapidly at pH 4–5, and decreased slowly at pH > 5. When the aqueous was in strong acid, the maximum adsorption capacity was 45.8 mg/g. The capacity decreased with increasing pH value, even down to 13.9 mg/g (Zhou et al., 2017). With a rise in pH (2–11), arsenic adsorption increased. Maximum arsenic adsorption (93–98%) was observed at slightly alkaline pH (8) (Sarwar et al., 2015).

Concentration of arsenic: The amount of arsenic that has been eliminated decreases as the starting concentration increases. The percentage will gradually decline as the arsenic content rises due to increased competition for active adsorption sites (Neisan et al., 2023). The inaccessible active sites to handle the significantly higher amount of arsenic available in the solution was the reason for an increase in arsenic absorption but a decrease in the percentage of adsorption (Lekić et al., 2013).

Adsorbent dosage: The absorbent is important in the arsenic removal adsorption process. As the dosage of the adsorbent was increased, so was the elimination of arsenic. More surface area adsorption sites are available, leading to an increasing adsorption with an increase in the adsorbent dose. (Babaee et al., 2018).

Temperature: In various temperature ranges, the effect on the elimination of As(III) has been examined. With increasing temperature, the % of Arsenic (III) removal increased. The increased adsorption capacity with increasing temperature is often referred to as an endothermic process, while a reduction in adsorption capacity with increasing solution temperature corresponds to an exothermic process (Afroze & Sen, 2018).

Revolution per minute (RPM): The proportion of As(V) elimination appears to have increased with decreasing stirring rate, according to research on the effect of different stirring rates at varied speeds. The surfaces of the adsorbate and biosorbents did not given enough time to interact, and an increasing stirring rate also caused loosely bounded adsorbates (Roy et al., 2017). On the other hand, adsorption rose once again when the rotor speed was increased (Hashim et al., 2019).

5.5 Adsorption Performances

5.5.1 Adsorption Kinetics

Adsorption kinetics research provides an explanation of the duration of adsorbate uptake, which controls the length of time that adsorbate uptake remains at the interface of the solid solution. Three types of adsorption kinetics are as follows, i.e. pseudo 1st order, pseudo 2nd order, and intraparticle diffusion.

Pseudo-first-order: The pseudo 1st order equation, presented by Lagergren in 1898, characterizes the adsorption of liquid–solid solutions based on solid capability (Rodríguez-Romero et al., 2020).

In Eq. 5.1, pseudo-first-order kinetics was defined by Lagergren.

$$\frac{da_e}{dt} = k_1(a_e - a_t) \tag{5.1}$$

Integration of the above equation yields the following solution:

$$\ln(a_e - a_t) = \ln(a_e - k_1 T) \tag{5.2}$$

where

a_e—(in equilibrium) arsenic adsorbed,
a_t—at time (t) arsenic adsorbed,
k_1—adsorption rate constant.

Pseudo-second-order kinetics: According to Ho and Mckay in 1988 introduced the second order kinetic model, chemisorption is the method used to adsorb the adsorbate (Bhan et al., 2021).

In Eq. 5.3, the pseudo-second-order equation are as follows.

$$\frac{da_t}{dt} = k_2(a_e - a_t)^2 \tag{5.3}$$

The equilibrium constant, k_2, is measured in g/mg min. After rearranging and integrating the previous formula

$$\frac{t}{a_t} = \frac{1}{k_2 a_e^2} + \frac{t}{a_e} \tag{5.4}$$

Intraparticle diffusion model:

The intraparticle diffusion (Weber-Morris) model was used to determine the kinetic data to ascertain whether intraparticle diffusion is a rate-controlling step in adsorption. The formulation of Weber and Morris' intraparticle diffusion model is as follows in Eq. 5.5:

$$C_t = k_i t^{1/2} + K \tag{5.5}$$

where

k_i—intraparticle diffusion rate constant,
K—thickness of the boundary layer (Sahu et al., 2021).

Kinetics applied in removal of arsenic from Wastewater: López-Luna et al. (2019) reported using manganese ferrite nanoparticles for arsenic removal and showed best fit of Pseudo II order kinetics. Mukherjee et al., (2021) stated that coal combustion was used for arsenic removal and followed pseudo–second order kinetics. Singh et al., (2020) reported using plastic waste char for efficient removal of arsenic and Pseudo-Second-Order rate effectively fitted well. Navarathna et al., (2019) using Douglas fir biochar for arsenic removal and showed pseudo second order model more suitable.

5.5.2 *Adsorption Isotherm*

Temperature is a sensitive factor that influences the adsorption process and the sorption capacity. Adsorption isotherms at constant temperature show how the adsorbent and adsorbates are in equilibrium.

Langmuir model: Irving Langmuir presented the Langmuir adsorption model in 1932, based on the following assumptions.

- The adsorption by which molecules adhere to a solid surface is described by the Langmuir model, a concept relevant to both surface chemistry and adsorption research.

- This model predicts that maximal surface coverage will occur at higher solute concentrations and linear adsorption at lower adsorption densities.
- The model is characterized by a monolayer adsorption capacity and an equilibrium constant relating to the adsorption and desorption rates.
- This model is widely used to understand and predict the behavior of gases and liquids interacting with solid surfaces in various industrial and scientific applications.

The Langmuir isotherm is expressed as follows in Eq. 5.6.

$$q_e = \frac{q_m K_L C_e}{1 + K_L C_e} \tag{5.6}$$

where

C_e (mg L^{-1})—equilibrium arsenic conc.,
q_e (mg/g)—amount of sorption,
q_m (mg g^{-1})—Adsorption capacity of substances.
K_L (L mg^{-1})—Langmuir constant (Rahdar et al., 2019).

Furthermore, this model is more versatile and easier to employ with computational simulations.

Freundlich adsorption isotherm model:

The Freundlich model is used to explain adsorption processes, particularly for heterogeneous surfaces where adsorbate molecules adhere to the surface of an adsorbent. It is named after the German chemist Herbert Freundlich, who proposed this model in 1909.

In their following curves Eq. 5.7 as a straight line.

$$\frac{x}{m} = K_p^{1/n} \tag{5.7}$$

$$\log\frac{x}{m} = \log k + \frac{1}{n}\log P \tag{5.8}$$

The Freundlich equation is typically expressed as:

$$q_e = K_e . C_e^{\frac{1}{n}} \tag{5.9}$$

where

1/n—heterogeneity factor,
K_F (mg/g)—rate constant (Cadaval et al., 2015; Grover et al., 2012; Wang et al., 2018).

$$q_e = \frac{q_m a_S C_e^{1/n}}{1 + a_S C_e^{1/n}} \tag{5.10}$$

Here

a_s—Sips constant (sorption energy (L/mg).

Temkin isotherm

Theoretically, adsorption on solid surfaces may be described by the Temkin isotherm model, especially in terms of physical adsorption. It is assumed that as the surface becomes more covered, the heat of molecule adsorption decreases linearly with coverage as the connections between the adsorbate and the adsorbent surface diminish.

Here are the key points of the Temkin isotherm model:

1. **Assumption of Linearity**: According to the model, the adsorption heat drops linearly as the adsorbate's surface coverage increases.
2. **Equation**: The Temkin equation is described as follows in Eq. 5.11 (Tanzifi, 2016).

$$q_e = B \ln(K_T) + B \ln C_e \tag{5.11}$$

where

KT—Temkin constant,
q_e—amount (at equilibrium).
C_e—Concentration (at equilibrium).
B—Constant (adsorption heat).

Isotherm applied in the removal of arsenic from Wastewater: Khan and Imteaz (2021) reported using different natural adsorbents for arsenic removal and showed the best fit of the Langmuir isotherm model. In their article, Chen et al., (2021) reported using waste wood generated from magnetic biochar for arsenic removal and showed best fit of the Langmuir isotherm model. Siddiqui et al., (2020) reported using starch-functionalized maghemite nanoadsorbents for water purification of As and showed the best fit of Freundlich isotherm model. Rahdar et al., (2019) reported using modified saxaul ash for arsenic adsorption from fluid solution and the most suitable Freundlich adsorption isotherm.

5.6 Conclusion and Future Scope

The goal of this chapter is to cover as many adsorbents and treatment techniques as possible that have been utilized to eliminate arsenic from wastewater. Arsenic water contamination is a big issue that exists globally. In the removal of last several years, the arsenic from fluid solutions has become more significant in the fight against water pollution. The application of inexpensive adsorbents has drawn a great deal of interest. To efficiently remove arsenic from water streams with high amounts of arsenic, a variety of methods are available, i.e. adsorption, precipitation, electrocoagulation, ion exchange, and reverse osmosis methods. With its straightforward design, low cost, and simple operating process, the adsorption technique is a very appealing option for future arsenic removal. In their, common examples of conventional adsorbents are biochar, activated charcoal, and iron-based sorbents; natural adsorbents include soils, clays, hydroxyl apatite, industrial and agricultural wastes, and bio sorbents used to treat arsenic. In summary, it should be highlighted that methods that do not lead to additional environmental issues use less energy, are more affordable, and are considered sustainable recycling technologies. These methods also require further research to become a more efficient means of reducing arsenic contamination. In ideal circumstances, the most effective methods can remove arsenic at a rate of more than 100%.

The potential of advanced novel functional materials such as graphite oxides, metal organic frameworks (porous materials linked with metal ion and an organic linker to form coordinated bonds), carbon nanotubes, and other nano- adsorbent materials are much better alternatives for the future development of arsenic removal technologies. At the same time, effects and threats implicated in the use of nanoadsorbents on the environment should also be committed. Although field trials are more costly and prone to failure, more field experience is necessary.

Acknowledgements The authors express my gratitude to the HBTU Chemical Engineering Department for producing the available Web of Science journal access functionality.

References

Afroze, S., & Sen, T. K. (2018). A review on heavy metal ions and dye adsorption from water by agricultural solid waste adsorbents. *Water, Air, and Soil Pollution, 229*, 1–50.

Alka, S., Shahir, S., Ibrahim, N., Ndejiko, M. J., Vo, D. V. N., & Abd Manan, F. (2021). Arsenic removal technologies and future trends: A mini review. *Journal of Cleaner Production, 278*, Article 123805.

Altowayti, W. A. H., Othman, N., Shahir, S., Alshalif, A. F., Al-Gheethi, A. A., Al-Towayti, F. A. H., Saleh, Z. M., & Haris, S. A. (2021). Removal of arsenic from wastewater by using different technologies and adsorbents: A review. *International Journal of Environmental Science and Technology*, 1–24.

Asere, T. G., Stevens, C. V., & Du Laing, G. (2019). Use of (modified) natural adsorbents for arsenic remediation: A review. *Science of the Total Environment, 676*, 706–720.

Babaee, Y., Mulligan, C. N., & Rahaman, M. S. (2018). Removal of arsenic (III) and arsenic (V) from aqueous solutions through adsorption by Fe/Cu nanoparticles. *Journal of Chemical Technology & Biotechnology, 93*(1), 63–71.

Bhan, C., Singh, J., & Sharma, Y. C. (2021). Development of adsorbent from Mentha plant ash and its application in fluoride adsorption from aqueous solution: A mechanism, isotherm, thermodynamic, and kinetics studies. *International Journal of Phytoremediation, 23*(11), 1113–1123.

Brahman, K. D., Kazi, T. G., Afridi, H. I., Baig, J. A., Abro, M. I., Arain, S. S., Ali, J., & Khan, S. (2016). Simultaneously removal of inorganic arsenic species from stored rainwater in arsenic endemic area by leaves of Tecomella undulata: A multivariate study. *Environmental Science and Pollution Research, 23*, 15149–15163.

Cadaval, T. R., Dotto, G. L., & Pinto, L. A. (2015). Equilibrium isotherms, thermodynamics, and kinetic studies for the adsorption of food azo dyes onto chitosan films. *Chemical Engineering Communications, 202*(10), 1316–1323.

Cantu, J., Gonzalez, L. E., Goodship, J., Contreras, M., Joseph, M., Garza, C., Eubanks, T. M., & Parsons, J. G. (2016). Removal of arsenic from water using synthetic Fe_7S_8 nanoparticles. *Chemical Engineering Journal, 290*, 428–437.

Chen, C. K., Chen, J. J., Nguyen, N. T., Le, T. T., Nguyen, N. C., & Chang, C. T. (2021). Specifically designed magnetic biochar from waste wood for arsenic removal. *Sustainable Environment Research, 31*, 1–16.

Chiavola, A., Amato, E. D., Stoller, M., Chianese, A., & Boni, M. R. (2016). Application of iron based nanoparticles as adsorbents for arsenic removal from water. *Chemical Engineering Transactions, 47*, 325–330.

da Silva, E. B., Mussoline, W. A., Wilkie, A. C., & Ma, L. Q. (2019). Anaerobic digestion to reduce biomass and remove arsenic from As-hyperaccumulator Pteris vittata. *Environmental Pollution, 250*, 23–28.

Darban, Z., Shahabuddin, S., Gaur, R., Ahmad, I., & Sridewi, N. (2022). Hydrogel-based adsorbent material for the effective removal of heavy metals from wastewater: A comprehensive review. *Gels, 8*(5), 263.

De Almeida Ohana, N., HM, L. F., & Luis, N. G. (2022). Adsorption of arsenic anions in water using modified lignocellulosic adsorbents. *Results in Engineering, 13*, 100340.

Feng, L., Cao, M., Ma, X., Zhu, Y., & Hu, C. (2012). Superparamagnetic high-surface-area Fe_3O_4 nanoparticles as adsorbents for arsenic removal. *Journal of Hazardous Materials, 217*, 439–446.

Grover, V. A., Hu, J., Engates, K. E., & Sipley, H. J. (2012). Adsorption and desorption of bivalent metals to hematite nanoparticles. *Environmental Toxicology and Chemistry, 31*, 86–92.

Gupta, A. R., Joshi, V. C., & Sharma, S. (2019). Remediation of arsenic from water by utilizing waste of Jatropha seed cake as low cost and ecofriendly bio-adsorbent. *International Journal of Scientific and Engineering Research, 10*, 1053–1077.

Hansdah, B. (2016). *Adsorption of methylene blue and phenol on activated carbon prepared from Fox Nutshell by K_2CO_3 activator* [Doctoral dissertation].

Hashim, M. A., Kundu, A., Mukherjee, S., Ng, Y. S., Mukhopadhyay, S., Redzwan, G., & Gupta, B. S. (2019). Arsenic removal by adsorption on activated carbon in a rotating packed bed. *Journal of Water Process Engineering, 30*, Article 100591.

Indurkar, P. D., Gupta, A. R., & Sharma, S. (2023). Iron oxide impregnated Pongamia pinnata microporous biochar composite as an effective bio-adsorbent for the remediation of arsenic ions from water. *Biomass Conversion and Biorefinery*, 1–18.

Khan, A. M., & Ganai, S. A. (2020). Removal and recovery of heavy metal ions using natural adsorbents. In *Modern age waste water problems: Solutions using applied nanotechnology* (pp. 251–260).

Khan, S. A., & Imteaz, M. A. (2021). Experimental studies on arsenic removal efficiencies through adsorption using different natural adsorbents. *Water, Air, & Soil Pollution, 232*, 1–10.

Lata, S., & Samadder, S. R. (2016). Removal of arsenic from water using nano adsorbents and challenges: A review. *Journal of Environmental Management, 166*, 387–406.

Lekić, B. M., Marković, D. D., Rajaković-Ognjanović, V. N., Đukić, A. R., & Rajaković, L. V. (2013). Arsenic removal from water using industrial by-products. *Journal of Chemistry, 2013*(1), Article 121024.

Lin, S., Yang, H., Na, Z., & Lin, K. (2018). A novel biodegradable arsenic adsorbent by immobilization of iron oxyhydroxide (FeOOH) on the root powder of long-root Eichhornia crassipes. *Chemosphere, 192*, 258–266.

Lin, Y., Jin, X., Khan, N. I., Owens, G., & Chen, Z. (2022). Bimetallic Fe/Ni nanoparticles derived from green synthesis for the removal of arsenic (V) in mine wastewater. *Journal of Environmental Management, 301*, Article 113838.

Liu, B., Kim, K. H., Kumar, V., & Kim, S. (2020). A review of functional sorbents for adsorptive removal of arsenic ions in aqueous systems. *Journal of Hazardous Materials, 388*, Article 121815.

López-Luna, J., Ramírez-Montes, L. E., Martinez-Vargas, S., Martínez, A. I., Mijangos-Ricardez, O. F., González-Chávez, M. D. C. A., Carrillo-González, R., Solís-Domínguez, F. A., Cuevas-Díaz, M. D. C., & Vázquez-Hipólito, V. (2019). Linear and nonlinear kinetic and isotherm adsorption models for arsenic removal by manganese ferrite nanoparticles. *SN Applied Sciences, 1*, 1–19.

Mohan, D., & Pittman, C. U., Jr. (2007). Arsenic removal from water/wastewater using adsorbents—A critical review. *Journal of Hazardous Materials, 142*(1–2), 1–53.

Mondal, M. K., & Garg, R. (2017). A comprehensive review on removal of arsenic using activated carbon prepared from easily available waste materials. *Environmental Science and Pollution Research, 24*, 13295–13306.

Mukherjee, S., Thakur, A. K., Goswami, R., Mazumder, P., Taki, K., Vithanage, M., & Kumar, M. (2021). Efficacy of agricultural waste derived biochar for arsenic removal: Tackling water quality in the Indo-Gangetic plain. *Journal of Environmental Management, 281*, Article 111814.

Navarathna, C. M., Karunanayake, A. G., Gunatilake, S. R., Pittman, C. U., Jr., Perez, F., Mohan, D., & Mlsna, T. (2019). Removal of Arsenic (III) from water using magnetite precipitated onto Douglas fir biochar. *Journal of Environmental Management, 250*, Article 109429.

Navarro, P., & Alguacil, F. J. (2002). Adsorption of antimony and arsenic from a copper electrorefining solution onto activated carbon. *Hydrometallurgy, 66*(1–3), 101–105.

Neisan, R. S., Saady, N. M. C., Bazan, C., Zendehboudi, S., Al-nayili, A., Abbassi, B., & Chatterjee, P. (2023). Arsenic removal by adsorbents from water for small communities' decentralized systems: Performance, characterization, and effective parameters. *Clean Technologies, 5*(1), 352–402.

Nguyen, T. H., Loganathan, P., Nguyen, T. V., Vigneswaran, S., Nguyen, T. H. H., Tran, H. N., & Nguyen, Q. B. (2022). Arsenic removal by pomelo peel biochar coated with iron. *Chemical Engineering Research and Design, 186*, 252–265.

Niazi, N. K., Bibi, I., Shahid, M., Ok, Y. S., Burton, E. D., Wang, H., Shaheen, S. M., Rinklebe, J., & Lüttge, A. (2018). Arsenic removal by perilla leaf biochar in aqueous solutions and groundwater: An integrated spectroscopic and microscopic examination. *Environmental Pollution, 232*, 31–41.

Nikić, J., Tubić, A., Watson, M., Maletić, S., Šolić, M., Majkić, T., & Agbaba, J. (2019). Arsenic removal from water by green synthesized magnetic nanoparticles. *Water, 11*(12), 2520.

Podder, M. S., & Majumder, C. B. (2016). Application of granular activated carbon/$MnFe_2O_4$ composite immobilized on C. glutamicum MTCC 2745 to remove As(III) and As(V): Kinetic, mechanistic and thermodynamic studies. *Spectrochimica Acta Part a: Molecular and Biomolecular Spectroscopy, 153*, 298–314.

Podgorski, J., & Berg, M. (2020). Global threat of arsenic in groundwater. *Science, 368*(6493), 845–850.

Rahdar, S., Taghavi, M., Khaksefidi, R., & Ahmadi, S. (2019). Adsorption of arsenic (V) from aqueous solution using modified saxaul ash: Isotherm and thermodynamic study. *Applied Water Science, 9*, 1–9.

Rajaković, L. V. (1992). The sorption of arsenic onto activated carbon impregnated with metallic silver and copper. *Separation Science and Technology, 27*(11), 1423–1433.

Rajendran, S., Priya, A. K., Kumar, P. S., Hoang, T. K., Sekar, K., Chong, K. Y., Khoo, K. S., Ng, H. S., & Show, P. L. (2022). A critical and recent developments on adsorption technique for removal of heavy metals from wastewater-A review. *Chemosphere, 303*, Article 135146.

Rathi, B. S., & Kumar, P. S. (2021). A review on sources, identification and treatment strategies for the removal of toxic Arsenic from water system. *Journal of Hazardous Materials, 418*, Article 126299.

Rodríguez-Romero, J. A., Mendoza-Castillo, D. I., Reynel-Ávila, H. E., de Haro-Del Rio, D. A., González-Rodríguez, L. M., Bonilla-Petriciolet, A., Duran-Valle, C. J., & Camacho-Aguilar, K. I. (2020). Preparation of a new adsorbent for the removal of arsenic and its simulation with artificial neural network-based adsorption models. *Journal of Environmental Chemical Engineering, 8*(4), Article 103928.

Roy, P., Dey, U., Chattoraj, S., Mukhopadhyay, D., & Mondal, N. K. (2017). Modeling of the adsorptive removal of arsenic (III) using plant biomass: a bioremedial approach. *Applied Water Science, 7*, 1307–1321.

Sahu, N., Bhan, C., & Singh, J. (2021). Removal of fluoride from an aqueous solution by batch and column process using activated carbon derived from iron infused Pisum sativum peel: Characterization, Isotherm, kinetics study. *Environmental Engineering Research, 26*(4).

Sahu, U. K., Sahu, S., Mahapatra, S. S., & Patel, R. K. (2019). Synthesis and characterization of magnetic bio-adsorbent developed from Aegle marmelos leaves for removal of As(V) from aqueous solutions. *Environmental Science and Pollution Research, 26*, 946–958.

Saif, S., Adil, S. F., Khan, M., Hatshan, M. R., Khan, M., & Bashir, F. (2021). Adsorption studies of arsenic (V) by CuO nanoparticles synthesized by Phyllanthus emblica leaf-extract-fueled solution combustion synthesis. *Sustainability, 13*(4), 2017.

Sarkar, A., & Paul, B. (2016). The global menace of arsenic and its conventional remediation-A critical review. *Chemosphere, 158*, 37–49.

Sarwar, A., Mahmood, Q., Bilal, M., Bhatti, Z. A., Pervez, A., Saqib, A. N. S., Khan, A. R., & Sultan, S. (2015). Investigation on Melia azedarach biomass for arsenic remediation from contaminated water. *Desalination and Water Treatment, 53*(6), 1632–1640.

Shakoor, M. B., Niazi, N. K., Bibi, I., Shahid, M., Sharif, F., Bashir, S., Shaheen, S. M., Wang, H., Tsang, D. C., Ok, Y. S., & Rinklebe, J. (2018). Arsenic removal by natural and chemically modified water melon rind in aqueous solutions and groundwater. *Science of the Total Environment, 645*, 1444–1455.

Shirin, S., & Yadav, A. K. (2014). Physico-chemical analysis of municipal wastewater discharge in Ganga River, Haridwar district of Uttarakhand, India. *Current World Environment, 9*(2), 536–543. https://doi.org/10.12944/CWE.9.2.39

Siddiqui, S. I., Singh, P. N., Tara, N., Pal, S., Chaudhry, S. A., & Sinha, I. (2020). Arsenic removal from water by starch functionalized maghemite nano-adsorbents: Thermodynamics and kinetics investigations. *Colloid and Interface Science Communications, 36*, Article 100263.

Singh, E., Kumar, A., Khapre, A., Saikia, P., Shukla, S. K., & Kumar, S. (2020). Efficient removal of arsenic using plastic waste char: Prevailing mechanism and sorption performance. *Journal of Water Process Engineering, 33*, Article 101095.

Tanzifi, M. (2016). Removal of Congo Red anionic dye from aqueous solution using polyaniline/TiO_2 and polypyrrole/TiO_2 nanocomposites: Isotherm, kinetic, and thermodynamic studies. *International Journal of Engineering, 29*(12), 1659–1669.

Vences-Alvarez, E., Chazaro-Ruiz, L. F., & Rangel-Mendez, J. R. (2022). New bimetallic adsorbent material based on cerium-iron nanoparticles highly selective and affine for arsenic (V). *Chemosphere, 297*, Article 134177.

Verma, L., Siddique, M. A., Singh, J., & Bharagava, R. N. (2019). As(III) and As(V) removal by using iron impregnated biosorbents derived from waste biomass of Citrus limmeta (peel and pulp) from the aqueous solution and ground water. *Journal of Environmental Management, 250*, Article 109452.

Verma, P., Agarwal, A., & Singh, V. K. (2014). Arsenic removal from water through adsorption-A Review. *Recent Research in Science and Technology, 6*(1), 541.

Wang, Y., Zhang, Y., Li, S. Y., Zhong, W. H., & Wei, W. (2018). Enhanced methylene blue adsorption onto activated reed-derived biochar by tannic acid. *Journal of Molecular Liquid, 268*, 658–666.

Weerasundara, L., Ok, Y. S., & Bundschuh, J. (2021). Selective removal of arsenic in water: A critical review. *Environmental Pollution, 268*, Article 115668.

Yao, R., & Yang, H. (2020). An overview of As(V) removal from water by adsorption technology. *Annals of Musculoskeletal Medicine, 4*(1), 015–020.

Yeo, K. F. H., Li, C., Zhang, H., Chen, J., Wang, W., & Dong, Y. (2021). Arsenic rliliemoval from contaminated water using natural adsorbents: A review. *Coatings, 11*(11), 1407.

Yin, H., Kong, M., Gu, X., & Chen, H. (2017). Removal of arsenic from water by porous charred granulated attapulgite-supported hydrated iron oxide in bath and column modes. *Journal of Cleaner Production, 166*, 88–97.

Yu, Y., Yu, L., Koh, K. Y., Wang, C., & Chen, J. P. (2018). Rare-earth metal based adsorbents for effective removal of arsenic from water: A critical review. *Critical Reviews in Environmental Science and Technology, 48*(22–24), 1127–1164.

Zhou, Z., Liu, Y. G., Liu, S. B., Liu, H. Y., Zeng, G. M., Tan, X. F., Yang, C. P., Ding, Y., Yan, Z. L., & Cai, X. X. (2017). Sorption performance and mechanisms of arsenic (V) removal by magnetic gelatin-modified biochar. *Chemical Engineering Journal, 314*, 223–231.

Ziarati, P., Far, B. F., Mashayekhi, E., & Sawicka, B. (2019). Removing arsenic by food-processing waste (Zizyphus jujuba seeds) and study on its adsorptive properties.

Chapter 6
Role of Adsorption on Arsenic Contamination in Groundwater

Rijin M. D. Tom, Thomas J. Kallupurakel, Neethu Shajan, and Vishnu Manirethan

Abstract Metalloid arsenic is one of the most infamous groundwater contaminants that puts millions of lives at risk every day. Anthropogenic factors such as mining and mineral processing, as well as natural ones such as leeching and rock weathering, all add to this rapidly deteriorating situation worldwide. Continued or acute exposure to such sources and doses of inorganic arsenic can cause a myriad of ailments, including but not limited to cardiac dysrhythmia, bloody diarrhea, hyper and hypo pigmentation, encephalopathy, and cancers, all of which fall under the umbrella term of arsenicosis, which eventually results death. The WHO suggests that the maximum amount of arsenic in potable water be limited to 10 μg/L because of these factors. Although it is a global issue, the least fortunate populations of the world are disproportionately affected by the problem of arsenic contamination in groundwater. In light of this, adsorption can be a very effective, rapidly deployable, and affordable solution that can be incorporated into both current water treatment systems and newly planned ones. This paper reviews the various such adsorption materials and methods that have been developed over the past decade. Adsorbents that incorporate oxides of metals like titanium and iron in various forms have shown a high affinity for inorganic arsenic species. Various zeolite-based adsorbents have also been developed that display promising results. With their adjustable surface characteristics and functional groups that may be designed for specific arsenic adsorption, modified organic adsorbents such as biochar offer a sustainable alternative. The key focus of this chapter is on the adsorption mechanisms, material synthesis, and performance evaluation of different adsorbents. Desorption and regeneration of the spent adsorbents will also be covered.

Keywords Arsenic adsorbents · Arsenic contamination · Inorganic arsenic · Water treatment

R. M. D. Tom · T. J. Kallupurakel · V. Manirethan (✉)
Department of Chemical Engineering, Amal Jyothi College of Engineering, Kanjirappally, India
e-mail: vishnumvishnu@gmail.com

N. Shajan
Chemical and Bioprocess Engineering, Hamburg University of Technology, Hamburg, Germany

A. K. Yadav and S. Shirin (eds.), *Emerging and Innovative Arsenic Removal Technologies for a Sustainable Future*, Water Science and Technology Library 131,
https://doi.org/10.1007/978-3-032-01176-3_6

6.1 Introduction

Arsenic, the twentieth most commonly occurring element in the Earth's crust, is notorious for its toxicity. Various arsenic minerals, including realgar, orpiment, and arsenopyrite, have been identified in different geological formations. These mineral rocks can erode and weather, releasing arsenic into the soil, water, and air. When minerals containing arsenic, such as arsenopyrite, weather, arsenic finds its way into ground water and surrounding environments. Volcanic activity is one of the most important natural sources of air pollution. During and between volcanic eruptions, many contaminants, including arsenic compounds, are released into the atmosphere. Arsenic concentrations rise as a result of these airborne materials being deposited on land and water. Arsenic contamination is largely caused by human activities such as industrial operations, waste disposals, and modern farming methods. Arsenic is often discovered as arsenopyrite (FeAsS) during gold mining. Arsenopyrite is released in the form of dust, contaminating air, land, and water. The wastewater from smelters and gold mines can also contribute to arsenic pollution. 300–1600 mg/L of arsenic is typical concentration of arsenic in such industries (Langsch et al., 2012).

A myriad of experiments and studies have focused on the mitigation of arsenic contamination of water. The main methods include membrane filtration, biological treatments, ion exchange, enhanced oxidation, flocculation and coagulation, adsorption, and constructed wetlands. The four main membrane filtration techniques for arsenic contamination are reverse osmosis (RO), nanofiltration (NF), ultrafiltration (UF), and microfiltration (MF). Biological treatments for the mitigation of arsenic have shown promising results. In one study, rice seedlings were hydroponically cultivated using oxygen nanobubbles (ONB). It was discovered that arsenic oxidizes and that ONB rupture produces hydroxyl radicals. This process increased iron plaque, which retained As(V), because arsenic and iron have a strong affinity for each other. As a result, arsenic uptake was inhibited and the arsenic content in rice roots and shoots decreased by 12.5% and 46.4%, respectively. All things considered, ONB reduced rice's vulnerability to arsenic toxicity, indicating potential applications in the future (Huang et al., 2023).

Several studies have demonstrated the efficacy of plant growth-promoting microorganisms (PGPM) in enhancing plant tolerance and development while lowering arsenic accumulation in plants. Several pollutants, including arsenic, can be effectively removed by constructed wetlands (CW), which are low-cost, natural systems. Constructed wetlands are deep, shallow basins with long hydraulic retention times (HRTs) that are covered in a waterproof sheath and seeded with macrophytes. The basins are then filled with sand, gravel and soil. An integrated strategy that takes into account the biological, physical and chemical interactions between plants, substrata, and soil is used to remediate wastewater in CWs. The highest percentage of arsenic removed (99%) was found in limestone wetland columns, followed by zeolite columns (92%) and gravel columns (43%). Under favorable environmental conditions, bacteria can also potentially mediate these processes and have a significant impact. Coagulation and flocculation are two established methods of treating water polluted with arsenic. Coagulants such as ferric sulfate, calcium hydroxide and polyaluminium chlorides (PACs) have shown good efficacy in removing arsenic, precipitated as insoluble compounds. Chemicals added during these processes help aggregate formations and further remove arsenic. The optimum pH range for these processes was found to be between 5 and 7 for better removal efficiency (Carotenuto et al., 2014). Another prospective technique for removing arsenic from water bodies can be found in adsorption. Conventional adsorbents include activated carbon and charcoal. Arsenic-contaminated water can be treated with modified adsorbents such as $FeCl_3$, poly-Fe-based, Fe-doped organic materials, and granular activated carbon (GAC).

6.2 Principles of Adsorption Technology

Adsorption describes the mass transfer operation that governs the deposition of materials at the interface between two phases. The substance adsorbed is known as the adsorbate, while the material that holds the adsorbate is called the adsorbent (Fig. 6.1). The two types of adsorption processes are physisorption and chemisorption. Physisorption, or physical adsorption, occurs when gas or liquid molecules attach themselves to a solid surface by weak van der Waals forces (intermolecular forces). This process is exothermic in nature and is preferred at low temperatures. As a result of the weak forces involved, physisorption can exhibit reversible behaviour. In contrast, chemisorption, or chemical adsorption, involves the chemical bonding of gas or liquid molecules with a solid surface. During this process, there is a strong interaction between the adsorbent and the adsorbate, often resembling covalent or ionic bonds. Adsorption technology is crucial to numerous industrial, environmental, and medical applications due to its efficacy and versatility.

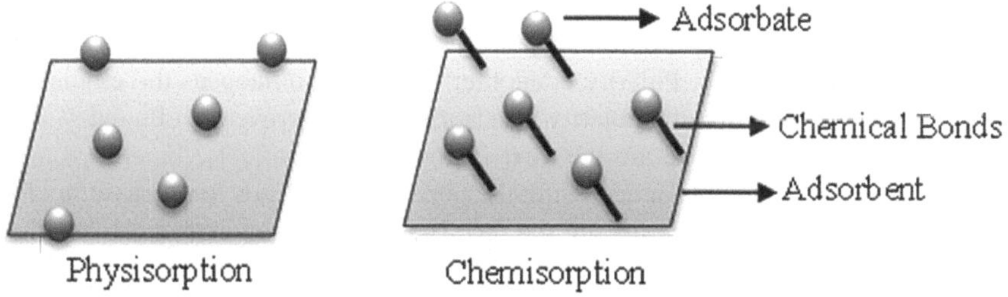

Fig. 6.1 Physisorption and chemisorption mechanism

6.3 Types of Absorbents

There are multiple classifications of adsorbents. On the basis of their source, they are classified as natural and synthetic adsorbents. Zeolite and clay are two well-known natural adsorbents. Some of the synthetic adsorbents are metal organic frameworks (MOFs), and synthetic adsorbents are designed and fabricated for special purposes. Adsorbents can be segregated as microporous, mesoporous, and macroporous on their pore size. If the pores present in the adsorbent have a size less than 2 nm, it is termed as a microporous structure. Similarly, if it is between 2 and 50 nm, it is termed as mesoporous types, and if it is greater than 50 nm, it is termed as macroporous type adsorbents. On the basis of the chemical nature, adsorbents are classified as carbonaceous, inorganic, and polymeric types. Various carbonaceous materials serve as effective adsorbents, including carbon nanotubes, graphene oxide, and activated carbon. Alumina and silica are commonly used inorganic adsorbents. Functionalized polymers are used in polymeric adsorbents, and these are designed for specific adsorption purposes. Based on the application field, adsorbents can be classified as water and wastewater treatment adsorbents, air purification adsorbents and reactive adsorbent.

6.4 Factors Influencing Adsorption Efficiency

Different physical and chemical parameters affect the efficiency of adsorption. The particle size and pore size are significant physical parameters influence the efficiency. The surface chemistry of adsorbents is crucial during chemisorption. If the surface area of the adsorbents is larger, these are generally preferred because of their greater adsorption capacity; however, for specific purposes, optimization is also carried out. The particle size and the pore size are aligned accordingly for the process to maximize efficiency. The presence of functional groups improves the adsorption efficiency. Throughout the process, the adsorbate concentration also influences the overall removal efficiency. Polarity is another factor that influences the efficiency, adsorbents are selected on the polarity of adsorbates to improve the efficiency. The molecular structure and size are the next factors that can affect efficiency. Large molecules are incompetent for accessing the pore, and this leads to decrease in efficiency. Temperature and pressure are the factors that have been extensively studied in adsorption. Increased temperatures can improve the adsorption process in most of the cases, but they may not be applicable in desorption. At higher pressures, gases adsorb more easily on solid surfaces. Finally, pH is an important factor; a change in pH can alter the degree of ionization of the participants, which can affect the efficiency (Shirin & Yadav, 2021; Wang et al., 2023).

6.5 Types of Adsorbents for Arsenic Removal

6.5.1 *Activated Carbon*

In the water filtration process activated carbon is the most used adsorbent. Activated carbon is made by heat treating carbonaceous biomass to create high-surface-area materials, followed by surface activation. These two-stage processes are termed carbonization and activation. In the carbonization stage, moisture is fist removed, and the remaining raw materials are processed in an elevated oxygen-free environment. In the next stage, the surface is activated by physical or chemical methods.

6.5.2 *Metal Oxides and Hydroxides*

Metal oxides consist of metal cations and oxide anions; in metal hydroxides the components are metal cations and hydroxide anions. For the elimination of arsenic in contaminated water, multiple studies have concluded that the metal oxides can be used effectively. Aluminum oxide (Al_2O_3), an inexpensive metal oxide, can effectively adsorb arsenic. A study concluded that Al_2O_3 can remove 95% of As(V) while maintaining a pH less than 7 (Jeong et al., 2007). Modifications to the surface and activation in the physical structure of Al_2O_3 have also been found to increase the removal efficiency. Thermal activation in this study was carried out at 500 °C and for chemical activation strong acids are used (Chanchal, 2018). Another potentially used in arsenic removal are copper oxides (CuO), Compared to Al_2O_3, CuO nanoparticles have higher affinity adsorbing arsenic. Studies proved that CuO nanoparticles rapidly adsorb both As(III) and As(V). A bottom-up approach is used for the synthesis of CuO nanoparticles. One of the commonly used processes involves microwaving the $CuCl_2.2H_2O$, NaOH and polyethylene glycol in ethanol, in specific composition (Martinson & Reddy, 2009; Reddy et al., 2013).

Another inexpensive and abundant adsorbent found in nature is manganese dioxide (MnO_2). MnO_2 can easily oxidize arsenite and adsorb arsenic. With a greater surface area and many active sites, MnO_2 can be promote the physisorption and chemisorption of arsenic from wastewater. There is an oxidative/reductive mechanism present in these removal (Manning et al., 2002). Iron oxides are promising materials for arsenic mitigation. Geothite (α-FeOOH), magnetite (Fe_3O_4), and hematite (Fe_2O_3), are some of the iron oxides used in studies. Natural and synthetic iron oxides show similar adsorption capacities. Hematite, a natural iron oxide, has shown the highest adsorption capacity for both trivalent and pentavalent arsenic species at optimized pH values (Giménez et al., 2007).

Titanium dioxide (TiO_2) a commonly used substance in water treatment for the elimination of quite a few contaminants including arsenic. TiO_2 has a strong adsorption capability and photocatalytic activity, which contribute significantly to the overall removal efficiency of arsenic in contaminated water. A study concluded that the adsorption capacity of amorphous TiO_2 is superior than that of crystalline. The TiO_2 nanoparticles synthesised in this study is by sol–gel method (Jegadeesan et al., 2010). Binary metal oxides are increasingly being used in water treatment due to their large surface areas, adsorption capabilities, and catalytic properties. Aluminium–Iron Oxide (Al–Fe_2O_3), Iron–Titanium Oxide (Fe–TiO_2), Iron–Manganese Oxide (Fe–MnO_2), are some of the binary metal oxides for arsenic adsorption.

6.5.3 Natural Adsorbents

Soils a readily available and inexpensive source is a potential adsorbent. Laterite soil, red soil, termite mound, etc. are used to reduce the arsenic levels. Various minerals, such as clay consisting of hydrous aluminium silicate, Hydroxylapatite ($Ca_5(PO_4)_3OH$, HAP), bentonites, montmorillonites, and zeolites, are also used to remediate arsenic. One of the sustainable technologies that has gained more interest for removing heavy metals such as arsenic and toxic ions from industrial effluents and natural water bodies is biosorption. Cellulose derived from natural materials treated for greater surface area is proven to be efficacious in their use as biosorbent for arsenic removal. Alginate, a polysaccharide extracted from algae, has been successfully tested for removal arsenic impurities from water. Some microbial species have also been used for mitigation of arsenic. These materials, coated with organic and inorganic materials, improve overall efficiency. Modifications in natural adsorbents have also been performed to improve adsorption in number of studies (Asere et al., 2019; Shirin et al., 2019).

6.5.4 Synthetic Adsorbents

Some of the synthetic adsorbents are modified metal oxides, ion exchange membranes, and nanomaterials. Modification in metal oxides can be done by different methods. Doping with significant adsorptive materials such as TiO_2 can improve adsorption process. Another method is surface functionalization for modifying metal oxides. For example, iron oxide nanomaterials attached by amine groups improve chemisorption. In addition, metal oxides can be nanostructured to improve surface area which then promote the efficiency. Ion-exchange resins are commonly used in water and wastewater treatment to remove ions from metals and other compounds. The different resins used in arsenic removal are strong base anion (SBA) resins and weak base anion (WBA) resins. Nanomaterials are potential adsorbents that can be used for the mitigation of arsenic. Several types of nanomaterial have demonstrated great promise in the efficient removal of arsenic, including metal oxide nanoparticles, carbon-based nanomaterials, MOFs, and polymer-based nanocomposites. Some research and development efforts have been made to optimize these materials for realistic and sustainable water treatment applications, despite obstacles such as aggregation, toxicity, and expense (Lata & Samadder, 2016).

6.6 Advancements in Adsorption Technologies

6.6.1 Nanotechnology in Arsenic Adsorption

Recent innovations in nanotechnology have significantly improved arsenic removal efficiency, offering promising solutions for cleaner water. Maity et al. (2021), provides a relatively recent comprehensive review of these advancements. A few examples of these innovations are detailed below. Yu et al. (2019), investigates the development and use of porous graphene aerogel impregnated with Fe_2O_3 nanocubes (PGA/P-Fe_2O_3). This novel adsorbent is prepared by mixing a graphene oxide solution with potassium ferrocyanide ($K_4Fe(CN)_6$) solution, which is then treated with ferric chloride ($FeCl_3$) and reduced with vitamin C. The mixture is then heated, dialyzed, freeze dried, and calcined to produce the required PGA/P-Fe_2O_3. SEM analysis carried out on the prepared adsorbent confirmed uniform suspension of the nanoparticles, with a bulk pore size ranging between 40 and 80 nm. Raman spectroscopy and XRD analyses confirmed the chemical composition and crystal structures. Batch adsorption experiments with initial trivalent and pentavalent arsenic concentrations of 5 mg/L revealed astonishing adsorption capacities of 172.27 and 217.34 mg/g, with equilibrium times of 30 and 5 min for trivalent and pentavalent arsenic, respectively. Reaction kinetics corresponded well with the pseudo-second order kinetic adsorption model, with adsorption conforming to the Freundlich isotherm model. As with other iron-based arsenic adsorbents, adsorption at a lower pH range of 2 to 4 was ideal for complete removal of arsenic. This strong adsorption capacity was attributed to the nanoporous Fe_3O_3 nanocube structure, which provided a specific surface area of 177.9 m^2/g. Furthermore, a high pH environment facilitated by NaOH, coupled with centrifugation, was found to be an excellent means of desorption and adsorbent regeneration with minimal loss in adsorption capacity.

Another interesting and novel adsorbent are magnetic nanoparticles modified with Fe–Mn binary oxide (MNp-FeMn) generated and studied by Nikić et al., (2021). The base magnetic nanoparticles were synthesized by coprecipitation of alkaline $FeCl_3$ and $FeSO_4$ and then dried. These nanoparticles were treated with $FeCl_3$ and $MnSO_4$ solutions in a 3:1 ratio in an alkaline environment facilitated by NaOH, washed, oxidized with an NaOCl solution, and dried to produce the final MNp-FeMn. BET analysis revealed a specific surface area of 109 m^2/g, while SEM analysis showed a highly porous structure with an average pore size of 8.07 nm. EDS and XRD analyzes confirmed the crystallography and chemical composition of the nanoparticles. Batch studies using initial trivalent and pentavalent arsenic concentrations of 10 mg/L revealed adsorption capacities of 56 mg/g and 54 mg/g for each of the species, respectively, with equilibrium being achieved in less than 420 min across a broad pH range of 3 to 9. The kinetics of the adsorption process aligned with the pseudo-second order kinetic model of adsorption, while the adsorption mode was chemisorption, which corresponds well with the Freundlich adsorption isotherm model. However, competition from other ionic species in contaminated water, such as PO_4^{3-}, SiO_3^{2-}, and CO_3^{2-}, was found to be detrimental to the sorption of trivalent and pentavalent arsenic. FTIR analysis confirmed that inner-sphere surface complex formation was the main driver for adsorption, occurring in two distinct steps. First, MnO_2 oxidizes As(III) in solution to As(V), and then it is adsorbed onto Fe_2O_3 in the second step. Similarly to the previous study, a strong alkaline environment facilitated by NaOCl was instrumental in desorbing arsenic from Fe and regenerating the adsorbent by rapidly oxidizing the Mn, bringing the adsorbent back to near full capacity (with only a capacity loss after five cycles). It is interesting to note that such iron-based nanoparticulate adsorbents do not suffer from the separation and recovery issues that most others of their kind face, as they can be separated magnetically due to their ferromagnetic nature.

6.6.2 Hybrid Adsorbents

Hybrid adsorbents, such as biochar and chitosan-based materials, combine multiple materials to improve arsenic removal efficiency, offering improved selectivity, capacity, and regeneration capabilities, making them a promising solution for effective water treatment.

Biochar-based hybrid adsorbents, in addition to being a cost-effective and sustainable alternative to conventional arsenic adsorbents, have the added benefit of being compatible with and, in most cases, beneficial for use in soil remediation. This is due to biochar's nature as a pH buffer and nutrient moderator. An example of such a hybrid biochar-based adsorbent was developed and studied by Lin et al. (2019). Their study focused on the development and efficacy of biochar impregnated with Fe–Mn oxides, which utilized biochar derived from powdered corn stem biomass. The base biomass was subjected to slow pyrolysis in an N_2 environment, after which it was treated with $FeSO_4$ and $KMnO_4$ solutions on a weight ratio basis. Batch adsorption studies revealed that the ideal weight ratio of biochar:Fe was 18:3:1. The BET surface area of the adsorbent was determined to be 7.53 m^2/g, with an impressive oxidation–reduction potential of 317.1 mV. The process of adsorption described by the pseudo-second-order kinetic model was found to be the best fit, while adsorption corresponded well with the Freundlich isotherm model, indicating a strong bias toward chemisorption and an apparent adsorption capacity of 8.80 mg/g. Subsequent batch adsorption studies determined that with a preliminary trivalent arsenic concentration of 30 mg/L at pH 3–7, equilibrium was achieved in less than 120 min, resulting in a 92% reduction in As(III) content. Further FTIR and XPS studies helped elucidate the mechanism of As(III) adsorption, which was determined to occur primarily through three means: oxidation of As(III) to As(V) by both Fe and, more prominently, Mn; chemisorption of As(V) onto Fe; and minimal but direct adsorption of As(III) onto Fe. This process was theorized to have been aided by the electron transfer capabilities of biochar.

Another class of hybrid adsorbents leverages the stability and robust framework of metallic oxide minerals, commonly combined with the biocompatibility and biodegradability of chitosan. This synergy enhances the adsorption efficiency, making these hybrid materials highly effective for environmental remediation applications. A recent and novel example of this is explained in the work of Han et al. (2019), who developed and studied the efficacy of using chitosan-coated Na–X zeolite (C-NX) for the adsorptive removal of As(V). The adsorbent was prepared by first homogenizing and calcining fly ash with NaOH, which formed the matrix for the adsorbent. It was then mixed into an acetic acid chitosan solution, filtered, washed, and dried. XRD and FTIR analyzes confirmed the effective bonding of chitosan on the surface of the Na-X zeolite. Batch adsorption studies conducted with a preliminary concentration of pentavalent arsenic of 5 mg/L at pH 2.1 resulted in the near complete elimination of pentavalent arsenic from the sample tested, with an equilibrium time of 2 h. With a notable reported capacity of 63.23 mg/g, the adsorption was shown to align with the Langmuir adsorption isotherm, suggesting the dominance of monolayer adsorption, and corresponds to the pseudo-second order kinetic model. XPS and FTIR analyses of the samples before and after adsorption helped determine the adsorption mechanism. The protonated amino surface groups ($-NH_3^+$) of the chitosan and Al–OH groups played the greatest role in As(V) adsorption through surface complex formation, due to the large positive charge they facilitate. It should be noted that this adsorbent functions effectively in very low pH ranges (pH 2.1–4.6), making it a perfect fit for the remediation of mining wastewater, which frequently requires adjustment of pH before remediation.

6.6.3 Innovations in Adsorbent Design

Defect engineering in adsorbent design involves the intentional introduction of vacancies, interstitials, or dislocations within the structure of a material to create additional active sites. These defects theoretically enhance the surface reactivity and adsorption capacity of the adsorbent, thereby improving the material's ability to selectively remove contaminants. Hou et al. (2022) explored one such adsorbent, namely goethite (α-FeO(OH)). Although the adsorptive properties of goethite have been studied in earlier works, this study focused on the influence of defects, specifically iron vacancy defects, on adsorption capacity for arsenic species. To investigate this, goethite with varying but consistent and reproducible degrees of defects was prepared by hydrothermal coprecipitation of $Fe_2(SO_4)_3$ and NaOH solutions for different periods. The longer the samples were allowed to age, the higher the degree of defects (1 day—defect-rich, 3 days—defect-medium, and 5 days—defect-poor).

Scanning-transmission electron microscopy (STEM) analysis of the three selected adsorbents revealed that all exhibited the classic needle-shaped structure, albeit with minor variations. However, closer inspection using high-angle annular dark-field

STEM (HAADF-STEM) techniques exposed the fact that the defect-rich and defect-medium samples had a prominently broken surface with shorter needles. The arrangement of the Fe atoms in the lattice also showed a decreasing degree of regularity from defect-poor to defect-rich, with varying Fe ratios. Batch adsorption studies conducted in the pH range of 3–7 revealed that the adsorption isotherm model conformed to the Langmuir adsorption isotherms. Defect-rich goethite demonstrated the highest observed capacity for arsenic adsorption, with 608.2 μmol/g towards As(V), which is 2.8 times and 10.2 times higher than that of defect-medium and with poor defects, respectively. Similarly, for As(III), the defect-rich goethite exhibited an adsorption capacity of 605.8 μmol/g, which is 4.8 times and 22.2 times higher than that of defect-medium and defect-poor samples, respectively.

6.7 Future Directions and Research Needs

Advancement in water and wastewater treatment technique has been gaining sharp popularity through the introduction of low-cost adsorbents. This upcoming progression toward these absorbents accounts for the persistent existing drawbacks including the limited surface area and stability, lowered mechanical properties and reusability confinements, which overall hinders the functionality of the conventional adsorbents (Eniola et al., 2019). Several studies illustrate the successful use of reformation or alteration on these prevailing adsorbents by co-operation with existing properties while maintaining cost effectiveness and usability. Pyrolysis self—activation (Gao et al., 2023) or co–pyrolysis (Wang et al., 2022a; Yoon et al., 2020), calcination (Shehzad et al., 2018; Xie et al., 2022), hybridisation (Basu et al., 2021; Das et al., 2020), chemical (Bhatti et al., 2020; Ochedi et al., 2020) and thermal activation (Ma et al., 2022; Nieto-Delgado et al., 2019) are some of the techniques implemented in augmenting these adsorbents that are further discussed below. In addition to improving the adsorption capacity, these methodologies were found to enhance the selectivity toward the arsenic variants.

6.7.1 Pyrolysis

Thermal decomposition implemented in the absence of oxygen results in the conversion of biomass into char, bio-oil, and gases, and describes pyrolysis (Aasberg-Petersen et al., 2004). Agricultural and other organic wastes are predominately used to produce the low-cost adsorbents for arsenic removal. Effective and cost efficient combined with eco-friendliness are the added-on characteristics of this approach. Self-activated pyrolysis of biowaste into activated carbon is among one such method. Gao et al. (2023) presented a technique to generate activated carbon through the self-activated pyrolysis of bamboo shoot shells. Figure 6.2 shows the pictorial representation of the system. In the absence of external activators or auxiliary gases,

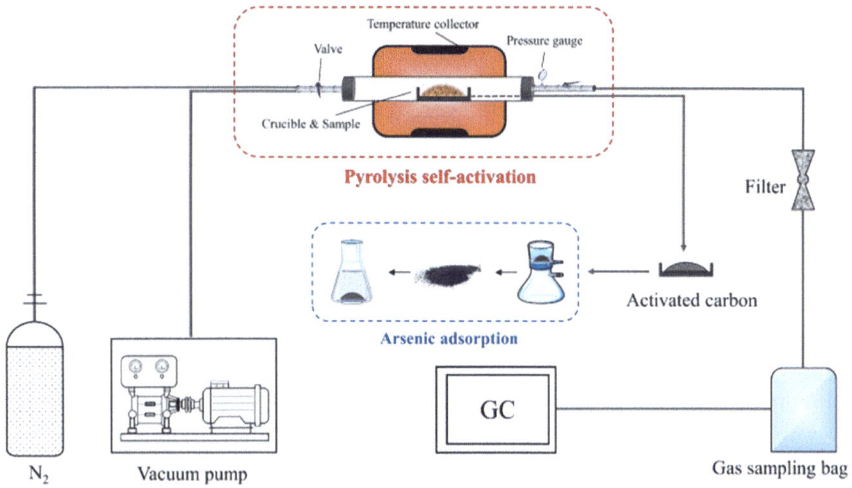

Fig. 6.2 Schematic representation of self-activation pyrolysis of bamboo shoot shells into activated carbon for arsenic removal (Gao et al., 2023)

self-activated pyrolysis uses H_2O and CO_2 that are produced as by-products of the process. These reagents have a higher heating value of 11.7 MJ/Nm^3. The activated carbon thus formed exhibited excellent arsenic adsorption capabilities, with maximum capacities of 10.9 mg/g for As(III) and 16.0 mg/g for As(V). The process was carried out at an activation temperature of 1000 °C for a residence period of 4 h. Reduction of residence time and utilization of combustible gases are the future focus points of this research, which could further uplift the overall effectiveness of the process.

The use of more than one material is used as feedstock compared to conventional pyrolysis, known as Copyrolysis (Prakash et al., 2018), is another method that is gaining popularity in the modification of low-cost adsorbents for the removal of arsenic. A study by Wang et al., (2022a, 2022b), incinerated sewage sludge ash (ISSA) from iron-containing solid waste from the combustion of dewatered sewage (Chen & Poon, 2017) was used as primary feedstock which is then copyrolyzed with peanut shells to produce an iron-based biochar. This biochar exhibited arsenic adsorption properties of 17.44 mg/g for As(III) and 57.6 mg/g for As(V). However, the process showed a drawback of high leaching of reactive silica from the biochar, resulting in the formation of corrosive products and hindering some active adsorption sites. Although the process was found to reduce production costs, more research is necessary to address this issue.

Microwave pyrolysis, a thermochemical process that utilizes microwave heating to achieve extensive decomposition, is another procedure that is gaining popularity in this field. The novel bagasse biochar/ferrite nanocomposite produced by this method demonstrated not only high performance as an adsorbent but also a low-cost methodology that widely utilizes readily available renewable biomaterials (Bai et al., 2022). Plastic pyrolysis, which involves the heating of plastic at extremely high temperatures in the absence of oxygen, is recently being utilized as an eco-friendly synthesis method. This process employs waste plastic, which reveals excellent surface chemistry and evolved morphology in activated char adsorbents.

6.7.2 Hybridisation

The process of enhancing adsorption efficiency can be accomplished through the incorporation of multiple low-cost adsorbents, which strengthens their physicochemical morphology (Ahmed et al., 2021; Eniola et al., 2020). The use of geologically accessible materials and industrial waste supports the reduction of hybrid adsorbent costs (Unuabonah et al., 2017). For example, (Basu et al., 2021) developed a composite adsorbent consisting of GO-MnO_2-Goethite core–shell microspheres impregnated onto calcium alginate to form hydrogel beads (Fig. 6.3). Their study details the reaction kinetics, thermodynamic parameters, and pH studies. The beads followed pseudo first-order kinetics with a Langmuir isotherm model and were found to be endothermic and spontaneous in nature. Parametrical changes in this study accounted for the enhancement of the rate of interparticle diffusion of arsenic groups within the modified bead adsorbent. Maximum uptake was observed at a pH of 4 to 6, with sorption capacities of 27.53 mg/g for As(III) and 34.17 mg/g for As(V). Similarly, a hybrid clay sorbent was synthesised by intercalating a surfactant cation into the interlayers of kaolin clay by a study by Mudzielwana et al. (2020). These adsorbents demonstrated regenerative properties for up to 7 cycles using 0.01 M HCl as the regenerant. They achieved maximum sorption capacities of 7.99 mg/g for As(III) and 7.32 mg/g for As(V) at room temperature.

Das et al. (2020) discovered an ultra-high-performance adsorbent through the deposition of Fe nanoparticles onto graphene oxide (GO), termed GFeN (Fig. 6.4). This adsorbent exhibited exceptionally high adsorption power by improving Fe efficiency and arsenic mobility. Electrostatic interactions and surface complexation contributed to this ultra-high adsorption capacity. The GO sheets acted as a reservoir for electron release due to the corrosion of FeNPs, with electrons being transferred to rejuvenate the oxidized platform, aiding in arsenic removal. Sorption capacities of 306 mg/g for As(III) and 431 mg/g for As(V) were achieved.

Despite these exceptional characteristics, there are insufficient data on the eradication of microbes by these materials, which is also crucial for the treatment of water and wastewater (Baig et al., 2021; Simon et al., 2024). Further research should focus on extending the applicability of these adsorbents to the elimination of microbial, organic, and inorganic residual impurities.

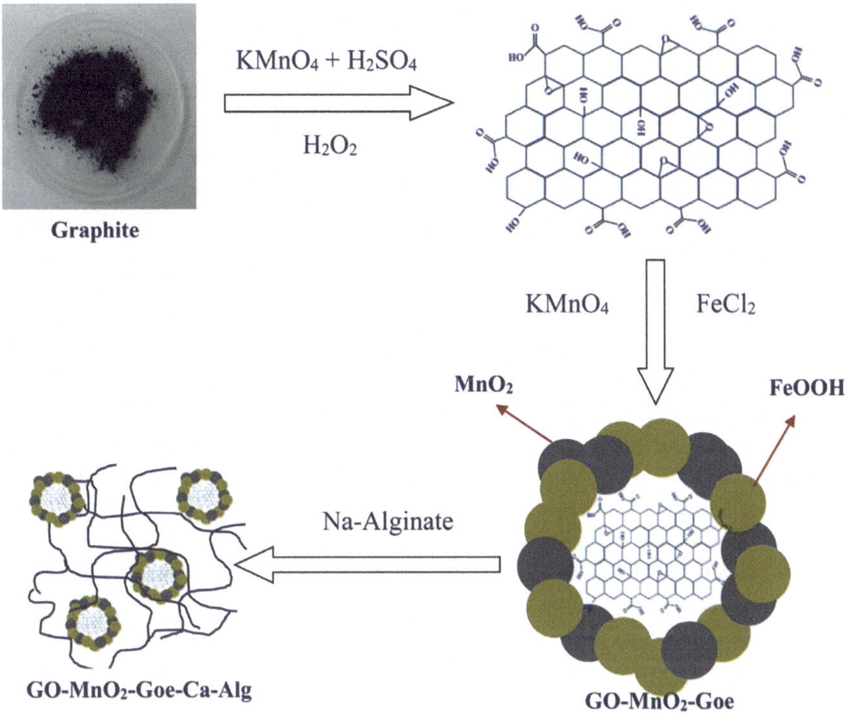

Fig. 6.3 Representation of synthesis of GO—MnO_2—Goethite—Ca—Alginate beads

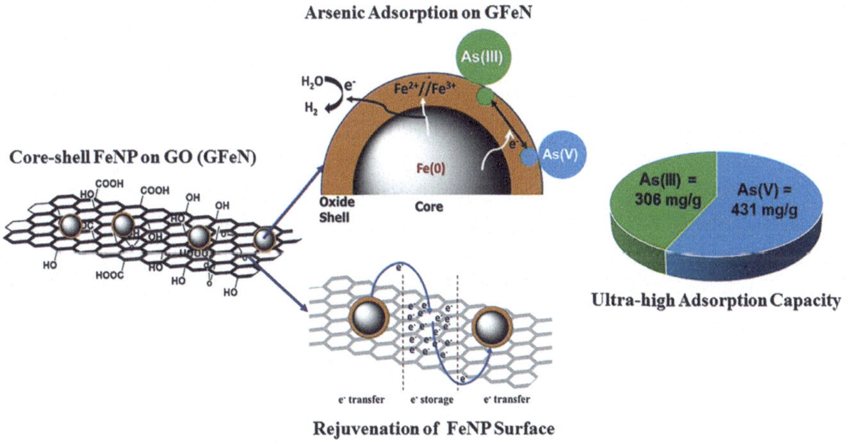

Fig. 6.4 Graphical representation of core–shell FeNP on GO for arsenic adsorption

6.7.3 Aerogel

Aerogels and their derivatives are emerging as promising adsorbents in the field of heavy metal remediation (Ihsanullah et al., 2022). Aerogels are characterized by a porous solid network that contains air pockets, which occupy the majority of the space within the material. Their remarkable properties, including exceptionally low density, immense active surface area, and low thermal conductivity, make them suitable for fabrication from various materials (Eniola et al., 2021). Traditionally, aerogels were associated with high costs due to the use of expensive nanomaterials and complex fabrication techniques (Partow et al., 2022). However, recent advances have led to the use of low-cost renewable materials, making aerogels more cost-effective and eco-friendlier (Ji et al., 2020). One of the latest studies by Li et al. (2021) developed a hierarchically monolithic porous iron metal organic gel/carbaliferial cellulose aerogel (Fe-MOG/BC) (Fig. 6.5). This aerogel is ultralight, has an abundant number of active sites, and is mechanically stable. It exhibited the highest adsorption capacity of 495 mg/g for arsenate. The advantages include ease of generation, low cost, high selectivity, scalability, and reusability, making it an excellent low-cost adsorbent. It effectively decontaminates 5 ppm of As(V) and shows potential for point-of-use and point-of-entry applications.

Similarly, Yu et al. (2019) developed a porous graphene aerogel impregnated with Fe_2O_3 nanocubes, which demonstrated maximum adsorption capacities of 172.27 mg/g for As(III) and 217.34 mg/g for As(V). This material showed ultra-fast removal capabilities, achieving reductions in highly concentrated solutions of As(III) and As(V) within just 60 and 5 min, respectively. Even after 8 cycles, the removal efficiency remained nearly unchanged. Aerogels show great promise as low-cost adsorbents because of their ultrafast and high adsorption power. Future research should focus on experimenting with various naturally occurring materials and wastes, as well as developing cost-efficient fabrication methods. This could herald a new era for adsorbents. Implementing supportive policies and regulatory frameworks is crucial to advance the use of these low-cost adsorbents. Regular monitoring through Life Cycle Assessment (LCA) and Health Risk Assessment is necessary to evaluate their impact. The regeneration and reusability of the adsorbents are crucial factors along with the development of novel adsorbents. Thus, encouraging thorough research in these areas and the implementation of effective policies are essential to establish efficient and effective solutions that in addition affect the overall cost of the adsorbent.

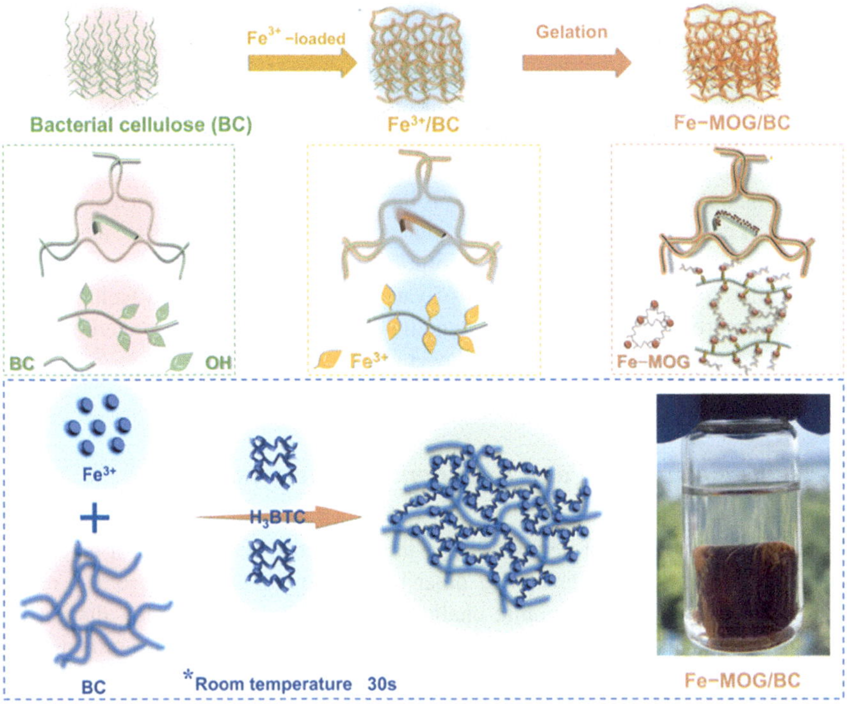

Fig. 6.5 Schematic representation of fabrication of Fe—MOG/BC aerogel (Li et al., 2021)

6.8 Challenges and Limitations

Remediation of arsenic-contaminated water and soil through adsorption is a quite promising technique among other methodologies, however, this approach is heavily impacted by various challenges and limitations. This section outlines numerous inadequacies of the prevailing adsorbents, their environmental impacts, and the associated costs, emphasizing the need for further research and development in order to address the mentioned issues.

6.8.1 Limitations of Current Adsorbents

Despite the capability of the adsorption technique, it is limited by various factors that hinder the true potential of the efficiency of the method. Specific parameters that include precise pH levels, temperatures, and arsenic speciation are necessary for the optimal functionality of the adsorbent. The difficulties in reusability and reproducibility of these adsorbents, because of their physiochemical properties, are making their economic viability questionable. Another issue faced is the scalability, that is, these adsorbents when used in bulk often display capacities lower than those of batch studies, due to issues such as channel formation, leaching, and inadequate detention time. The reuse and regeneration of adsorbents is one of the best quality of these adsorbents, but it is also a major concern in this domain. The results of regeneration techniques, such as thermal, redox, and eluent desorption, show promise. Among them, eluent-based desorption, which utilizes pH fluctuations and the polarizing characteristics of acids and alkalis, is the simplest and least expensive method. However, it presents challenges, such as managing and disposing of toxic desorption solutions and the gradual decline in adsorption capacity over multiple regeneration cycles. Therefore, further exploration of various other methods and techniques is needed to improve these processes, extend the life of the adsorbent, and address the disposal of regeneration solutions (Yadav et al., 2022).

6.8.2 Environmental and Economic Challenges

The reliability and ease of deployment of the adsorptive remediation of arsenic is one of the attractive features among other removal methods, but these also come with environmental and economic challenges. Arsenic contamination disproportionately affects less affluent or developing regions, where continuous remediation and responsible waste disposal may be economically unfeasible. Most developing countries are forced to dispose of spent adsorbents in surface waterways or landfills without adequate toxicity evaluation due to inadequate reusability or disposal methods. Studies handling the toxic characteristic leaching procedure (TCLP) have shown that arsenic concentrations in leachate often exceed WHO recommended levels (Yadav et al., 2022), which also resulted into groundwater contamination and bioaccumulation in aquatic and soil environments, posing significant health risks to local populations (Al-Makishah et al., 2020; Yadav et al., 2022; Zhang et al., 2022).

Costs related to the acquisition of adsorbents, the maintenance of the system, and the disposal of spent adsorbents are some of the pressing issues in the implementation of arsenic remediation systems in isolated communities. However, research into low-cost, high-capacity adsorbents, innovative regeneration methods, and sustainable disposal solutions offers hope to address these challenges (Yadav et al., 2022).

6.8.3 Addressing Adsorbent Disposal Issues

The disposal of spent adsorbents, particularly those laden with arsenic, poses a substantial threat. Safe disposal methods vary worldwide, where developed countries typically follow regulated dumping practices, while less developed regions often resort to open dumping, which can lead to environmental harm and health problems. Incineration or mixing of immobilized arsenic adsorbents with livestock feed are some alternate disposal methods investigated that could be adopted. However, these methods are not completely safe or sustainable (Yadav et al., 2022). The necessity for better disposal methods is crucial. Rathore and Mondal (2017) explored an innovative approach through the combination of stabilized spent adsorbents with clay aggregate to manufacture sustainable bricks or mortars, in accordance with Indian standards. Adsorbents such as thermally treated laterite (TTL), acid–base treated laterite (ABTL), and aluminum oxide/hydroxide nanoparticles (AHNP) were used, and the resulting bricks demonstrated satisfactory physical stability and low arsenic leachability. However, high adsorption capacities and small particle sizes of nanoparticle adsorbents showed reduced binding to clay aggregate, affecting compressive strength and cumulative arsenic leachate concentration. This method has potential for developing safe and affordable housing materials in less developed regions, although its applicability is limited to inorganic Fe, Al, and Si-based adsorbents available in sufficient quantities. In conclusion, addressing the limitations of current adsorbents, their environmental impact, and disposal issues is essential for optimizing the effectiveness and sustainability of arsenic remediation. Persistent research and development are critical to overcoming these challenges and improving arsenic removal technologies.

6.9 Conclusion

Arsenic, even at low concentrations, is extremely toxic to living forms in the environment. Among the various removal techniques, adsorption is extensively used because of its ease of application, cost-effectiveness, and efficiency. The adsorption process is enhanced through the regeneration of adsorbents. To achieve optimized functionality, the development of low-cost yet highly effective adsorbents remains an ongoing area of advancement. Recent research has highlighted several promising adsorbents, including bioadsorbents, agricultural and industrial waste, metal-doped oxides, composites, and hybrid materials. These adsorbents offer unique benefits that contribute to the effective treatment of arsenic contamination. The development and modification of these adsorbents show promise in the treatment of contamination issues. Emerging low-cost adsorbents, such as hybrid materials and aerogels, and modifications of existing methodologies have demonstrated significant potential. However, these materials and fabrication techniques often have certain drawbacks. Therefore, extensive research and innovation are essential to optimize performance, usability, scalability, and environmental friendliness, thus ensuring effective reduction of arsenic contamination.

Acknowledgements The authors express their gratitude to the reviewers and the editor for dedicating their valuable time to enhancing this work.

References

Aasberg-Petersen, K., Christensen, T. S., Dybkjær, I., Sehested, J., Østberg, M., Coertzen, R. M., Keyser, M. J., & Steynberg, A. P. (2004). Synthesis gas production for FT synthesis. In *Studies in surface science and catalysis* (Vol. 152, pp. 258–405). https://doi.org/10.1016/s0167-2991(04)80461-0

Ahmed, S. F., Mofijur, M., Nuzhat, S., Chowdhury, A. T., Rafa, N., Uddin, M. A., Inayat, A., Mahlia, T. M. I., Ong, H. C., Chia, W. Y., & Show, P. L. (2021). Recent developments in physical, biological, chemical, and hybrid treatment techniques for removing emerging contaminants from wastewater. *Journal of Hazardous Materials, 416*, Article 125912. https://doi.org/10.1016/j.jhazmat.2021.125912

Al-Makishah, N. H., Taleb, M. A., & Barakat, M. A. (2020). Arsenic bioaccumulation in arsenic-contaminated soil: a review. *Chemical Papers, 74*(9), 2743–2757. https://doi.org/10.1007/S11696-020-01122-4

Asere, T. G., Stevens, C. V, & Du Laing, G. (2019). Use of (modified) natural adsorbents for arsenic remediation: A review. *Science of the Total Environment, 676*, 706–720. https://doi.org/10.1016/j.scitotenv.2019.04.237

Bai, L., Huang, A., Feng, J., Su, X., Mo, W., Ma, S., & Lin, H. (2022). One-step microwave pyrolysis synthesis of bagasse biochar/ ferrites nanocomposite and synergistic effect on As(V) adsorption in water. *Materials Chemistry and Physics, 283*, Article 126035. https://doi.org/10.1016/j.matchemphys.2022.126035

Baig, U., Faizan, M., & Sajid, M. (2021). Effective removal of hazardous pollutants from water and deactivation of water-borne pathogens using multifunctional synthetic adsorbent materials: A review. *Journal of Cleaner Production, 302*, Article 126735. https://doi.org/10.1016/j.jclepro.2021.126735

Basu, H., Singh, S., Venkatesh, M., Pimple, M. V., & Singhal, R. K. (2021). Graphene oxide-MnO_2-goethite microsphere impregnated alginate: A novel hybrid nanosorbent for As(III) and As(V) removal from groundwater. *Journal of Water Process Engineering, 42*, Article 102129. https://doi.org/10.1016/j.jwpe.2021.102129

Bhatti, Z. A., Qureshi, K., Maitlo, G., & Ahmed, S. (2020). Study of PAN fiber and iron ore adsorbents for arsenic removal. *Civil Engineering Journal (Iran), 6*(3), 548–562. https://doi.org/10.28991/cej-2020-03091491

Carotenuto, M., Lofrano, G., & Sharma, S. K. (2014). Arsenic contamination: An overview. In S. Sharma (Ed.), *Heavy metals in water: Presence, removal and safety*. The Royal Society of Chemistry. https://doi.org/10.1039/9781782620174-00086

Chanchal, M. (2018). Arsenic(V) removal using activated alumina: Kinetics and modeling by response surface. *Journal of Environmental Engineering, 144*(3), 4017115. https://doi.org/10.1061/(ASCE)EE.1943-7870.0001337

Chen, Z., & Poon, C. S. (2017). Comparing the use of sewage sludge ash and glass powder in cement mortars. *Environmental Technology, 38*(11), 1390–1398. https://doi.org/10.1080/09593330.2016.1230652

Das, T. K., Sakthivel, T. S., Jeyaranjan, A., Seal, S., & Bezbaruah, A. N. (2020). Ultra-high arsenic adsorption by graphene oxide iron nanohybrid: Removal mechanisms and potential applications. *Chemosphere, 253*, Article 126702. https://doi.org/10.1016/j.chemosphere.2020.126702

Eniola, J. O., Ansari, M. O., Barakat, M. A., & Kumar, R. (2021). Aerogels in photocatalysis. In *Advances in aerogel composites for environmental remediation* (pp. 87–108). Elsevier. https://doi.org/10.1016/B978-0-12-820732-1.00006-0

Eniola, J. O., Kumar, R., & Barakat, M. A. (2019). Adsorptive removal of antibiotics from water over natural and modified adsorbents. *Environmental Science and Pollution Research, 26*(34), 34775–34788. https://doi.org/10.1007/s11356-019-06641-6

Eniola, J. O., Kumar, R., Barakat, M.A. (2020). Nanomaterials for the adsorptive removal of antibiotics from aqueous solution: A minireview. *General Chemistry, 6*(2), 190018–190018. https://doi.org/10.21127/yaoyigc20190018

Gao, Q., Feng, Z., He, Y., Hou, Y., Ren, H., Su, M., Ni, L., & Liu, Z. (2023). Pyrolysis self-activation: An environmentally friendly method to transform biowaste into activated carbon for arsenic removal. *Bioresource Technology, 368*, Article 128353. https://doi.org/10.1016/j.biortech.2022.128353

Giménez, J., Martínez, M., de Pablo, J., Rovira, M., & Duro, L. (2007). Arsenic sorption onto natural hematite, magnetite, and goethite. *Journal of Hazardous Materials, 141*(3), 575–580. https://doi.org/10.1016/j.jhazmat.2006.07.020

Han, C., Yang, T., Liu, H., Yang, L., & Luo, Y. (2019). Characterizations and mechanisms for synthesis of chitosan-coated Na–X zeolite from fly ash and As(V) adsorption study. *Environmental Science and Pollution Research, 26*(10), 10106–10116. https://doi.org/10.1007/S11356-019-04466-X

Hou, J., Tan, X., Xiang, Y., Zheng, Q., Chen, C., Sha, Z., Ren, L., Wang, M., & Tan, W. (2022). Insights into the underlying effect of Fe vacancy defects on the adsorption affinity of goethite for arsenic immobilization. *Environmental Pollution, 314*(July), Article 120268. https://doi.org/10.1016/j.envpol.2022.120268

Huang, M., Nhung, N. T. H., Dodbiba, G., & Fujita, T. (2023). Mitigation of arsenic accumulation in rice (Oryza sativa L.) seedlings by oxygen nanobubbles in hydroponic cultures. *Ecotoxicology and Environmental Safety, 268*, 115700. https://doi.org/10.1016/j.ecoenv.2023.115700

Ihsanullah, I., Sajid, M., Khan, S., & Bilal, M. (2022). Aerogel-based adsorbents as emerging materials for the removal of heavy metals from water: Progress, challenges, and prospects. In *Separation and purification technology* (Vol. 291, p. 120923). https://doi.org/10.1016/j.seppur.2022.120923

Jegadeesan, G., Al-Abed, S. R., Sundaram, V., Choi, H., Scheckel, K. G., & Dionysiou, D. D. (2010). Arsenic sorption on TiO_2 nanoparticles: Size and crystallinity effects. *Water Research, 44*(3), 965–973. https://doi.org/10.1016/j.watres.2009.10.047

Jeong, Y., Fan, M., Singh, S., Chuang, C.-L., Saha, B., & Hans van Leeuwen, J. (2007). Evaluation of iron oxide and aluminum oxide as potential arsenic(V) adsorbents. *Chemical Engineering and Processing—Process Intensification, 46*(10), 1030–1039. https://doi.org/10.1016/j.cep.2007.05.004

Ji, Y., Wen, Y., Wang, Z., Zhang, S., & Guo, M. (2020). Eco-friendly fabrication of a cost-effective cellulose nanofiber-based aerogel for multifunctional applications in Cu(II) and organic pollutants removal. *Journal of Cleaner Production, 255*, Article 120276. https://doi.org/10.1016/j.jclepro.2020.120276

Langsch, J. E., Costa, M., Moore, L., Morais, P., Bellezza, A., & Falcão, S. (2012). New technology for arsenic removal from mining effluents. *Journal of Materials Research and Technology, 1*(3), 178–181. https://doi.org/10.1016/S2238-7854(12)70030-3

Lata, S., & Samadder, S. R. (2016). Removal of arsenic from water using nano adsorbents and challenges: A review. *Journal of Environmental Management, 166*, 387–406.

Li, H., Ye, M., Zhang, X., Zhang, H., Wang, G., & Zhang, Y. (2021). Hierarchical porous iron metal-organic gel/bacterial cellulose aerogel: Ultrafast, scalable, room-temperature aqueous synthesis, and efficient arsenate removal. *ACS Applied Materials and Interfaces, 13*(40), 47684–47695. https://doi.org/10.1021/acsami.1c14938

Lin, L., Song, Z., Huang, Y., Khan, Z. H., & Qiu, W. (2019). Removal and oxidation of arsenic from aqueous solution by biochar impregnated with Fe–Mn oxides. *Water, Air, and Soil Pollution, 230*(5). https://doi.org/10.1007/s11270-019-4146-5

Ma, Z., Li, J. shan, Xue, Q., Zhan, B., Chen, X., Wan, Y., Zhao, Y., Sun, Y., & Poon, C. S. (2022). Deep insight on mechanism and contribution of As(V) removal by thermal modification waste concrete powder. *Science of the Total Environment, 807*, 150764. https://doi.org/10.1016/j.scitotenv.2021.150764

Maity, J. P., Chen, C. Y., Bhattacharya, P., Sharma, R. K., Ahmad, A., Patnaik, S., & Bundschuh, J. (2021). Advanced application of nano-technological and biological processes as well as mitigation options for arsenic removal. *Journal of Hazardous Materials, 405*, Article 123885. https://doi.org/10.1016/j.jhazmat.2020.123885

Manning, B. A., Fendorf, S. E., Bostick, B., & Suarez, D. L. (2002). Arsenic(III) oxidation and Arsenic(V) adsorption reactions on synthetic birnessite. *Environmental Science & Technology, 36*(5), 976–981. https://doi.org/10.1021/es0110170

Martinson, C. A., & Reddy, K. J. (2009). Adsorption of arsenic(III) and arsenic(V) by cupric oxide nanoparticles. *Journal of Colloid and Interface Science, 336*(2), 406–411. https://doi.org/10.1016/j.jcis.2009.04.075

Mudzielwana, R., Gitari, M. W., & Ndungu, P. (2020). Enhanced As(III) and As(V) adsorption from aqueous solution by a clay based hybrid sorbent. *Frontiers in Chemistry*. https://doi.org/10.3389/fchem.2019.00913

Nieto-Delgado, C., Gutiérrez-Martínez, J., & Rangel-Méndez, J. R. (2019). Modified activated carbon with interconnected fibrils of iron-oxyhydroxides using Mn^{2+} as morphology regulator, for a superior arsenic removal from water. *Journal of Environmental Sciences, 76*, 403–414. https://doi.org/10.1016/j.jes.2018.06.002

Nikić, J., Watson, M. A., Isakovski, M. K., Tubić, A., Šolić, M., Kordić, B., & Agbaba, J. (2021). Synthesis, characterization and application of magnetic nanoparticles modified with Fe-Mn binary oxide for enhanced removal of As(III) and As(V). *Environmental Technology (United Kingdom), 42*(16), 2527–2539. https://doi.org/10.1080/09593330.2019.1705919

Ochedi, F. O., Liu, Y., & Hussain, A. (2020). A review on coal fly ash-based adsorbents for mercury and arsenic removal. *Journal of Cleaner Production, 267*, Article 122143. https://doi.org/10.1016/j.jclepro.2020.122143

Partow, A. J., Meng, S., Wong, A. J., Savin, D. A., & Tong, Z. (2022). Recyclable & highly porous organo-aerogel adsorbents from biowaste for organic contaminants' removal. *Science of the Total Environment, 827*, Article 154051. https://doi.org/10.1016/j.scitotenv.2022.154051

Prakash, R., Siddharth, R., & Gunasekar, N. (2018). Cracking of toxic waste. In *Energy from toxic organic waste for heat and power generation* (pp. 139–149). Elsevier. https://doi.org/10.1016/B978-0-08-102528-4.00010-9

Rathore, V. K., & Mondal, P. (2017). Stabilization of arsenic and fluoride bearing spent adsorbent in clay bricks: Preparation, characterization and leaching studies. *Journal of Environmental Management, 200*, 160–169. https://doi.org/10.1016/j.jenvman.2017.05.081

Reddy, K. J., McDonald, K. J., & King, H. (2013). A novel arsenic removal process for water using cupric oxide nanoparticles. *Journal of Colloid and Interface Science, 397*, 96–102. https://doi.org/10.1016/j.jcis.2013.01.041

Shehzad, K., Xie, C., He, J., Cai, X., Xu, W., & Liu, J. (2018). Facile synthesis of novel calcined magnetic orange peel composites for efficient removal of arsenite through simultaneous oxidation and adsorption. *Journal of Colloid and Interface Science, 511*, 155–164. https://doi.org/10.1016/j.jcis.2017.09.110

Shirin, S., Jamal, A., Ranjan, P., & Yadav, A. K. (2019). Study on assessment of slope stability and mixed disposal of overburden in voids of Singrauli Coalfield. *Environmental Quality Management, 28*(3), 131–139. https://doi.org/10.1002/tqem.21616

Shirin, S., & Yadav, A. K. (2021). Assessment and suitability for the dumping of flyash in an abandoned mine. *Environmental Challenges, 4*, Article 100111. https://doi.org/10.1016/j.envc.2021.100111

Simon, M., Joshi, H., Yadav, A. K., & Hopke, P. K. (2024). Simultaneous nitrogen and phosphorus removal from polluted water using a novel bacterial consortium (BACON-38): Promising application for wastewater remediation. *Journal of Water Process Engineering, 66*, Article 106044. https://doi.org/10.1016/j.jwpe.2024.106044

Unuabonah, E. I., Adewuyi, A., Kolawole, M. O., Omorogie, M. O., Olatunde, O. C., Fayemi, S. O., Günter, C., Okoli, C. P., Agunbiade, F. O., & Taubert, A. (2017). Disinfection of water with new chitosan-modified hybrid clay composite adsorbent. *Heliyon, 3*(8), Article e00379. https://doi.org/10.1016/j.heliyon.2017.e00379

Wang, Q., Li, J., & Poon, C. S. (2022a). An iron-biochar composite from co-pyrolysis of incinerated sewage sludge ash and peanut shell for arsenic removal: Role of silica. *Environmental Pollution, 313*, 120115. https://doi.org/10.1016/j.envpol.2022.120115

Wang, W., Wang, P., Cao, Z., Cao, Y., Wu, Z., & Su, B. (2023). Study on adsorption-diffusion-seepage behavior of oxygen in coal gangue under the coupling of temperature and pressure. *Materials Today Communications, 36*, 106817. https://doi.org/10.1016/j.mtcomm.2023.106817

Wang, Y., Liu, Q., Wong, N. H., Sunarso, J., Huang, J., Dai, G., Hou, X., & Li, X. (2022b). Near-infrared (NIR) light responsiveness of CuS/S–C_3N_4 heterojunction photocatalyst with enhanced tetracycline degradation activity. *Ceramics International, 48*(2), 2459–2469. https://doi.org/10.1016/j.ceramint.2021.10.027

Xie, X., Lu, C., Xu, R., Yang, X., Yan, L., & Su, C. (2022). Arsenic removal by manganese-doped mesoporous iron oxides from groundwater: Performance and mechanism. *Science of the Total Environment, 806*, Article 150615. https://doi.org/10.1016/j.scitotenv.2021.150615

Yadav, M. K., Saidulu, D., Ghosal, P. S., Mukherjee, A., & Gupta, A. K. (2022). A review on the management of arsenic-laden spent adsorbent: Insights of global practices, process criticality, and sustainable solutions. *Environmental Technology & Innovation, 27*, Article 102500. https://doi.org/10.1016/J.ETI.2022.102500

Yoon, K., Cho, D.-W., Bhatnagar, A., & Song, H. (2020). Adsorption of As(V) and Ni(II) by Fe-Biochar composite fabricated by co-pyrolysis of orange peel and red mud. *Environmental Research, 188*, Article 109809. https://doi.org/10.1016/j.envres.2020.109809

Yu, X., Wei, Y., Liu, C., Ma, J., Liu, H., Wei, S., Deng, W., Xiang, J., & Luo, S. (2019). Ultrafast and deep removal of arsenic in high-concentration wastewater: A superior bulk adsorbent of porous Fe_2O_3 nanocubes-impregnated graphene aerogel. *Chemosphere, 222*, 258–266. https://doi.org/10.1016/j.chemosphere.2019.01.130

Zhang, W., Miao, A. J., Wang, N. X., Li, C., Sha, J., Jia, J., Alessi, D. S., Yan, B., & Ok, Y. S. (2022). Arsenic bioaccumulation and biotransformation in aquatic organisms. *Environment International, 163*, Article 107221. https://doi.org/10.1016/J.ENVINT.2022.107221

Chapter 7
Electrochemical Processes for Arsenic Mitigation

Ram Raj Meena, Hemangini Rathva, and Pramod Soni

Abstract The recognition of arsenic as a hazardous environmental contaminant has notably increased, driving interest toward more effective remediation technologies, among which electrochemical (EC) methods have garnered significant attention. These methods are preferred because of their ability to perform in situ oxidation, produce minimal sludge, and achieve high removal efficiencies at lower operational costs compared to conventional approaches. They are adept at eliminating arsenite (As(III)) and arsenate (As(V)) from aqueous matrices. Among the various electrochemical techniques, electrocoagulation stands out, consistently achieving arsenic removal efficiencies greater than 90%. On the other hand, capacitive deionization is effective mainly at lower arsenic concentrations, with removal efficiencies approximating 40% for arsenite and 45% for arsenate. Transformation of arsenite to arsenate, typically through oxidation processes, is a prerequisite for its efficient elimination. Indirect electrochemical oxidation has been proven to be effective for this oxidation, although its applicability is generally restricted to small-scale operations. Integrating anodic oxidation with EC can result in nearly complete arsenic removal, effectively combining oxidation and precipitation processes. Furthermore, bioelectro-Fenton and electro-Fenton processes are emerging significant sustainable and economically favorable advanced oxidation processes, facilitating substantial reductions in arsenite toxicity and its removal from water systems.

Keywords Electrochemical techniques · Arsenic removal efficiency · Electrocoagulation · Oxidative techniques · Advanced oxidation processes

---

R. R. Meena (✉)
Department of Civil Engineering, Motilal Nehru National Institute of Technology Allahabad, Prayagraj, India
e-mail: ram.2020rce10@mnnit.ac.in

H. Rathva
Department of Horticulture, B. A. College of Agriculture, Anand Agricultural University, Anand, India

P. Soni
Department of Civil Engineering, Indian Institute of Technology (BHU) Varanasi, Varanasi, India

A. K. Yadav and S. Shirin (eds.), *Emerging and Innovative Arsenic Removal Technologies for a Sustainable Future*, Water Science and Technology Library 131,
https://doi.org/10.1007/978-3-032-01176-3_7

7.1 Introduction

Pollution of water is the contamination of aquatic systems, stemming from both natural phenomena and anthropogenic activities. The global crisis of water scarcity has increased the urgency of addressing water resource management. In response, groundwater extraction has become a prevalent solution to meet municipal, agricultural, and industrial water demands. Arsenic is discharged into the environment specifically into water, air, and soil, through natural processes and human actions such as cement production, extraction, combustion of fossil fuels, and paper manufacturing (Islam, 2023; Shirin & Yadav, 2014, 2021). This dispersal leads to the contamination of groundwater with arsenic, thus increasing both its concentration and the duration of human exposure. Prolonged acquaintance with increased concentrations of As in intake water can result in significant adverse health effects (Babu et al., 2019). The World Health Organization (WHO) stipulates a maximum permissible arsenic level of 0.010 mg/L in drinking water (WHO, 2012). On the contrary, the permissible limit in India and Bangladesh is set at 0.050 mg/L, as specified by the Bureau of Indian Standards (2012).

Upon entry into the human body, arsenic profoundly disrupts physiological processes, negatively affecting multiple organ systems, including the respiratory, gastrointestinal, cardiovascular, immune, reproductive, and central nervous systems (Hong et al., 2014). In the respiratory system, exposure to arsenic trioxide (AsO_3) can cause hemorrhage within the mucosal, submucosal, and alveolar regions and is accompanied by a great risk of pulmonary tuberculosis and lung cancer (Smith et al., 2011). Cardiovascular symptoms show the indication of hypertension (Abhyankar et al., 2012), hyperlipidemia, coronary heart disease (Hu et al., 2012) and vascular damage, which results in fatal outcomes (Chen et al., 2023). Hematological disorders such as anemia, suppressed erythropoiesis, and leucopenia (Muzaffar et al., 2023). Gastrointestinal impacts are noticeable by impaired liver function, which causes conditions such as cirrhosis, jaundice, and a high risk of gastrointestinal cancers, coupled with damage to the gastrointestinal epithelium (Prasad & Sahni, 2023). Arsenic's severe health effects, such as carcinogenic, mutagenic, enzyme inhibitory and genotoxic characteristics, highlight the need for its effective removal from aquatic environments to reduce public health risks. Recently, research efforts are dedicated to arsenic from aqueous solutions, utilizing various techniques such as coagulation, adsorption (Ali et al., 2014; Beolchini et al., 2007), ion exchange, flocculation, precipitation (Rashidi Nodeh et al., 2016), membrane filtration and bioremediation (Hesami et al., 2012; Lakshmanan et al., 2010; Rashidi Nodeh et al., 2016; Simon et al., 2024; Yogalakshmi et al., 2020), as well as advanced treatment techniques like ozonation and EC methods (Taha et al., 2022).

Hossain et al., (2005) performed the As removal in the West Bengal province, thoroughly monitoring 18 water treatment plants for 2 years from September 2001 to September 2003. These plants work with a combination of methods, including oxidation, adsorption, ion exchange, and catalytic precipitation, followed by coagulation/precipitation and filtration. Electrochemical techniques have arisen as alternatives to conventional treatment methods for efficient removal of arsenic from aqueous environments. These methods use electrons as reagents, thereby avoiding the need for additional chemicals and streamlining the purification process (Nidheesh & Singh, 2017). The electrochemical method generally proposes wide utility in curative functions (elimination, recovery, recycling) and precautionary applications such as electrosynthesis, supporting its developing adoption. The electron-driven oxidation procedure is particularly advantageous due to its extensive applicability, high selectivity, and lower propensity for generating secondary pollutants in comparison to old technologies. Additionally, the ease and cost effectiveness of the required design and operations, when properly engineered, offer important advantages over conventional technologies (Rai et al., 2021).

In trivalent form As often resists the effective removal using standard treatment methodologies. However, electrochemical techniques have demonstrated robust capabilities in removing both arsenate and arsenite from water systems. This chapter further describes the invaluable tool for scientists, providing thorough insights into various electrochemical methods for As removal, including indirect electrochemical oxidation (IEO), electrocoagulation (EC), anodic oxidation, capacitive deionization (CD), and electro-Fenton/bio-electro-Fenton processes (BEF), whether applied individually or in combined systems. Additionally, this chapter thoroughly assesses the impact of factors such as pH, initial As level, voltage, and the presence of various dissolved ions on the effectiveness of these electrochemical strategies.

7.2 Electrochemical Methods Used for Removal of Arsenic

Electrochemical treatment methodologies have been subjected to extensive consideration due to environmental compatibility and adaptability. These methods present an intriguing alternative as they enable simultaneous electrochemical and chemical reactions, facilitating the treatment of liquids, gases, and solids. In these methods, the electron acts as the primary reagent, which avoids the need for additional chemicals (Rai et al., 2021). Electrochemical techniques can generally be divided into two main types: separation techniques and oxidation techniques, as illustrated in Fig. 7.1. Within the separation category, the methods are further broken down into electro-sorption, electrodialysis, and electrocoagulation.

Electrosorption is an advanced adsorption technique in which contaminants are removed from a solution by adsorbing onto electrified electrodes through an applied electric current. In this process, an electrochemical double layer is formed, significantly enhance contaminant removal efficiency. Bain et al., (2010) focuses on the effectiveness of electrosorption for arsenic removal, reported remarkably improved removal rates at applied voltages between 1.0 and 1.5 V and an electrolyte pH of 4.5. In this study, it was found that 100 μg/L initial As concentration of 100 g/L, electro-sorption, outperformed traditional adsorption methods. Electrodialysis

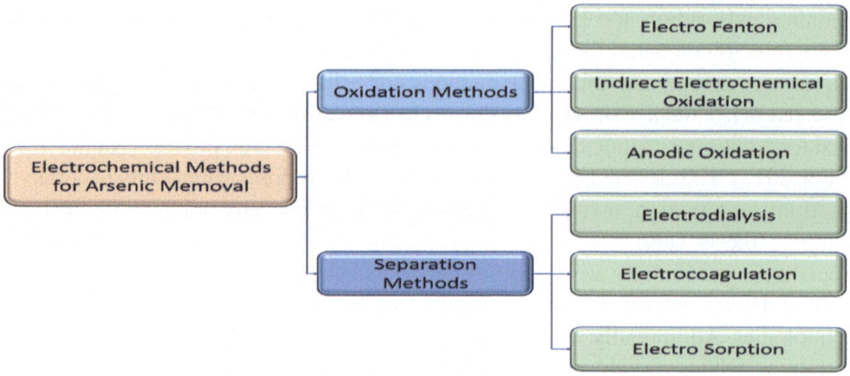

Fig. 7.1 Electrochemical methods for arsenic removal

is a membrane-based separation technique under the influence of an electric field that removes As ions through ion exchange or semipermeable membranes. In this process, an electric current is applied through the electrodes, which causes reduction and oxidation reactions, aiding the removal of oxidized ions across the membranes. This setup allows the transfer of counterions while restricted the passage of coions. Ribeiro et al., (2000) performed electrodialysis for As removal with an electrode of platinumized titanium and reached a removal efficiency over 30 days.

The EC process uses electrical energy instead of chemical coagulants to facilitate the removal of contaminants. The anode oxidizes when a current is applied; metal ions are released from the electrodes into the solution. In addition to the formation of metal oxides by these metal ions, arsenic is efficiently adsorbed and removed from the aqueous solution. Various studies performed how effective electrocoagulation is in removing arsenic (Kobya et al., 2022). The degree of oxidation of As has a major effect on the efficient removal of it from water. Arsenate can be removed more efficiently using technologies than arsenite (Kumar et al., 2004). This effectiveness difference is due to variances in their ionic charges; arsenite remains unchanged at pH less than 9.2 (Gutiérrez et al., 2022). Therefore, trivalent arsenic is less susceptible to removal by conventional techniques like precipitation, adsorption, and ion exchange (Kim & Nriagu, 2000). To remove As, peroxidation of arsenite to arsenate is necessary. Oxidation methods such as anodic oxidation (AO), indirect EC oxidation, and E–F are used to convert arsenite to arsenate. These methodologies support the oxidation of arsenite, which makes it more favorable to removal by subsequent separation methods. Further sections of this chapter will provide a comprehensive analysis of oxidation techniques and their role in arsenic removal strategies.

7.3 Separation Methods for Arsenic Removal

7.3.1 Electrocoagulation for Arsenic Removal

Electrocoagulation is an advanced electrochemical method in comparison to traditional chemical coagulation-flocculation processes for efficient removal of As from drinking water (Nidheesh & Singh, 2017). This method bypasses the external chemical need from conventional coagulation by producing coagulants directly by electrodissolution of substitute anodes in the water, such as iron or aluminum electrodes (Babu et al., 2019; Lacasa et al., 2013). EC offers multiple benefits, including reduced sludge production, enhanced flocculation capabilities, and superior separation efficiency (Mollah et al., 2001). The electrocoagulation process is elaborated in Fig. 7.2.

Electrocoagulation methodology involves anodic dissolution of the metal (Fe or Al), releasing Fe^{2+} ions into the solution in the case of iron (Eq. 7.1). When water is reduced in the cathode, hydroxide ions are created, which mix with these ions (see Eq. 7.2), leading to the formation of $Fe(OH)_2$. This compound then oxidizes to $Fe(OH)_3$ (see Eq. 7.3). The coagulants generated in the electrocoagulation process are effective in adsorbing and removing As(V) from the solution (illustrated in Eq. 7.4). Moreover, as a byproduct hydrogen gas is released, contributing in the removal of arsenic via electroflotation (depicted in Eq. 7.5).

$$Fe_{(S)} \rightarrow Fe^{2+} + 2e^{-} \tag{7.1}$$

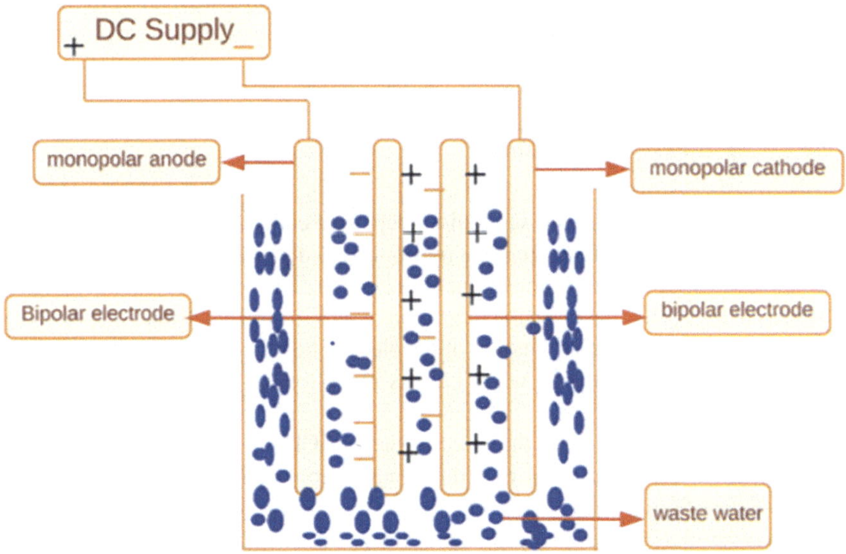

Fig. 7.2 Electrocoagulation process

$$2H_2O + 2e^- \rightarrow 2OH^- + 2H^+ \tag{7.2}$$

$$Fe^{2+} + 2OH^- \rightarrow Fe(OH)^{-2} + Fe(OH)_3 \tag{7.3}$$

$$As(V) + Fe(OH)_3 \rightarrow FeAsO_4 \tag{7.4}$$

$$2H^+ + 2e^- \rightarrow H_2 \uparrow \tag{7.5}$$

Lacasa et al., (2013) performed a relative study on the efficiency of coagulation versus electrocoagulation for the removal of arsenate from high-arsenic groundwater. Their findings indicated that electrocoagulation, using Fe and Al anodes as shown in Fig. 7.2, was superior in reducing arsenic levels below allowable limits, even though in the most polluted zones. The study revealed that electrocoagulation achieved almost complete arsenic removal, while conventional coagulation reached a maximum efficiency of 80% with Fe electrodes and 70% with Al electrodes. The proficiency of both methods was influenced by pH variations, with optimal arsenic removal observed at pH 8 to 10 for aluminum and pH 2 to 12 for iron, where specific hydroxide and arsenate species predominate.

Furthermore, Mohora et al., (2018) explored the use of Fe bipolar anodes for a horizontal continuous electrocoagulation technique to treat As in groundwater. At an initial As concentration of 38.15 μg/L, current density of 1.98 A/m^2 and a flow rate of 12 L/h, the removal efficiency of up to 96% was reported. Hence, electrocoagulation represents a highly effective and efficient method for arsenic removal, leveraging in situ generated coagulants and minimizing chemical usage, thus providing a superior alternative to conventional treatment methods.

7.3.2 Capacitive Deionization

A complex electric field-driven separation technique called capacitive deionization (CDI) is intended to effectively remove ions or charged particles from aqueous solutions, even at small concentrations (Fan et al., 2016). The core mechanism of CDI, illustrated in Fig. 7.3, is based on the migration of ions toward electrodes that bear the opposite charge due to the application of the electric field. Anions are drawn toward the positively charged cathode, while cations move toward the negatively charged anode.

Fan et al., (2016) estimated the effectiveness of CDI in removing arsenic. The CDI setup included a couple of electrodes made from activated carbon equipped with an electrosorption unit. These electrodes were produced by mixing activated carbon powder in a 9: 1 weight ratio with polyvinylidene fluoride (PVDF) and using N, N-dimethylacetamide as the solvent to create a suspension. This slurry was then

Fig. 7.3 Capacitive deionization

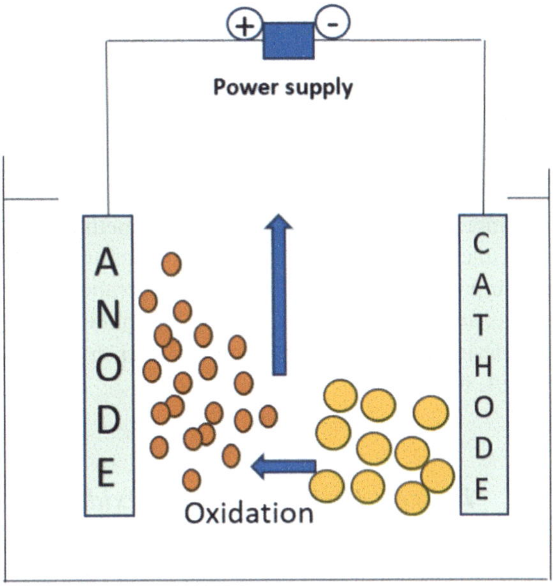

spread onto a titanium plate, resulting in electrodes with a dry weight of 0.5 g of activated carbon each. Positioned with a 2 mm gap between them to allow for fluid flow, for 120 min, the system was tested with an initial As of 0.2 mg/L and a flow rate of 5 ml/min. The performance of CDI system was assessed by varying both the voltage (0.0 to 1.2 V) and initial concentrations of As (0.01–200 mg/L). At elevated voltages, the adsorption capacities of arsenite and arsenate on the electrodes were found to be 1.37×10^{-2} mg/g and 2.47×10^{-2} mg/g, respectively. Arsenate enhanced sorption was observed at pH levels around 5.5 to 6, attributed to its negative charge within this pH range, facilitating its electrostatic adsorption on the carbon electrode. On the contrary, arsenite, which remains uncharged between pH 6.0 and 7.6, demonstrated a lower sorption efficiency. Consequently, negatively charged arsenate complexes were preferentially adsorbed onto the anode in the CDI system. In a concentration range of 0.01 to 200 mg/L, the adsorption efficiency of arsenate improved from 0.0116 to 10.31 mg/g, while arsenite increased from 0.0077 to 7.57 mg/g. Complete arsenate adsorption was observed at lower concentrations; however, the efficiency decreased at higher concentrations because of electrode saturation. The comparatively lower adsorption efficiency for arsenite is due to the limited oxidation of As(III) to As(V) under experimental conditions described (Fan et al., 2016). Therefore, the CDI presents an efficient method for the removal of As, using electrostatic interactions between charged arsenic species and activated carbon electrodes, with efficiency influenced by factors such as voltage, pH, and initial concentration of arsenic.

7.4 Oxidation Technologies for the Removal of Arsenic

Oxidation technologies have been shown to be effective for the removal of arsenic, and since arsenic (III) is difficult to remove by conventional adsorption or filtration techniques, oxidation shows an important role in converting arsenite to arsenate, which is removed through processes such as coagulation, precipitation, adsorption, or membrane filtration. Various oxidation methods, such as chemical oxidation, electrochemical oxidation, and advanced oxidation processes (AOPs), are employed for this purpose. Careful optimization of operational parameters such as pH, oxidant dosage, and contact time is essential to avoid incomplete oxidation or the formation of undesirable by-products, allowing safer drinking water standards (WHO limit) 10 μg/L for arsenic in drinking water.

7.4.1 Indirect Electrochemical Oxidation

Oxidizing arsenite to arsenate is essential to achieve maximum removal efficiency in arsenic treatment processes. Electrochemical oxidation is a superior method compared to chemical oxidation because it facilitates pollutant oxidation directly on the surface of the anode. The efficacy of EC arsenite oxidation lies in various key factors, including solution pH, current density, and electrode material.

Perez et al., (2007) examined the use of porous reticulated vitreous carbon (RVC) as an anode and cathode of graphite in an EC reactor for the electrochemical oxidation of arsenite (seen in Fig. 7.4), the reactor's design included a cationic ion exchange ultrafiltration membrane to divide the anode and cathode compartments. An electrolyte solution containing 0.10 M NaCl and 10 mg/L of arsenic was fed into the anodic chamber, while only 0.10 M NaCl was present in the cathodic chamber. To inhibit chemical oxidation, oxygen-free nitrogen gas was continuously purged through the anodic solution during the experiments. The study demonstrated enhanced removal of As efficiency at a voltage of 3 V, attributed to an increase in the oxidation rate at higher voltages. In the presence of 0.10 M NaCl at 3 V, oxidation and reduction occur. This process highlights the importance of optimizing voltage and supporting electrolyte conditions to maximize the electrochemical oxidation efficiency of arsenite.

7.4.2 Anodic Oxidation

Anodic oxidation is a refined EAOP which breaks down the various contaminants by creating hydroxyl radicals on the surface of the anode (Rabaaoui et al., 2013). To generate hydroxyl radicals in aqueous environments, there are two methods: DAO and IAO (Feng et al., 2016; Rahmani et al., 2015). Pollutant degradation occurs in

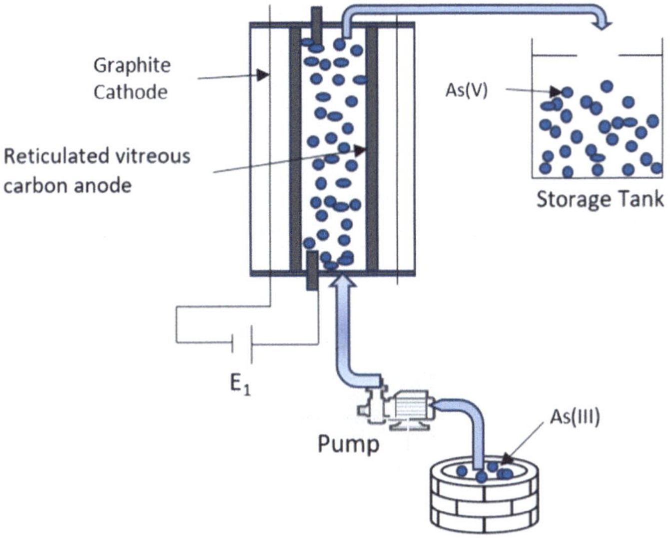

Fig. 7.4 Indirect electrochemical reactor design

the DAO process upon adsorption onto the anode surface, with the electron being the sole reactant. In contrast, IAO includes the participation of additional elements, such as oxygen evolution, to facilitate the oxidation process, but DAO does not include (Feng et al., 2016). The efficacy of anodic oxidation is based on the anode material (Panizza & Martinez-Huitle, 2013). Anode materials are characterized in two groups based on their simplicity and oxygen evolution capacity: active anodes and non-active anodes. Active anodes, such as IrO_2, PtO_2, and RuO_2, exhibit relatively low oxygen evolution overpotentials, which encourage contaminants to oxidize partially and selectively. While, high oxygen evolution overpotentials including boron-doped diamond (BDD), PbO_2, and SnO_2, allow for the total and non-selective oxidation of pollutants in non-active anodes (Feng et al., 2016). Figure 7.5 shows the mechanism of the elimination of pollution based on anodic oxidation. Extremely reactive Hydroxyl radicals (OH), and these also able to decomposing a wide range of organic and inorganic pollutants that are produced during this mechanism. The efficacy of removal of pollutant is highly dependent on the electrochemical properties of the anode material, current density, and the reactor's operating conditions. Optimal selection of anode materials and process parameters can achieve high removal efficiencies for various contaminants.

Zhao et al., (2010) inspected the dimensionally stable anodes in the arsenite oxidation and then electrocoagulation for its removal. The electrochemical reactor used in this study featured five electrodes: three Fe electrodes in the middle and two dimensionally stable anodes at each end, arranged parallel and spaced 2 cm apart. The experiments were carried out at a pH of 8 with a primary As level of 1 mg/L. The research showed that the dimensionally stable anodes effectively oxidized

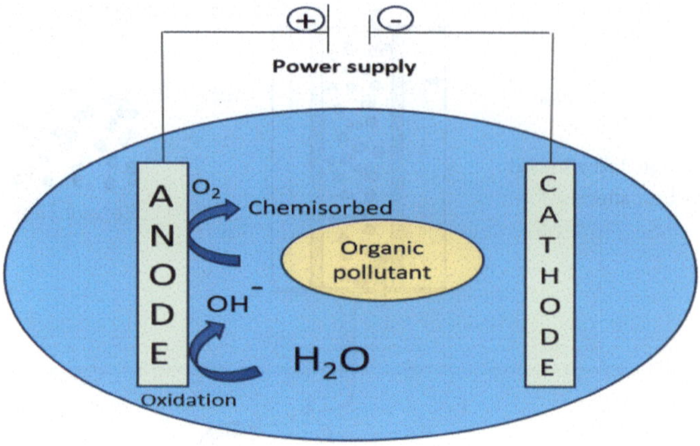

Fig. 7.5 Anodic oxidation process

arsenite, which was then efficiently removed as arsenate via electrocoagulation. It was found that the removal efficiency was pH-sensitive, suggesting that even small changes in pH would result in a reduction in efficiency because less metal arsenite/ arsenate complexes would form. The solubility of As particles is influenced by pH of solution, which is related to this effect. The research also evaluated the impact of interference from other metals such as bicarbonate, calcium, chloride, phosphate, silicate, and magnesium at various levels. Phosphate was identified as having the most substantial negative effect on the process, likely due to its structural resemblance to arsenite (Zhao et al., 2010). The sequence of anion interference on arsenic removal was ranked as phosphate > chloride > silicate > bicarbonate. On the contrary, cations such as magnesium and calcium were found to enhance arsenic removal, according to findings by Kumar and Goel (2010).

Zhang et al. (2015) performed removal of As from GW by a sequential process of anodic oxidation followed by electrocoagulation. The experimental setup utilized a composite anode made from Sb_2O_5 and SnO_2 coated on Ti/TiO_2 nanotubes, with the cathode of Fe. The operating parameters included initial arsenite = 6.67 μg/L, pH = 7, and a current = 50 mA. The procedure lasted 70 min, with the first 60 min devoted to the anodic oxidation of arsenite and the final 10 min focused on electrocoagulation, achieved by using the iron electrode as a sacrificial anode and reversing the polarity. Higher currents were found to enhance the formation of iron precipitates, which is why the effectiveness of arsenate removal increased with current. The entire amount of arsenic in groundwater was successfully removed using this integrated method. Additionally, the study found that bicarbonate and sulfate had a negative impact on the oxidation of arsenite, while chloride notably improved the oxidation process.

7.4.3 Bio-electro-Fenton (BEF)

BEF techniques leverage electrochemical processes to generate Fenton's reagents on site, thereby enhancing the oxidation and removal of pollutants. In this method, an electrical generator increases the production of H_2O_2 at the cathode through the reduction of dissolved O_2. Subsequently, this hydrogen peroxide interacts with iron ions, initiating the Fenton reaction to produce highly reactive OH (Brillas et al., 2009; Nidheesh & Gandhimathi, 2015). Parallelly, the oxidation of H_2O occurs at anode.

Because OH is continuously produced in situ in bio-electro-Fenton systems, contaminants can be persistently oxidized and removed. As described in Eq. (7.6) electrons are generated through biological processes in the anodic region. An electrical circuit carries these electrons to the cathode. Equations (7.7) and (7.8) state that Fe^{2+} is generated in situ by the cathodic iron material's corrosion, while H_2O_2 is continuously generated at the cathode by oxygen reduction of oxygen (Kahoush et al., 2018). The interaction of Fe^{2+} with H_2O_2 results in the production of extremely reactive OH.

$$CH_3COO^- + 2H_2O \rightarrow^{microbial} 2CO_2 + 7H^+ + 8e^- \quad (7.6)$$

$$O_2 + 2H^+ + 2e^- \rightarrow H_2O_2 \quad (7.7)$$

$$Fe^{3+} + e^- \rightarrow Fe^{2+} \quad (7.8)$$

Equation (7.9) describes how the Shewanella putrefaciens SP200 strain oxidizes lactate, producing electrons at the anode that are then transferred to the cathodic chamber in the bioelectro-Fenton process. Equations (7.10) and (7.11) describe how these electrons in the cathode help reduce oxygen to H_2O_2 and Fe^{3+} to Fe^{2+}, the latter of which occurs through solid-phase reduction before desorbing into the solution. By reducing O_2 and γ-FeOOH simultaneously, H_2O_2 and Fe^{2+} are produced, which subsequently undergo the Fenton reaction to generate hydroxyl radicals.

Hydroxyl radicals, along with the H_2O_2 and Fe^{2+}–Fe^{3+} intermediates, actively oxidize arsenite present in the solid and aqueous phases (as shown in Eqs. (7.12–7.14)).

$$CH_3CH(OH)COO^- \rightarrow_{S.putrefacines\ \mathrm{sp.}\ 200} HCO_3^- + H^+ \quad (7.9)$$

$$Fe(III)_{(S)} \rightarrow^{Solid\ phase\ reduction} Fe(II)_{(S)} \quad (7.10)$$

$$Fe(II)_{(S)} \rightarrow^{Desorbtion} Fe(II)_{(aq)} \quad (7.11)$$

$$As(III)_{(S)} \rightarrow^{Oxidise} As(V)_{(S)} \quad (7.12)$$

$$As(III)_{(aq)} \rightarrow^{H_2O_2} As(V)_{(aq)} \quad (7.13)$$

$$As(III)_{(aq)} \rightarrow^{Fenton(.OH)} As(V)_{(aq)} \quad (7.14)$$

This indicates that the BEF process integrates microbial fuel cell technology by electro-Fenton reaction, achieving efficient oxidation and removal through in situ generation of hydroxyl radicals and optimized pollutant degradation.

7.4.4 Other Methods

Under alkaline conditions in an electrolytic cell, oxygen is reduced in the cathode to produce active oxidants (Mao et al., 2012). For electrolytic elimination of arsenic using a 30 cm long and 5.0 cm diameter electrolytic column. This column is divided into three sections, each of which has electrodes that serve as both anodes and cathodes (Qian et al., 2015). The electrodes are coated with titanium diamond mesh and are composed of IrO_2 which is mixed with Ta_2O_5. These electrodes are arranged sequentially upwards to foster an alkaline environment around the cathode. Figure 7.6 shows the electrolyte mechanism for arsenic removal and reactions at electrodes.

At the bottom of reactor arsenite was added to the electrolyte solution, which had a pH of 7, at a starting concentration of 6.67 μM. When an applied current of 30 mA was provided at the cathode, OH generation caused the pH to increase to 11, while

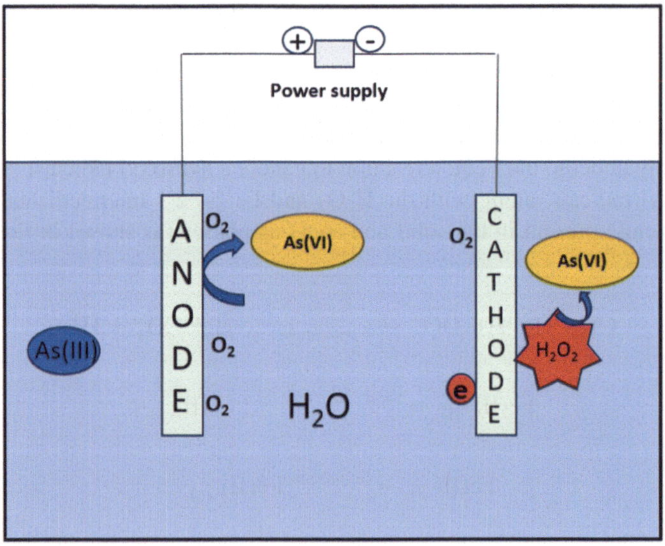

Fig. 7.6 Electrolyte method for the arsenic removal

H^+ ion production caused the pH to decrease to approximately 2.5 near the anode. H^+ ions were found to migrate 1.76 times faster than OH^+ ions.

Arsenic concentration decreased by 28% at the anode and at cathode by 23%, accompanied by increased levels of arsenate. A minor overall reduction in total arsenic was also observed, likely because at higher pH values adsorption was found. The O_2 produced in the anode traveled to the cathode, where it was converted to active oxidants like H_2O_2, resulting in a 94% oxidation of arsenite within 60 min. The presence of chloride (Cl^-) and bicarbonate (HCO^{3-}) in the groundwater and soil positively influenced arsenite oxidation. Bicarbonate expedited the oxidation process by forming CO_3^{2-}-As(III) compounds. The oxidation efficiency increased from 65 to 72% at the cathode with a bicarbonate concentration of 2 mM. However, an increase to 5 mM bicarbonate reduced the oxidation efficiency to 45%, due to a corresponding decrease in pH. Cathodic oxidation is notably sensitive to pH changes, and at a concentration of 4 mM, the occurrence of Cl^- increased the anodic oxidation of arsenite by 96%. Additionally, an increase from 36 to 43% has been found in cathodic oxidation rates. Calcium (Ca^{2+}) and magnesium (Mg^{2+}) ions also facilitated the oxidation process by forming calcium and magnesium precipitates (Qian et al., 2015).

7.5 Comparison of Electrochemical Techniques

Separation methods such as electrocoagulation (EC) and capacitive deionization (CDI) have been shown to be very effective in eliminating arsenate in a broad range of concentrations, outperforming traditional methods like adsorption technologies (Ociński et al., 2014) and chemical coagulation (Lakshmanan et al., 2010). However, electrocoagulation typically struggles to effectively remove arsenite, and several studies highlighting its limitations. Nonetheless, some reports indicate successful arsenite removal using EC treatments, often credited to the pre-oxidation of arsenite to more readily removable forms (Flores et al., 2014). Before utilizing separation technologies for arsenic removal, converting As(V) to As(III) is crucial. Indirect electrochemical oxidation effectively achieves near-complete arsenite oxidation by using reticulated vitreous carbon anodes with a porosity of 100 pores per inch. Anodic oxidation, an advanced oxidation process, employs dimensionally stable anodes that produce hydroxyl radicals with high oxidative potential. These radicals efficiently transform arsenite into arsenate, facilitating its subsequent removal by electrocoagulation. Furthermore, the bio-electro-Fenton method, which oxidizes approximately 54.7% of arsenite, depends on the production of active OH by the interaction between H_2O_2 and Fe^{2+} (Wang et al., 2014).

7.6 Parametric Effects

7.6.1 pH

The highest removal efficiencies for arsenite using the electrocoagulation process are achieved within a pH between 5 and 8, whereas for arsenate, optimum pH range varies from 5 to 7.5 (Lakshmanan et al., 2010). The speciation and ionic charges of arsenic and metal ions impact the effectiveness of electrocoagulation across different pH levels. Arsenite exists mainly as a neutral species between pH 3 and 9, becoming negatively charged at pH levels from 9 to 11. In contrast, arsenate generally carries a negative charge throughout the pH range of 3 to 11 (Gupta & Chatterjee, 2017; Nidheesh & Singh, 2017). Arsenic elimination is enhanced by the generation of positively charged ferrous and ferric hydroxides, which are prevalent between pH 5 and 8. Similarly, Al polymer elements that form at pH 5 to 7 are effective in removing arsenic (Lacasa et al., 2013; Mechelhoff et al., 2013).

Capacitive deionization (CDI) demonstrates a higher removal efficiency for arsenate compared to that for arsenite. The most effective removal of arsenate occurs at a pH of 5.5–6, where arsenite remains uncharged and arsenate is negatively charged. This charge difference leads to a stronger electrostatic adsorption of arsenate onto the anode compared to that of arsenite. Further research using real groundwater indicated that these efficiency levels remained consistent, showing no significant variations (Fan et al., 2017). Sequential anodic oxidation coupled with electrocoagulation achieves the maximum removal of As at an optimal pH level of 8. Deviations from this pH result in decreased removal proficiency due to changes in the generation rates of arsenite/metal-arsenate complexes, which are influenced by their solubility (Zhang & Jiao, 2018; Zhao et al., 2010). Furthermore, the production of hydrogen peroxide and the reduction of O_2 in the cathode under acidic conditions introduce a new approach to arsenite formation (Qian et al., 2015). When the pH of the arsenite was initially 7.5, almost 64% of it oxidized in 60 min. With a current of 30 mA, the pH near the cathode increased to approximately 11.0. (Qian et al., 2015). The E–F process is generally conducted at a low pH of 3. The production of OH is crucial for the process but becomes less effective as the pH increases. However, the BEF process has been successfully adapted to work at neutral pH levels for the oxidation of arsenic. In this variation, the pH in the anode chamber decreases to approximately 6.5, while it slightly increases in the cathode chamber due to the combination of protons and the consumption of the fuel (Wang et al., 2014). Therefore, optimizing pH is important for the efficacy of the EC and E–F methods in the removal of As. The selection of pH, current, and understanding of the behavior of coexisting ions play vital roles in maximizing the effectiveness of these advanced oxidation and separation techniques.

7.6.2 Voltage

The applied voltage is important in the EC technique, as it directly influences the current density. This, in turn, influences factors such as electrode dissolution, floc size, bubble production rate, and growth dynamics; these factors are vital to determining the removal efficiency. Elevating the voltage from lower to higher levels significantly improves arsenic removal efficiency, reduces the required operation time, and helps ensure that the residual arsenic concentration adheres to permissible limits (Ali et al., 2014; Kobya et al., 2022). Similarly, As removal efficiency in capacitive deionization heavily depends on the voltage applied. An increase in voltage corresponds to an enhanced rate of arsenic adsorption. For instance, an increment in the voltage from 0.4 to 0.8 V has not been effective in comprehensive arsenate elimination, but a further increase to 1.2 V led to total arsenate elimination. At a voltage of 1.2 V and pH levels between 5.5 and 6.5, arsenite removal was less effective due to its uncharged state under these conditions (Fan et al., 2016; Xu et al., 2019). Additionally, the applied cell potential significantly impacts the indirect electrochemical reactions that convert arsenite to arsenate. At cell voltages of 2.0, 2.5, 3.0, and 3.5 V, oxidation percentages were recorded at 33%, 40%, 59% and 85%, respectively, over a reaction time of 20 min (Perez et al., 2007). Anodic oxidation, crucial for arsenic removal, is a current-dependent process that involves both indirect and direct oxidation mechanisms, and indirect oxidation takes precedence at currents ranging from 30 to 80 mA, aligning with findings from other studies (Carter & Farrell, 2009).

7.6.3 Dissolved Ions

The effectiveness of EC methods such as electrocoagulation for As removal is significantly affected by dissolved ions in drinking and groundwater. For instance, phosphate ions markedly reduce arsenic removal efficacy due to competitive adsorption with arsenic components on metal hydroxides (Hong et al., 2014). Conversely, silica typically does not impact arsenic removal via EC due to its lower affinity for metal hydroxides than arsenic, though at higher concentrations, silica can inhibit arsenate adsorption. During the EC process, initially, it is found that phosphate and arsenate are adsorbed, followed by silica and arsenite as the generation of metal hydroxides increases (Li et al., 2012). It is widely recognized that calcium ions enhance the removal of As from both Fe/Al based EC systems. However, fluoride ions considerably reduce the removal efficiency, while sulfate ions have little to no impact. The presence of chloride ions improves efficiency by improving anode dissolution through pitting corrosion and preventing passivation (Bayramoglu et al., 2007; Colli et al., 2017). On the contrary, due to passivation effects, carbonate ions inhibit arsenic removal at concentrations greater than 5 mg/L. A significant increase in initial fluoride level between 0 and 5 mg/L gradually decreases arsenate removal due to the competitive interaction between fluoride and arsenate ions (Thakur & Mondal, 2017; Vasudevan et al., 2010).

In capacitive deionization, ions such as magnesium (Mg^{2+}), chloride (Cl^-), nitrate (NO_3^-), sulfate (SO_4^{2-}), fluoride (F^-), calcium (Ca^{2+}), potassium (K^+), and sodium (Na^+) are typically removed alongside arsenic without significantly impacting its removal efficiency (Fan et al., 2017). The preference for electrosorption of anions follows the order: nitrate > sulfate > fluoride > chloride > arsenic, and for cations, the order is: calcium > magnesium > sodium > potassium (Fan et al., 2017). However, high concentrations of sodium chloride (20 mM) can reduce sorption capacity by replacing arsenic ions with chloride ions on the anode surface. Furthermore, the presence of humic acid adversely affects the removal of arsenic in CDI systems (Xu et al., 2019). Table 7.1 represents the effective parameters of various electrochemical methods.

Table 7.1 Properties of electrochemical methods

S. No.	Method	Advantages	Disadvantage	References
1	Bio electro Fenton	Cathodic formation of hydrogen peroxide and in-situ formation of Fenton reagents, Cost-effective	High sludge formation, highly dependent on pH, time-consuming process, complex to operate	Feng et al. (2010), Syam Babu and Nidheesh (2021), Xu et al. (2013)
2	Anodic oxidation	Degrade pollutants, no dependency on pH, no chemical use, so sludge generation	Expensive, generates toxic pollutants, and causes electrode fouling	Qian et al. (2015), Zhang et al. (2014)
3	Capacitive deionization	Can be used for large volumes and salty water, cost-effective, energy-efficient	Highly dependent on pH, electrode fouling, and complex electrode design	Fan et al., (2016, 2017), Syam Babu and Nidheesh (2021)
4	Indirect electrochemical methods	A high oxidation rate can be used in low concentrations to eliminate additional chemicals	High maintenance, complex electrode design	Rahmani et al. (2015), Syam Babu and Nidheesh (2021)
5	Electrocoagulation	Easy to operate, low cost, no chemicals used, wide application, hydrogen gas generation at cathode improves efficiency	Depending on the electrolytic conductivity, the requirement of regular replacement of sacrificial anode	Li et al. (2012), Nidheesh and Singh (2017), Wang et al. (2022)

7.7 Conclusion

In this chapter, we explore various electrochemical methods for treating arsenic-contaminated water. Electrocoagulation (EC) is a robust method capable of treating various types of water types and effectively removing both organic and inorganic contaminants without generating secondary waste. The efficiency of EC is enhanced by hydrogen evolution, influenced by the electrode overpotential and cathode reaction zone size. Nevertheless, its effectiveness depends significantly on the conductivity of the electrolytes. The formation of an oxide layer on the anode causes electrode passivation, a significant limitation of EC. This problem can be mitigated by using AC instead of DC and by increasing aeration during the process (Islam, 2023; Singh et al., 2013). CDI is notable for its low-pressure requirements and low capital costs, consumes approximately one-third the energy of reverse osmosis and is broadly applicable. However, electrode fouling remains a challenge in CDI, addressed through techniques such as alternating electrode potentials or applying pulsed electric fields (Fan et al., 2017; Xu et al., 2019). The indirect electrochemical technique has been proficient in degrading organic compounds and removing inorganic compounds, even at very low concentrations. Although it is cost-effective for removing musty odors, it is generally suitable only for low-volume applications. Anodic oxidation, a traditional method used for surface modification, utilizes dimensionally stable anodes with a high oxidation potential for pollutant degradation to produce hydroxyl radicals. Its primary benefits include the capability to operate atmospheric pressure and room temperature, zero sludge generation, and non-selective and full degradation of pollutants. The E–F and bio-E–F processes are recognized to be among the most eco-friendly and cost-effective advanced oxidation processes compared to traditional Fenton methods. These techniques are particularly effective for treating low concentrations of persistent micropollutants, although the bioelectro-Fenton process is slower and requires meticulous maintenance of microbial conditions. These electrochemical methods represent practical and effective technologies to address arsenic contamination across diverse regions, because of their cost-effectiveness and environmentally benign operations. However, challenges such as electrode stability and the maintenance of optimal reaction conditions over prolonged periods persist.

Acknowledgements The authors express their gratitude to the reviewers and the editor for dedicating their valuable time to enhancing this work.

References

Abhyankar, L. N., Jones, M. R., Guallar, E., & Navas-Acien, A. (2012). Arsenic exposure and hypertension: A systematic review. *Environmental Health Perspectives*. https://doi.org/10.1289/ehp.1103988

Ali, I., Al-Othman, Z. A., Alwarthan, A., Asim, M., & Khan, T. A. (2014). Removal of arsenic species from water by batch and column operations on bagasse fly ash. *Environmental Science and Pollution Research*. https://doi.org/10.1007/s11356-013-2235-3

Babu, D. S., Srivastava, V., Nidheesh, P. V., & Kumar, M. S. (2019). Detoxification of water and wastewater by advanced oxidation processes. *Science of the Total Environment*. https://doi.org/10.1016/j.scitotenv.2019.133961

Bain, E. J., Calo, J. M., Spitz-Steinberg, R., Kirchner, J., & Axén, J. (2010). Electrosorption/electrodesorption of arsenic on a granular activated carbon in the presence of other heavy metals. *Energy and Fuels*. https://doi.org/10.1021/ef901542q

Bayramoglu, M., Eyvaz, M., & Kobya, M. (2007). Treatment of the textile wastewater by electrocoagulation. Economical evaluation. *Chemical Engineering Journal*. https://doi.org/10.1016/j.cej.2006.10.008

Beolchini, F., Pagnanelli, F., De Michelis, I., & Vegliò, F. (2007). Treatment of concentrated arsenic (V) solutions by micellar enhanced ultrafiltration with high molecular weight cut-off membrane. *Journal of hazardous materials, 148*(1–2), 116-121. https://doi.org/10.1016/j.jhazmat.2007.02.031

BIS. (2012). Indian Standard Drinking Water Specification (Second Revision). India: Bureau of Indian Standards.

Brillas, E., Sirés, I., & Oturan, M. A. (2009). Electro-fenton process and related electrochemical technologies based on fenton's reaction chemistry. *Chemical Reviews*. https://doi.org/10.1021/cr900136g

Carter, K. E., & Farrell, J. (2009). Electrochemical oxidation of trichloroethylene using boron-doped diamond film electrodes. *Environmental Science and Technology*. https://doi.org/10.1021/es9017738

Chen, M., Hu, H., Chen, M., Wang, C., Wang, Q., Zeng, C., Shi, Q., Song, W., Li, X., & Zhang, Q. (2023). In-situ production of iron flocculation and reactive oxygen species by electrochemically decomposing siderite: An innovative Fe-EC route to remove trivalent arsenic. *Journal of Hazardous Materials*. https://doi.org/10.1016/j.jhazmat.2022.129884

Colli, A. N., Toelzer, R., Bergmann, M. E. H., Bisang, J. M., Zhang, T., Shao, T., Zhang, Y., Pan, D., Huang, W., Li, B., Hakizimana, J. N., Gourich, B., Chafi, M., Stiriba, Y., Vial, C., Drogui, P., Naja, J., Yoon, S. K., Fichtl, W., & Kenis, P. J. A. (2017). Electrocoagulation process in water treatment: A review of electrocoagulation modeling approaches. *Desalination, 404*(August 2021), 1–21. https://doi.org/10.1016/j.desal.2016.10.011

Fan, C. S., Liou, S. Y. H., & Hou, C. H. (2017). Capacitive deionization of arsenic-contaminated groundwater in a single-pass mode. *Chemosphere*. https://doi.org/10.1016/j.chemosphere.2017.06.068

Fan, C. S., Tseng, S. C., Li, K. C., & Hou, C. H. (2016). Electro-removal of arsenic(III) and arsenic(V) from aqueous solutions by capacitive deionization. *Journal of Hazardous Materials*. https://doi.org/10.1016/j.jhazmat.2016.03.055

Feng, C. H., Li, F. B., Mai, H. J., & Li, X. Z. (2010). Bio-electro-fenton process driven by microbial fuel cell for wastewater treatment. *Environmental Science and Technology*. https://doi.org/10.1021/es9032925

Feng, Y., Yang, L., Liu, J., & Logan, B. E. (2016). Electrochemical technologies for wastewater treatment and resource reclamation. *Environmental Science: Water Research and Technology*. https://doi.org/10.1039/c5ew00289c

Flores, O. J., Nava, J. L., & Carreño, G. (2014). Arsenic removal from groundwater by electrocoagulation process in a filter-press-type FM01-LC reactor. *International Journal of Electrochemical Science*. https://doi.org/10.1016/s1452-3981(23)10918-7

Gupta, D. K., & Chatterjee, S. (2017). Arsenic contamination in the environment: The issues and solutions. In *Arsenic contamination in the environment: The issues and solutions*. https://doi.org/10.1007/978-3-319-54356-7

Gutiérrez, Á., Rodríguez, J. F., Castañeda, L. F., Nava, J. L., Coreño, O., & Carreño, G. (2022). Abatement of As and hydrated silica from natural groundwater by electrocoagulation in a continuous plant having an electrolyzer and a flocculator-settler. *Separation and Purification Technology, 281*. https://doi.org/10.1016/j.seppur.2021.119895

Hesami, F., Bina, B., Ebrahimi, A., & Amin, M. M. (2012). Arsenic removal by coagulation using ferric chloride and chitosan from water. *International Journal of Environmental Health Engineering*. https://doi.org/10.4103/2277-9183.110170

Hong, Y. S., Song, K. H., & Chung, J. Y. (2014). Health effects of chronic arsenic exposure. *Journal of Preventive Medicine and Public Health*. https://doi.org/10.3961/jpmph.14.035

Hossain, M. A., Sengupta, M. K., Ahamed, S., Rahman, M. M., Mondal, D., Lodh, D., Das, B., Nayak, B., Roy, B. K., Mukherjee, A., & Chakraborti, D. (2005). *Ineffectiveness and poor reliability of arsenic removal plants in West Bengal*. Environmental Science and Technology. https://doi.org/10.1021/es048703u

Hu, C., Liu, H., Chen, G., & Qu, J. (2012). Effect of aluminum speciation on arsenic removal during coagulation process. *Separation and Purification Technology*. https://doi.org/10.1016/j.seppur.2011.10.017

Islam, S. M. D. U. (2023). Electrochemical remediation of arsenic and fluoride from water : A review of the current state and future prospects. *Environmental Technology and Innovation, 31*. https://doi.org/10.1016/j.eti.2023.103148

Kahoush, M., Behary, N., Cayla, A., & Nierstrasz, V. (2018). Bio-Fenton and Bio-electro-Fenton as sustainable methods for degrading organic pollutants in wastewater. *Process Biochemistry*. https://doi.org/10.1016/j.procbio.2017.10.003

Kim, M. J., & Nriagu, J. (2000). Oxidation of arsenite in groundwater using ozone and oxygen. *Science of the Total Environment*. https://doi.org/10.1016/S0048-9697(99)00470-2

Kobya, M., Dolaz, M., Özaydın-Şenol, B., & Goren, A. Y. (2022). Removal of arsenic in groundwater from western Anatolia, Turkey using an electrocoagulation reactor with different types of iron anodes. *Heliyon, 8*(9). https://doi.org/10.1016/j.heliyon.2022.e10489

Kumar, N. S., & Goel, S. (2010). Factors influencing arsenic and nitrate removal from drinking water in a continuous flow electrocoagulation (EC) process. *Journal of Hazardous Materials*. https://doi.org/10.1016/j.jhazmat.2009.08.117

Kumar, P. R., Chaudhari, S., Khilar, K. C., & Mahajan, S. P. (2004). Removal of arsenic from water by electrocoagulation. *Chemosphere*. https://doi.org/10.1016/j.chemosphere.2003.12.025

Lacasa, E., Sáez, C., Cañizares, P., Fernández, F. J., & Rodrigo, M. A. (2013). Arsenic removal from high-arsenic water sources by coagulation and electrocoagulation. *Separation Science and Technology (Philadelphia)*. https://doi.org/10.1080/01496395.2012.690806

Lakshmanan, D., Clifford, D. A., & Samanta, G. (2010). Comparative study of arsenic removal by iron using electrocoagulation and chemical coagulation. *Water Research*. https://doi.org/10.1016/j.watres.2010.06.018

Li, L., Van Genuchten, C. M., Addy, S. E. A., Yao, J., Gao, N., & Gadgil, A. J. (2012). Modeling As(III) oxidation and removal with iron electrocoagulation in groundwater. *Environmental Science and Technology*. https://doi.org/10.1021/es302456b

Mao, X., Yuan, S., Fallahpour, N., Ciblak, A., Howard, J., Padilla, I., Loch-Caruso, R., & Alshawabkeh, A. N. (2012). Electrochemically induced dual reactive barriers for transformation of TCE and mixture of contaminants in groundwater. *Environmental Science and Technology*. https://doi.org/10.1021/es301711a

Mechelhoff, M., Kelsall, G. H., & Graham, N. J. D. (2013). Electrochemical behaviour of aluminium in electrocoagulation processes. *Chemical Engineering Science*. https://doi.org/10.1016/j.ces.2013.03.010

Mohora, E., Rončević, S., Agbaba, J., Zrnić, K., Tubić, A., & Dalmacija, B. (2018). Arsenic removal from groundwater by horizontal-flow continuous electrocoagulation (EC) as a standalone process. *Journal of Environmental Chemical Engineering*. https://doi.org/10.1016/j.jece.2017.12.042

Mollah, M. Y. A., Schennach, R., Parga, J. R., & Cocke, D. L. (2001). Electrocoagulation (EC)—Science and applications. *84*, 29–41.

Muzaffar, S., Khan, J., Srivastava, R., Gorbatyuk, M. S., & Athar, M. (2023). Mechanistic understanding of the toxic effects of arsenic and warfare arsenicals on human health and environment. *Cell Biology and Toxicology*. https://doi.org/10.1007/s10565-022-09710-8

Nidheesh, P. V., & Gandhimathi, R. (2015). Electro Fenton oxidation for the removal of Rhodamine B from aqueous solution in a bubble column reactor under continuous mode. *Desalination and Water Treatment*. https://doi.org/10.1080/19443994.2014.913266

Nidheesh, P. V., & Singh, T. S. A. (2017). Arsenic removal by electrocoagulation process: Recent trends and removal mechanism. *Chemosphere*. https://doi.org/10.1016/j.chemosphere.2017.04.082

Ociński, D., Jacukowicz-Sobala, I., & Kociołek-Balawejder, E. (2014). Oxidation and adsorption of arsenic species by means of hybrid polymer containing manganese oxides. *Journal of Applied Polymer Science*. https://doi.org/10.1002/app.39489

Panizza, M., & Martinez-Huitle, C. A. (2013). Role of electrode materials for the anodic oxidation of a real landfill leachate—Comparison between Ti-Ru-Sn ternary oxide, PbO_2 and boron-doped diamond anode. *Chemosphere*. https://doi.org/10.1016/j.chemosphere.2012.09.006

Perez, S., Lin Ho, S. W., & Felix, R. M. (2007). Electrochemical oxidation of arsenites, by an anode of reticulated glassy carbon as previous step for removal. *ECS Transactions. Doi, 10*(1149/1), 2806951.

Prasad, B. D., & Sahni, S. (2023). Mitigation of arsenic contamination through biotechnological approaches in rice. *Journal of Experimental Agriculture International*. https://doi.org/10.9734/jeai/2023/v45i122276

Qian, A., Yuan, S., Zhang, P., & Tong, M. (2015). A new mechanism in electrochemical process for arsenic oxidation: Production of H_2O_2 from anodic O_2 reduction on the cathode under automatically developed alkaline conditions. *Environmental Science and Technology*. https://doi.org/10.1021/acs.est.5b00808

Rabaaoui, N., Saad, M. E. K., Moussaoui, Y., Allagui, M. S., Bedoui, A., & Elaloui, E. (2013). Anodic oxidation of o-nitrophenol on BDD electrode: Variable effects and mechanisms of degradation. *Journal of Hazardous Materials*. https://doi.org/10.1016/j.jhazmat.2013.02.027

Rahmani, A. R., Godini, K., Nematollahi, D., & Azarian, G. (2015). Electrochemical oxidation of activated sludge by using direct and indirect anodic oxidation. *Desalination and Water Treatment*. https://doi.org/10.1080/19443994.2014.958761

Rai, V., Liu, D., Xia, D., Jayaraman, Y., & Gabriel, J. C. P. (2021). Electrochemical approaches for the recovery of metals from electronic waste: A critical review. *Recycling*. https://doi.org/10.3390/recycling6030053

Rashidi Nodeh, H., Wan Ibrahim, W. A., Ali, I., & Sanagi, M. M. (2016). Development of magnetic graphene oxide adsorbent for the removal and preconcentration of As(III) and As(V) species from environmental water samples. *Environmental Science and Pollution Research*. https://doi.org/10.1007/s11356-016-6137-z

Ribeiro, A. B., Mateus, E. P., Ottosen, L. M., & Bech-Nielsen, G. (2000). Electrodialytic removal of Cu, Cr, and As from chromated copper arsenate-treated timber waste. *Environmental Science and Technology*. https://doi.org/10.1021/es990442e

Shirin, S., & Yadav, A. K. (2014). Physico-chemical analysis of municipal wastewater discharge in Ganga river, Haridwar district of Uttarakhand, India. *Current World Environment, 9*(2), 536–543. https://doi.org/10.12944/CWE.9.2.39

Shirin, S., & Yadav, A. K. (2021). Assessment and suitability for the dumping of flyash in an abandoned mine. *Environmental Challenges, 4*, Article 100111. https://doi.org/10.1016/j.envc.2021.100111

Simon, M., Joshi, H., Yadav, A. K., & Hopke, P. K. (2024). Simultaneous nitrogen and phosphorus removal from polluted water using a novel bacterial consortium (BACON-38): Promising application for wastewater remediation. *Journal of Water Process Engineering, 66*, Article 106044. https://doi.org/10.1016/j.jwpe.2024.106044

Singh, K., Lataye, D. H., Wasewar, K. L., & Yoo, C. K. (2013). Removal of fluoride from aqueous solution: Status and techniques. *Desalination and Water Treatment*. https://doi.org/10.1080/19443994.2012.749036

Smith, A. H., Marshall, G., Yuan, Y., Liaw, J., Ferreccio, C., & Steinmaus, C. (2011). Evidence from chile that arsenic in drinking water may increase mortality from pulmonary tuberculosis. *American Journal of Epidemiology*. https://doi.org/10.1093/aje/kwq383

Syam Babu, D., & Nidheesh, P. V. (2021). A review on electrochemical treatment of arsenic from aqueous medium. *Chemical Engineering Communications*. https://doi.org/10.1080/00986445.2020.1715956

Taha, T. R., Castañeda, L. F., Coreño, O., Nava, J. L., Kobya, M., Dolaz, M., Özaydın-Şenol, B., Goren, A. Y., Chandraker, N., Chaudhari, P. K., Jyoti, G., Thakur, R. S., Oladipo, A. A., Mustafa, F. S., Ezugwu, O. N., Gazi, M., Enzyme, M. A., Copy, E., Version, E., … Griffioen, J. (2022). Removal of arsenic in groundwater from western Anatolia, Turkey using an electrocoagulation reactor with different types of iron anodes. *Journal of Environmental Chemical Engineering, 9*(5), 104–116. https://doi.org/10.1007/s10311-015-0496-4

Thakur, L. S., & Mondal, P. (2017). Simultaneous arsenic and fluoride removal from synthetic and real groundwater by electrocoagulation process: Parametric and cost evaluation. *Journal of Environmental Management, 190*, 102–112. https://doi.org/10.1016/j.jenvman.2016.12.053

Vasudevan, S., Sheela, S. M., Lakshmi, J., & Sozhan, G. (2010). Optimization of the process parameters for the removal of boron from drinking water by electrocoagulation—A clean technology. *Journal of Chemical Technology and Biotechnology, 85*(7), 926–933. https://doi.org/10.1002/jctb.2382

Wang, X. Q., Liu, C. P., Yuan, Y., Li, F., & Bai. (2014). Arsenite oxidation and removal driven by a bio-electro-Fenton process under neutral pH conditions. *Journal of Hazardous Materials*. https://doi.org/10.1016/j.jhazmat.2014.05.003

Wang, Y., Liu, L., Yang, X., Suib, S. L., & Qiu, G. (2022). Removal of As(V) from wastewaters using magnetic iron oxides formed by zero-valent iron electrocoagulation. *Journal of Environmental Management, 307*(January). https://doi.org/10.1016/j.jenvman.2022.114519

World Health Organization, World Health Organization, & World Health Organization. (2012). Regional Office for Europe (WHO, 2012). *Environmental health inequalities in Europe. Copenhagen.*

Xu, N., Zhang, Y., Tao, H., Zhou, S., & Zeng, Y. (2013). Bio-electro-Fenton system for enhanced estrogens degradation. *Bioresource Technology*. https://doi.org/10.1016/j.biortech.2013.03.157

Xu, X., Tan, H., Wang, Z., Wang, C., Pan, L., Kaneti, Y. V., Yang, T., & Yamauchi, Y. (2019). Extraordinary capacitive deionization performance of highly-ordered mesoporous carbon nanopolyhedra for brackish water desalination. *Environmental Science: Nano*. https://doi.org/10.1039/c9en00017h

Yogalakshmi, K. N., Sivashanmugam, P., Kavitha, S., Varjani, S., Adishkumar, S., Kumar, G., J, Mostafa, M., Uddin, N., Nannou, T., Bhushan, S., Rana, M. S., Nandan, N., Prajapati, S. K., Yadav, M., Rengasamy, R. S., Gupta, D., Elwakeel, K. Z., et al. (2020). Hydrogen-rich gas production from steam co-gasification of banana peel with agricultural residues and woody biomass. *Bioresource Technology Reports, 7*(4), 204–214. https://doi.org/10.1016/j.wasman.2021.02.042

Zhang, C., Tan, S., Niu, X., & Su, P. (2015). Treatment of geothermal water with high fluoride content by electrocoagulation. *Desalination and Water Treatment, 54*(8), 2223–2227. https://doi.org/10.1080/19443994.2014.900727

Zhang, G., & Jiao, K. (2018). Multi-phase models for water and thermal management of proton exchange membrane fuel cell: A review. *Journal of Power Sources*. https://doi.org/10.1016/j.jpowsour.2018.04.071

Zhang, P., Tong, M., Yuan, S., & Liao, P. (2014). Transformation and removal of arsenic in groundwater by sequential anodic oxidation and electrocoagulation. *Journal of Contaminant Hydrology*. https://doi.org/10.1016/j.jconhyd.2014.06.009

Zhao, X., Zhang, B., Liu, H., & Qu, J. (2010). Removal of arsenite by simultaneous electro-oxidation and electro-coagulation process. *Journal of Hazardous Materials*. https://doi.org/10.1016/j.jhazmat.2010.08.058

Chapter 8
Arsenic Removal from the Environment Using Biological Methods

Laxmi Kant Bhardwaj, Prangya Rath, Divya Upadhyay, and Vartika Singh

Abstract Arsenic (As) is a toxic metalloid that is distributed globally. It has two oxidation states, trivalent As and pentavalent As. As(III) is more toxic than As(V). As is mostly released into the environment by natural sources and man-made causes. It is the most significant environmental pollutant due to human activities. According to the World Health Organization (WHO) report, the presence of a high concentration of As in drinking water is the leading cause of serious health problems and the WHO recommends a maximum concentration of As(10 μg/L) in drinking water. It is incredibly difficult to remove As from the environment. There are various methods for removing As from contaminated soil and water. The effectiveness and waste associated with each method depend on criteria such as speciation, starting adsorbate concentration, absorbent types, and interfering species. As removal from the environment may be done efficiently and sustainably via biological methods. Microorganisms remove As from water, soil and sediments by metabolism, detoxification, oxidation–reduction, bioadsorption, bioprecipitation, and volatilization processes. The most effective approach to cleaning up polluted areas may require the use of hybrid (biological, chemical, and physical) and best-available technologies (BAT). This chapter focuses on the biological removal of As from the environment.

Keywords Arsenic · Heavy metals · Environment · Biological treatment

---

L. K. Bhardwaj (✉)
Amity Institute of Environmental Toxicology, Safety and Management, Amity University, Noida, Uttar Pradesh, India
e-mail: bhardwaj.laxmikant@gmail.com

P. Rath · D. Upadhyay
Amity Institute of Environmental Sciences, Amity University, Noida, Uttar Pradesh, India

V. Singh
Amity Institute of Global Warming and Ecological Study, Amity University, Noida, Uttar Pradesh, India

A. K. Yadav and S. Shirin (eds.), *Emerging and Innovative Arsenic Removal Technologies for a Sustainable Future*, Water Science and Technology Library 131,
https://doi.org/10.1007/978-3-032-01176-3_8

8.1 Introduction

Arsenic (As) is a harmful substance that can enter the environment through natural processes or human activity (Mairaj et al., 2022). Inorganic and organic forms of As exist in the environment. Common versions of As are arsenite and arsenate oxyanions (Shakya & Ghosh, 2018). In naturally occurring water, arsenate is more prevalent and persistent in aerobic environments rich in oxygen (O_2), while arsenite is more prevalent in anaerobic environments, such as groundwater (Uppal et al., 2019). Organic As comes in several forms, such as dimethyl arsenic acid, monomethyl arsenic acid, dimethyl arsenic acid, and monomethyl arsenic acid as examples (Meharg & Hartley-Whitaker, 2002). In nature, As is frequently found in mixtures with iron (Fe), O_2, and sulfur (S) (Jain & Ali, 2000). The primary source of As in natural surface environments is likely to be volcanic activities, followed by the weathering of rocks. These sources contribute significantly to the spread of As in subsurface systems (Osuna-Martínez et al., 2021).

The presence and spread of As are influenced by various factors such as organic matter dispersion, levels of oxygenation, and microbial flora. According to Sakhar et al. (2018), variations in pH and redox potential can lead to an increase in the spread of As throughout water systems. Doherty et al. (2017) claimed that the mobilization of As species is influenced by pH, total Fe, and free iron oxides (Fe_2O_3) in sediments. The toxicity of different forms of As is largely determined by their oxidative states and chemical compositions (Fatoki & Badmus, 2022). As pollution is an important environmental and public health hazard internationally, particularly in South Asia (Raju, 2022).

The main source of groundwater contamination in the Indo-Gangetic plain is the substance found in soil and sediments in various forms (Patel et al., 2022; Verma et al., 2023). The extensive industrial and agricultural use of As has led to significant groundwater and sediment contamination. This contamination, although smaller in effect compared to natural sources, has become a global issue, affecting countries such as Bangladesh, India, Chile, and Taiwan (Tareq et al., 2003). Natural geological sources remain a primary cause of As contamination (Ratnaike, 2003). Furthermore, soils and groundwater may include As mobility, particularly in the anaerobic atmosphere (Polya & Middleton, 2017). As and other minerals become more soluble through oxidation–reduction processes, increasing their mobility in the environment via water systems (Jain & Ali, 2000). Factors that influence the mobility of As include grain size, organic content, oxidation reduction, adsorption–desorption, precipitation-dissolution, and biological activity (Podgorski & Berg, 2020; Shirin & Yadav, 2021).

Mohan and Pittman (2007) reported a high amount of As in groundwater in West Bengal, Bihar, Uttar Pradesh and Assam and stated that As levels exceed safe drinking water limits and put millions of people's health at risk. By 2012, 19 provinces in China reported As concentrations in drinking water exceeding safe levels, with Inner Mongolia, Xinjiang, and Shanxi being notable hotspots (He & Charlet, 2013). In

Bangladesh and West Bengal, groundwater contamination in the Ganges-Meghna-Brahmaputra deltaic plain poses a significant environmental threat, primarily due to rock weathering and sediment deposition from the Himalayas (Tareq et al., 2003). Efforts to provide pathogen-free water through well construction inadvertently introduced As contamination (Oremland & Stolz, 2005). In Nepal, especially in the highly populated Terai districts, As pollution is a serious problem (Thakur et al., 2010).

A review of 25,058 samples from 20 districts showed that 23% had As levels between 10 and 50 µg/L, and 8% had levels higher than that. In 737,009 samples, recent studies revealed contamination of 2.3% and 7.9% at 10 to 50 µg/L and > 50 µg/L, respectively (Thakur et al., 2010). Other affected regions include the southwest coast of Taiwan; Antofagasta in Chile; regions in North Mexico; Argentina; Utah in the United States; Nova Scotia in Canada; and several provinces in China (Mandal & Suzuki, 2002). Zheng (2020) stressed the urgent need to test groundwater worldwide for As, with a focus on high-risk locations determined by models. Although many nations have effectively switched out dangerous homes. As with safer alternatives, more work has to be done to lower exposure, especially among scattered rural communities. Treatment for As is expensive and time-consuming, even for wealthy rural households. Reducing exposure is essential since well water contains both natural and man-made As (Zheng, 2020). This chapter focuses on the usage, toxicity, and methods for removing As from sediments, water, and soil. This chapter also suggests some recommendations for the removal of As from the environment.

8.2 Usage of Arsenic

The use of As has a long and diverse history. In medicine, it was first used to treat respiratory issues and to enhance appearance and plumpness. The application of As in medication made significant progress with the creation of arsphenamine, also referred to as Salvarsan, in 1908 (Paul et al., 2023). The first antimicrobial medication to be utilized for treating infectious diseases like sleeping sickness and syphilis (Wang & Rossman, 2023). Arsenic trioxide (As_2O_3) is another well-known form of As that is widely used in agriculture, manufacturing, and medicine for the treatment of leukemia with acute promyelocytic leukemia (Yang et al., 2024). Its negative history is equally significant. Due to its tasteless, colorless, as well as highly toxic qualities, it was used in homicides and suicides and thus was referred to as "inheritance powder" (Bentley & Chasteen, 2002).

It moved into the agricultural industry during the 1970s. It acted as an important component in preserving woods, cotton desiccants, as well as insecticides. One of the main reasons for its wide usage was that As was affordable and gave the desired output (Basu et al., 2014). As is used as an ingredient in paint pigments, food additives, pharmaceuticals, as well as glass–ceramics industries. A study has shown that to promote growth, 4-aminobenzenearsenic acid or *p*-arsenilic acid was used as one of the supplements in the diet of grilled chickens (Yang et al., 2024). Another application of As is increased usage in electronic industries in the contemporary period.

This is because gallium arsenide (GaAs) is a critical semiconductor material of solar cells and integrated circuits (Papež et al., 2021). Besides, As-containing chemicals are explored for application in the latest cancer therapies that more precisely target cancerous cells with minimal damage to healthy cells (Zhang et al., 2024). Nevertheless, there exist critical environmental issues related to the widespread application of As. Advanced As cleanup measures have been necessitated by the occurrence of As in water bodies from industrial and natural geological activities (Table 8.1).

Table 8.1 Historical and modern uses of arsenic

S. No.	Use case	Description	References
1	Medication has been used since ancient times	Used by females to improve, alleviate respiratory issues in living organisms	Zhang et al. (2024)
2	Used as an antimicrobial agent	The first antimicrobial drug used (salvarsan) for the treatment of syphilis, as well as sleeping sickness	Yang et al. (2024)
3	Medical treatment	Arsenic trioxide was used to treat acute promyelocytic leukemia and some other cancers	
4	Agricultural usage	Used as preservatives for woods, cotton desiccants, and pesticide components	Basu et al. (2014)
5	Industrial applications	Used in paint pigments, the production of glass, ceramics	Chen et al. (2016)
6	Animal feed	*p*-arsenilic acid utilised for grill chicken feed	Yang et al. (2024)
7	Electronics	Used in solar cells, semiconductors	Papež et al. (2021)
8	Homicides and suicides	Arsenic trioxide is poisonous, colorless, and tastes bad; it is sometimes employed in illegal activities	Bentley and Chasteen (2002)
9	Environmental contaminations	Global pollution problems are associated with industrial and natural activities	Wang and Rossman (2023)

8.3 Toxicity of Arsenic (As) to Human Health and Environment

As is a significant carcinogen linked to serious diseases by damaging genetic material as well as human cells. Several researchers studied the toxicity of As and stated that numerous health issues, such as cancers, pulmonary, and cardiovascular problems, are present. As poisoning is associated with gastrointestinal difficulties, neurological diseases, liver disease, immunological and endocrine abnormalities, reproductive health concerns, skin infections, and genotoxicity (Liu et al., 2018). Inorganic As compounds cause toxicity by reacting with sulfhydryl protein groups and replacing phosphate groups with arsenate. Furthermore, methylation of inorganic As can produce hazardous pathways that encourage detrimental biological processes, leading to gastrointestinal problems, skin conditions, and malignancies (Fan et al., 2019).

Consuming As contaminated drinking water can cause severe health issues, including As dermatosis, various cancers (lung, liver, uterus, skin) and bladder and hepatocellular carcinoma (Alam et al., 2023). In Southwestern Taiwan, consumption of As-laden artesian well water has caused blackfoot disease, a severe peripheral vascular disease (Tseng, 2005). Research has shown that even at low concentrations (< 50 mg/L), exposure can cause birth problems and significantly affect neurodevelopment, especially if exposure occurs early in life (Zheng, 2020). Changes in diabetes-related gene expression pathways have been associated with prenatal As poisoning, increasing the likelihood of developing diabetes as an adult (Navas-Acien et al., 2019). According to a DNA methylation study across the genome, a study conducted on 396 people as part of the Health Effects of Arsenic Longitudinal Study (HEALS), as toxicity is significantly influenced by the epigenome (Demanelis et al., 2019). As values greater than 1.0 μg/g in hair, 20 to 130 μg/g in nails, and more than 100 μg daily in urine are indicative of As poisoning. As the levels in the hair, toenails, and urine samples had significant connections with each other (Goto et al., 2013). A meta-analysis indicates that children between the ages of 5 and 15 have a 0.4-point drop in IQ for every 50% increase in urine As levels (Rodríguez-Barranco et al., 2013).

Acute As poisoning can occur when ingesting pesticides or insecticides containing As. If doses of 100 to 300 mg are taken, acute As poisoning can be lethal. Acute poisoning can include cramps, diarrhea, vomiting, and abdominal discomfort. The progression of these symptoms may include pulmonary edema, respiratory failure, hematological abnormalities (such as anemia and leukemia), renal failure, shock, coma, and death (Duarte et al., 2009). Chronic exposure to As can cause a number of major health problems, including skin lesions, malignancies, cardiovascular diseases, and neurological consequences (Ma et al., 2001). For example, investigations in West Bengal have found that long-term exposure to As-contaminated water has resulted in widespread health problems such as skin cancer and other significant disorders. Industrial As contamination has led to chronic poisoning, manifesting itself as skin disorders in 19.9% of the population in Ron Phibun, Thailand (Mandal & Suzuki, 2002).

Consumption of As-contaminated water doubles the risk of diabetes mellitus. In the U.S., higher urine As concentrations correlate with an increased prevalence of diabetes (Maull et al., 2012). Food and beverages contaminated with As have caused poisoning in Japan, England, Germany, and China (Mandal & Suzuki, 2002). A significant number of samples, including tuna 90%, salmon 48%, mullet 31%, and octopus 100%, surpassed the highest level of As allowed by European standards (Morgano et al., 2014). In Bangladesh, high levels of As have been found in rice (Li et al., 2009). Phosphate fertilization can reduce arsenate uptake in plants; however, in flood-prone environments, As is available as arsenite, which can enhance As mobility and does not compete with phosphate (Sahoo & Kim, 2013).

8.4 Biological Treatment Approaches for Arsenic

Remediation is a crucial process that focuses on resolving issues and restoring something to its proper condition (Song et al., 2022). Many remediation technologies have been developed for As and are mostly based on physical or chemical methods and a few biological methods (Alka et al., 2021). Traditional As removal technologies, such as coagulation, precipitation, and adsorption, have severe limitations, including high operating costs and the formation of hazardous waste. Both organic and inorganic contaminants are of special concern (Vandenbossche et al., 2014). Due to their detrimental ecological effects such as resistance to degradation, bioaccumulation, and biomagnification, heavy metals and metalloids (such as As, Cd, Cr, Cu, Hg, Ni, Pb, and Zn) have drawn a lot of attention in the multidisciplinary geosciences and environmental sciences (Bhardwaj & Sharma, 2021; Bhardwaj et al., 2023a).

According to Fls et al. (2017), microorganisms are common in polluted media and have evolved a variety of defense mechanisms against the stress and toxicity caused by various heavy metals. Microorganisms have the ability to mineralize organic contaminants and produce goods. (e.g., CO_2 and H_2O), or to produce metabolic intermediates that act as the primary building blocks for the proliferation of cells (Dixit et al., 2015). It is impossible for inorganic pollutants, including heavy metals, to break down directly into innocuous substances. However, through growth, metabolism, and metabolic products, bacteria can alter the chemical form, mobility, toxicity, and bioavailability of heavy metals. Biosorption, bioaccumulation, bioassimilation, bioprecipitation, bioleaching, biodegradation/biosynthesis, and biotransformation are the primary mechanisms through which microbial organisms interact with heavy metals.

According to Ahemad (2014), biosorption is the process by which soluble heavy metals bind to the surface of cells by complexation (such as electrostatic, covalent, and exopolysaccharides), chelation/coordination, reduction, precipitation, and cation/anion exchange. By means of microbial metabolism, soluble species are converted into insoluble hydroxides, carbonates, phosphates, and sulfides. For example, the bioprecipitation of heavy metals using microbiologically produced sulfides in the bioprecipitation process. An effective way to immobilize heavy metals is by bioprecipitation of heavy metals using microbiologically generated sulfides, such as SRB (Tabak et al., 2005). The process of dissolving metallic minerals and releasing related metals through the action of microorganisms is known as bioleaching. The most well-known strains of Fe/S-oxidizing bacteria, such as *Thiobacillus* and *Leptospirillum ferrooxidans* are capable of oxidizing iron and sulfide, releasing related heavy metals into an aqueous solution and generating sulfuric acid (Akcil et al., 2015). Large-scale operations have recovered metals from ores using this method.

Biological treatment approaches use inherent metabolic activities to convert As into less harmful or more readily removed forms (Srivastava et al., 2011). Biological treatment techniques are a viable path for As cleanup, with possible benefits in terms of cost, sustainability, and low secondary contamination. Due to their cost-effectiveness and suitability for the environment, biological processes have been extensively studied for As removal, and several possible uses have been proposed to clean up As-contaminated ecosystems (Sher & Rehman, 2019). However, problems remain, such as the need for precise control of microbial activity and the risk of partial removal of As under suboptimal circumstances (Jiang et al., 2013). These approaches may be divided into two categories: biooxidation and bioreduction.

The bio-oxidation process may be incorporated into existing water treatment systems, providing a sustainable and cost-effective way to increase the efficiency of As removal. Certain bacteria in biooxidation can transform the more poisonous arsenite (As(III)) into the less toxic and more readily removed arsenate (As(V)). Bio-reduction is the employment of microorganisms that can convert arsenate (As(V)) to arsenite (As(III)) or elemental As in anaerobic settings. *Geobacter* and *Shewanella* bacteria help to promote this transition.

Remediating environmental systems polluted with As also requires consideration of the interactions between various As species and microbes (Huang, 2014). Bacterial species of the genera *Thiomonas* and *Pseudomonas* help to assist the treatment process. Arsenate, which is less hazardous and more easily adsorbed onto solid surfaces, may be efficiently removed from water using the following treatment procedures, such as coagulation, precipitation, or adsorption (Santini et al., 2002). *Thiomonas arsenivorans*, for example, have been shown to successfully oxidize arsenite under aerobic conditions, greatly increasing the efficacy of subsequent As removal methods (Simeonova & Lievremont, 2004). *Geobacter sulfurreducens*, for example, may reduce arsenate to arsenite under anaerobic circumstances, which can subsequently precipitate as As sulfide or adsorb on iron minerals, making it easier to remove from polluted water or soil (Islam et al., 2005).

Microbial activities that mobilize and immobilize through sorption and biomethylation include oxidation and reduction. Currently, different combinations of microbes are used to increase crop yields and address issues related to heavy metals. For instance, different types of microorganisms display diverse responses to As, such as chelation, compartmentalization, exclusion, and immobilization (Tsai et al., 2009), showcasing the complex range of interactions that microorganisms can have in heavy metal remediation. Each species of microorganism interacts uniquely with the soil and with different elements. For instance, the fungal yeast *Saccharomyces cerevisiae* has been found to selectively bind with monovalent metal ions and exhibit resistance to high levels of As. Certain genes in *S. cerevisiae* have been linked to resistance to arsenate and arsenite, and removing these genes increased arsenite sensitivities in yeast (Clay, 2012), suggesting that although *S. cerevisiae* might not accumulate heavy metals, it could potentially aid in the growth by surviving in soils with high As content.

To improve the native microbial flora for bioremediation, stimulants are introduced into the system as part of the biostimulation process. These stimulants may include S, water, nutrients, and/or O_2. Numerous additional stimulants, including ethanol (C_2H_5OH), sodium acetate (CH_3COONa), phosphorus (P) and nitrogen (N), have also been studied (Lopes et al., 2022). Introducing microbial strains that have been tested in laboratories into a synthetic environment is a step in the bioaugmentation process. In such a procedure, bioremediation rates are investigated in small-scale laboratory settings before the most productive microbial communities are added to a larger system. It is also possible to generate and employ very effective strains that are genetically modified. These bacteria can be identified and used for bioaugmentation, according to recent research on arsenite-oxidizing bacteria on the Indian subcontinent (Sarkar et al., 2022).

Several kinds of green plants are used in phytoremediation (Sabreena et al., 2022). Since the rhizosphere of the amount of O_2 present in the plant species affects transportation from the soil to the roots of the plant. The redox state of the soil has a significant impact on speciation (Tripathi et al., 2007). According to Vithanage et al. (2012), arsenate is more readily accessible in aerobic conditions, and arsenite is more prevalent in anaerobic conditions (Vithanage et al., 2012). Through their roots, plants draw toxins from groundwater, transport them through xylem vessels, volatilize them, and release them during transpiration (Awa & Hadibarata, 2022).

High biomass aquatic plants such as *Ceratophyllum demersu*, *Hydrilla verticillata* (Nigam et al., 2012), *Lemna gibba*, *Lemna minor*, *Azolla caroliniana*, *Pistia stratiotes* (Farnese et al., 2014), *Salvinia natans* (Rahman, 2009) and *Eichhornia crassipes* may be able to help remediate contaminated water bodies. It has been shown that *Lemna gibba* may acquire As up to 1022 mg/kg dry biomass in 21 days from surface water that is polluted and contains 41.37 to 47 μg/L of As. It was discovered that *L. gibba* had a maximum biomass accumulation and As removal potential of 73.6 t/ha/y and 752 kg As/ha/y, respectively. In another investigation, it was shown that *E. crassipes* reduced the original As concentration by 83% and accumulated around 498 mg As/kg dry weight from a solution of 0.5 mg/L As in 10 days (De Souza et al., 2018). Islam et al. (2015) evaluated the utilization of water fern, *Mircanthemum umbrosum*, in the remediation of As and Cd. Studies on phytoremediation have also shown the use of emerging aquatic plants such as *Cyperus vaginatus* and *Vetiveria zizanioides* (Aryal et al., 2016). *Jatropha curcas* (Yadav et al., 2009), shrub willow (*Salix spp.*), sunflower (*Helianthus annuus*) and Indian mustard (*Brassica juncea*) (Srivastava et al., 2009) are a few high biomass plants with strong potential for As accumulation.

A small number of plant species acquire metals in their roots and move them to the surface when growing in polluted sediments (Srivastava et al., 2021).

In Bihar, community-based biofiltration devices have been installed in numerous villages, utilizing locally produced materials and As oxidizing bacteria to offer clean drinking water at a reduced cost. Similarly, in Assam, research initiatives have investigated the use of locally available plants for phytoremediation, demonstrating that plants such as *Colocasia esculenta* can absorb high levels of As from polluted water. These biological approaches provide a possible answer to India's As-contaminated areas.

8.5 Conclusions and Recommendations

It poses a severe threat to the environment and public health due to its toxic nature and widespread presence in soil, water, and sediments. Acute poisoning and chronic health problems such as cancer, heart disease, and neurological disorders are among the well-documented drawbacks of exposure to As. The importance of creating effective and durable As removal methods cannot be overstated. Although they typically result in hazardous waste creation and significant operating costs, traditional As removal techniques, including coagulation, precipitation, and adsorption, are generally ineffective. On the other hand, biological therapy alternatives offer affordable, practical, and environmentally safe alternatives. Through processes like biooxidation, bioreduction, biosorption, and precipitation, these methods use the natural metabolic activity of microorganisms to change As into less dangerous or easier-to-remove forms.

The sustainable management of the biological and geochemical cycles of As depends on the understanding of the distribution of various As species in major water systems. With this information, the geochemical mechanisms causing rising concentrations in different hydrogeological settings will become clearer. *Thiomonas* and *pseudomonas* are two examples of bacteria that support biooxidation, which changes the more toxic arsenite (As(III)) into the less dangerous arsenate (As(V)), which can subsequently be removed by other treatment methods. On the other hand, in bioreduction, arsenate is broken down in anaerobic settings by bacteria such as *Shewanella* and *Geobacter* into elemental arsenic or arsenite, which then precipitates or adsorbs onto solid surfaces.

Several studies have also demonstrated the potential of phytoremediation, a cleaning technique that is both environmentally benign and sustainable. It works by using plants such as *Colocasia esculenta*, *Jatropha curcas*, and Indian mustard to collect As from contaminated soil and water. Similarly, soils can be immobilized and their leakage into groundwater can be prevented by using bioremediation approaches that include microbial consortia. There are still difficulties in spite of these developments. The effectiveness of biological treatments can be affected by variables like pH, redox potential, and the presence of other contaminants. Furthermore, to guarantee reliable and comprehensive As removal, accurate control of microbial activity is necessary. To maximize these biological activities and incorporate them into all-encompassing approaches for managing As, more investigation and advancement are required.

The combination of low secondary contamination, sustainability, and cost-effectiveness makes biological treatment approaches for As removal a viable choice. However, to implement in a widespread way, we need to overcome technological barriers, improve our knowledge of how microbes and plants interact with As, and create scalable solutions that are suited to particular environmental circumstances. Collaboration between scientists, legislators, and communities is crucial to advance these creative solutions and safeguard the environment and the general public's health against As poisoning. Furthermore, because As in groundwater is distributed unevenly geographically, a targeted testing program is required for high As wells. The development of reliable, affordable, and user-friendly detection techniques, ideally with quick on-site monitoring capabilities, is necessary to improve the screening and identification of people exposed to inorganic arsenic in water.

Acknowledgements The authors are deeply grateful to Amity University, Noida, for providing the platform to do this study.

References

Ahemad, M. (2014). Remediation of metalliferous soils through the potential of microbial bio-agents. *Environmental Science and Pollution Research, 21*(17), 10183–10194.

Akcil, A., Koldas, S., & Donati, E. R. (2015). Bioremediation of cyanide contaminated mining wastewater by Scenedesmus obliquus. *Journal of Hazardous Materials, 293*, 97–104.

Alam, S., Bhardwaj, L. K., Mallick, R., & Rai, S. (2023). Estimation of heavy metals and fluoride ion in vegetables grown nearby the stretch of River Yamuna, Delhi (NCR). *India. Indian Journal of Environmental Protection, 43*(1), 64–73.

Alka, S., Shahir, S., Ibrahim, N., Ndejiko, M. J., Vo, D. V. N., & Abd Manan, F. (2021). Arsenic removal technologies and future trends: A mini review. *Journal of Cleaner Production, 278*, Article 123805.

Aryal, R., Nirola, R., Beecham, S., & Kamruzzaman, M. (2016). Impact of elemental uptake in the root chemistry of wetland plants. *International Journal of Phytoremediation, 18*(10), 936–942.

Awa, S. H., & Hadibarata, T. (2022). Microbial degradation of polycyclic aromatic hydrocarbons (PAHs) in the environment: A review. *Environmental Science and Pollution Research, 29*(1), 658–680.

Basu, A., Saha, D., Saha, R., Ghosh, T., & Saha, B. (2014). A review on sources, toxicity and remediation technologies for removing arsenic from drinking water. *Research on Chemical Intermediates, 40*, 447–485.

Bentley, R., & Chasteen, T. G. (2002). Arsenic curiosa and humanity. *The Chemical Educator, 7*, 51–60.

Bhardwaj, L. K., & Sharma, A. (2021). Estimation of physico-chemical, trace metals, microbiological and phthalate in PET bottled water. *Chemistry Africa, 4*(4), 981–991.

Bhardwaj, L. K., Kumar, D., & Kumar, A. (2023a). Phytoremediation potential of Ocimum sanctum: A sustainable approach for remediation of heavy metals. In *Phytoremediation potential of medicinal and aromatic plants* (pp. 46–57). CRC Press.

Bhardwaj, L. K., Rath, P., & Tokas, R. (2024). E-waste management in developing countries: Current practices, challenges, disposal, and impact on human health and environment. In *Sustainable solutions for e-waste and development* (pp. 90–102).

Bhardwaj, L. K., Rath, P., Jain, H., & Choudhury, M. (2025). Exploring the effects of e-waste on soil, water quality and human health. *Discover Civil Engineering, 2*(1), 1–16.

Bhardwaj, L. K., Sharma, S., & Jindal, T. (2023b). Estimation of physico-chemical and heavy metals in the lakes of Grovnes & Broknes Peninsula, Larsemann Hill, East Antarctica. *Chemistry Africa, 6*(5), 2677–2694.

Chen, W. Q., Shi, Y. L., Wu, S. L., & Zhu, Y. G. (2016). Anthropogenic arsenic cycles: A research framework and features. *Journal of Cleaner Production, 139*, 328–336.

Clay, S. E. (2012). *Removing arsenic and lead from soils using a bioremediation approach: Design, implementation and analysis of a feasibility study on Vashon Island, Washington* [Doctoral dissertation, Evergreen State College].

de Souza, T. D., Borges, A. C., Teixeira de Matos, A., Veloso, R. W., & Braga, A. F. (2018). Optimization of arsenic phytoremediation using Eichhornia crassipes. *International Journal of Phytoremediation, 20*(11), 1129–1135.

Demanelis, K., Argos, M., Tong, L., Shinkle, J., Sabarinathan, M., Rakibuz-Zaman, M., Sarwar, G., Shahriar, H., Islam, T., Rahman, M., Yunus, M., & Pierce, B. L. (2019). Association of arsenic exposure with whole blood DNA methylation: An epigenome-wide study of Bangladeshi adults. *Environmental Health Perspectives, 127*(5), 057011.

Dixit, R., Wasiullah, X., Malaviya, D., Pandiyan, K., Singh, U. B., Sahu, A., Shukla, R., Singh, B. P., Rai, J. P., Sharma, P. K., & Lade, H. (2015). Bioremediation of heavy metals from soil and aquatic environment: An overview of principles and criteria of fundamental processes. *Sustainability, 7*(2), 2189–2212.

Doherty, S. J., Tighe, M. K., & Wilson, S. C. (2017). Evaluation of amendments to reduce arsenic and antimony leaching from co-contaminated soils. *Chemosphere, 174*, 208–217.

Duarte, A. A., Cardoso, S. J., & Alçada, A. J. (2009). Emerging and innovative techniques for arsenic removal applied to a small water supply system. *Sustainability, 1*(4), 1288–1304.

Fan, X. Y., Liu, Y. J., Cai, Y. M., Wang, A. D., Xia, Y. Z., Hu, Y. J., Jiang, F. L., & Liu, Y. (2019). A mitochondria-targeted organic arsenical accelerates mitochondrial metabolic disorder and function injury. *Bioorganic & Medicinal Chemistry, 27*(5), 760–768.

Farnese, F., Oliveira, J., Lima, F., Leão, G., Gusman, G., & Silva, L. (2014). Evaluation of the potential of Pistia stratiotes L. (water lettuce) for bioindication and phytoremediation of aquatic environments contaminated with arsenic. *Brazilian Journal of Biology, 74*(Suppl 1), S108–S112.

Fatoki, J. O. & Badmus, J. A. (2022). Arsenic as an environmental and human health antagonist: A review of its toxicity and disease initiation. *Journal of Hazardous Materials Advances*. sciencedirect.com.

Fls, S. H., Shen, Z., Kuppusamy, S., & Bolan, N. (2017). Comparative remediation of potentially toxic elements by biochar and compost in soil and water systems. *Environmental Pollution, 229*, 405–415.

Goto, R., Okamoto, T., Ishikawa, H., Hamamura, Y., & Kato, M. (2013). Molecular phylogeny of echiuran worms (Phylum: Annelida) reveals evolutionary pattern of feeding mode and sexual dimorphism. *PLoS ONE, 8*(2), Article e56809.

Gupta, U., Shukla, S., Bhardwaj, L. K., Rath, P., Upadhyay, D., Sharma, B., Rupam, Asthana, H., Jindal, T., & Singh, H. (2024). Unraveling urban plant strategies for heavy metal uptake and detoxification. In *Urban forests, climate change and environmental pollution* (pp. 93–119). Springer.

He, J., & Charlet, L. (2013). A review of arsenic presence in China drinking water. *Journal of Hydrology, 492*, 79–88.

Huang, J. H. (2014). Impact of microorganisms on arsenic biogeochemistry: A review. *Water, Air, & Soil Pollution, 225*(2), 1–25.

Islam, F. S., Boothman, C., Gault, A. G., Polya, D. A., & Lloyd, J. R. (2005). Potential role of the Fe (III)-reducing bacteria Geobacter and Geothrix in controlling arsenic solubility in Bengal delta sediments. *Mineralogical Magazine, 69*(5), 865–875.

Islam, M. S., Saito, T., & Kurasaki, M. (2015). Phytofiltration of arsenic and cadmium by using an aquatic plant, Micranthemum umbrosum: Phytotoxicity, uptake kinetics, and mechanism. *Ecotoxicology and Environmental Safety, 112*, 193–200.

Jain, C. K., & Ali, I. (2000). Arsenic: Occurrence, toxicity and speciation techniques. *Water Research, 34*(17), 4304–4312.

Jiang, D., Zeng, G., Huang, D., Chen, M., Zhang, C., Huang, C., Wang, L., et al. (2013). Remediation of heavy metal contaminated sediments using bioremediation: A review. *Environmental Science & Technology, 43*(9), 2124–2132.

Li, R. Y., Stroud, J. L., Ma, J. F., McGrath, S. P., & Zhao, F. J. (2009). Mitigation of arsenic accumulation in rice with water management and silicon fertilization. *Environmental Science & Technology, 43*(10), 3778–3783.

Liu, Q., Leslie, E. M., Moe, B., Zhang, H., Douglas, D. N., Kneteman, N. M., & Le, X. C. (2018). Metabolism of a phenylarsenical in human hepatic cells and identification of a new arsenic metabolite. *Environmental Science & Technology, 52*(3), 1386–1392.

Lopes, C. B., Couto, R. P., Barrocas Dias, A., & Pereira, E. (2022). The role of metal-resistant bacteria in the bioremediation of polluted environments. *Environmental Pollution, 295*, Article 118710.

Ma, L. Q., Komar, K. M., Tu, C., Zhang, W., Cai, Y., & Kennelley, E. D. (2001). A fern that hyperaccumulates arsenic. *Nature, 409*(6820), 579.

Mairaj, S., Nagar, R. D., Bhardwaj, L. K., Rehman, F., Punnakal, A., Chauhan, A., & Jindal, T. (2022). Antioxidant responses of arsenite-induced oxidative stress in rice (Oryza sativa L.) and its modulation by eugenol (extracted from Ocimum sanctum). *Indian Journal of Agricultural Research, 56*(4), 396–400.

Mandal, B. K., & Suzuki, K. T. (2002). Arsenic round the world: A review. *Talanta, 58*(1), 201–235.

Maull, E. A., Ahsan, H., Edwards, J., Longnecker, M. P., Navas-Acien, A., Pi, J., Silbergeld, E. K., Styblo, M., Tseng, C. H., Thayer, K. A., & Loomis, D. (2012). Evaluation of the association between arsenic and diabetes: a National Toxicology Program workshop review. *Environmental Health Perspectives, 120*(12), 1658–1670.

Meharg, A. A., & Hartley-Whitaker, J. (2002). Arsenic uptake and metabolism in arsenic resistant and nonresistant plant species. *New Phytologist, 154*(1), 29–43.

Mohan, D., & Pittman, C. U. (2007). Arsenic removal from water/wastewater using adsorbents—A critical review. *Journal of Hazardous Materials, 142*(1–2), 1–53.

Morgano, M. A., Rabonato, L. C., Milani, R. F., Miyagusku, L., & Quintaes, K. D. (2014). As, Cd, Cr, Pb and Hg in seafood species used for sashimi and evaluation of dietary exposure. *Food Control, 36*(1), 24–29.

Navas-Acien, A., Spratlen, M. J., Abuawad, A., LoIacono, N. J., Bozack, A. K., & Gamble, M. V. (2019). Early-life arsenic exposure, nutritional status, and adult diabetes risk. *Current Diabetes Reports, 19*, 1–8.

Nigam, S., Gopal, K., & Vankar, P. S. (2012). Biosorption of arsenic in drinking water by submerged plant: Hydrilla verticillata. *Environmental Science and Pollution Research, 20*(6), 4000–4008.

Oremland, R. S., & Stolz, J. F. (2005). Arsenic, microbes and contaminated aquifers. *Trends in Microbiology, 13*(2), 45–49.

Osuna-Martínez, C. C., Armienta, M. A., Bergés-Tiznado, M. E., & Páez-Osuna, F. (2021). Arsenic in waters, soils, sediments, and biota from Mexico: An environmental review. *Science of the Total Environment, 752*, Article 142062.

Papež, N., Dallaev, R., Ţălu, Ş, & Kaštyl, J. (2021). Overview of the current state of gallium arsenide-based solar cells. *Materials, 14*(11), 3075.

Patel, D., Pamidimukkala, P., Chakraborty, D., & Yadav, A. (2022). Bharuch district, Gujarat, India: Factor analysis and geographical distribution of water quality characteristics. *Environmental Nanotechnology, Monitoring & Management, 18*, Article 100732. https://doi.org/10.1016/j.enmm.2022.100732

Paul, N. P., Galván, A. E., Yoshinaga-Sakurai, K., Rosen, B. P., & Yoshinaga, M. (2023). Arsenic in medicine: Past, present and future. *BioMetals, 36*(2), 283–301.

Podgorski, J., & Berg, M. (2020). Global threat of arsenic in groundwater. *Science, 368*(6493), 845–850.

Polya, D. A., & Middleton, D. R. (2017). Arsenic in drinking water: Sources & human exposure. In *Best practice guide on the control of arsenic in drinking water* (p. 15).

Rahman, K. Z. (2009). Treatment of arsenic containing artificial wastewater in different laboratory-scale constructed wetlands. Oldenbourg Industrieverl.

Raju, N. J. (2022). Arsenic in the geo-environment: A review of sources, geochemical processes, toxicity and removal technologies. *Environmental Research.*

Ratnaike, R. N. (2003). Acute and chronic arsenic toxicity. *Postgraduate Medical Journal, 79*(933), 391–396.

Rizvi, A., Zaidi, A., Ameen, F., Ahmed, B., AlKahtani, M. D., & Khan, M. S. (2020). Heavy metal induced stress on wheat: Phytotoxicity and microbiological management. *RSC Advances, 10*(63), 38379–38403.

Rodríguez-Barranco, M., Lacasaña, M., Aguilar-Garduño, C., Alguacil, J., Gil, F., González-Alzaga, B., & Rojas-García, A. (2013). Association of arsenic, cadmium and manganese exposure with neurodevelopment and behavioural disorders in children: A systematic review and meta-analysis. *Science of the Total Environment, 454*, 562–577.

Sabreena, Hassan, S., Bhat, S. A., Kumar, V., Ganai, B. A., & Ameen, F. (2022). Phytoremediation of heavy metals: An indispensable contrivance in green remediation technology. *Plants, 11*(9), 1255.

Sahoo, P. K., & Kim, K. (2013). A review of the arsenic concentration in paddy rice from the perspective of geoscience. *Geosciences Journal, 17*, 107–122.

Sakhar, W., Mehmood, K., Abbasi, H. N., & Kausar, S. (2018). Heavy metal remediation potential of microbial biofilms. *Environmental Science and Pollution Research, 25*(12), 12078–12089.

Santini, J. M., Sly, L. I., Schnagl, R. D., & Macy, J. M. (2002). A new chemolithoautotrophic arsenite-oxidizing bacterium isolated from a gold mine: Phylogenetic, physiological, and preliminary biochemical studies. *Applied and Environmental Microbiology, 66*(1), 92–97.

Sarkar, D., Mandal, S., Ghosh, S., & Ghosh, D. (2022). Enhanced phytoremediation of heavy metals using transgenic plants: Recent developments and future perspectives. *Environmental Science and Pollution Research, 29*(12), 18356–18374.

Shakya, A. K., & Ghosh, P. K. (2018). Simultaneous removal of arsenic and nitrate in absence of iron in an attached growth bioreactor to meet drinking water standards: Importance of sulphate and empty bed contact time. *Journal of Cleaner Production, 186*, 304–312.

Sher, S., & Rehman, A. (2019). Use of heavy metals resistant bacteria—A strategy for arsenic bioremediation. *Applied Microbiology and Biotechnology, 103*, 6007–6021.

Shirin, S., & Yadav, A. K. (2021). Assessment and suitability for the dumping of flyash in an abandoned mine. *Environmental Challenges, 4*, Article 100111. https://doi.org/10.1016/j.envc.2021.100111

Simeonova, D. D., & Lievremont, D. (2004). Microbial arsenite oxidation: Formation of arsenate, In *Arsenic in groundwater: Geochemistry and occurrence* (pp. 439–454). Springer.

Song, Y., Jiang, J., Liu, X., & Zhang, W. (2022). Mechanisms and applications of biochar for the remediation of soil pollutants: A critical review. *Environmental Pollution, 297*, Article 118739.

Srivastava, M., Ma, L. Q., & Singh, N. (2011). Phytoremediation of arsenic-contaminated sites using Pteris vittata L.: Role of arsenic-resistant bacteria. *Chemosphere, 83*(5), 583–589.

Srivastava, S., Shukla, A., Rajput, V.D., Kumar, K., Minkina, T., Mandzhieva, S., Shmaraeva, A., & Suprasanna, P., (2021). Arsenic remediation through sustainable phytoremediation approaches. *Minerals, 11*(9), 936.

Srivastava, S., Srivastava, A. K., Suprasanna, P., & D'souza, S. F. (2009). Comparative biochemical and transcriptional profiling of two contrasting varieties of Brassica juncea L. in response to arsenic exposure reveals mechanisms of stress perception and tolerance. *Journal of Experimental Botany, 60*(11), 3419–3431.

Tabak, H. H., Lens, P., van Hullebusch, E. D., & Dejonghe, W. (2005). Developments in bioremediation of soils and sediments polluted with metals and radionuclides–1. Microbial processes and mechanisms affecting bioremediation of metal contamination and influencing metal toxicity and transport. *Reviews in Environmental Science and Bio/Technology, 4*(2), 115–156.

Tareq, S. M., Safiullah, S., Anawar, H. M., Rahman, M. M., & Ishizuka, T. (2003). Arsenic pollution in groundwater: A self-organizing complex geochemical process in the deltaic sedimentary environment. *Bangladesh. Science of the Total Environment, 313*(1–3), 213–226.

Thakur, J. K., Thakur, R. K., Ramanathan, A. L., Kumar, M., & Singh, S. K. (2010). Arsenic contamination of groundwater in Nepal—An overview. *Water, 3*(1), 1–20.

Tripathi, R. D., Srivastava, S., Mishra, S., Singh, N., Tuli, R., Gupta, D. K., & Maathuis, F. J. (2007). Arsenic hazards: Strategies for tolerance and remediation by plants. *Trends in Biotechnology, 25*(4), 158–165.

Tsai, S. L., Singh, S., & Chen, W. (2009). Arsenic metabolism by microbes in nature and the impact on arsenic remediation. *Current Opinion in Biotechnology, 20*(6), 659–667.

Tseng, C. H. (2005). Blackfoot disease and arsenic: A never-ending story. *Journal of Environmental Science and Health, 23*(1), 55–74.

Uppal, H., Chawla, S., Joshi, A. G., Haranath, D., Vijayan, N., & Singh, N. (2019). Facile chemical synthesis and novel application of zinc oxysulfide nanomaterial for instant and superior adsorption of arsenic from water. *Journal of Cleaner Production, 208*, 458–469.

Vandenbossche, V., Verstraete, W., & Boon, N. (2014). Microbial electrosynthesis: An emerging strategy for CO_2 bioconversion to value-added products. *Trends in Biotechnology, 32*(3), 124–136.

Verma, A., Sharma, A., Kumar, R., & Sharma, P. (2023). Nitrate contamination in groundwater and associated health risk assessment for Indo-Gangetic Plain, India. *Groundwater for Sustainable Development, 23*, Article 100978.

Vithanage, M., Dabrowska, B. B., Mukherjee, A. B., Sandhi, A., & Bhattacharya, P. (2012). Arsenic uptake by plants and possible phytoremediation applications: A brief overview. *Environmental Chemistry Letters, 10*(3), 217–224.

Wang, Z., & Rossman, T. G. (2023). The carcinogenicity of arsenic. In *Toxicology of metals* (Vol. I, pp. 221–229). CRC Press.

Yadav, S. K., Juwarkar, A. A., Phani Kumar, G., Thawale, P. R., Singh, S. K., & Chakrabarti, T. (2009). Bioaccumulation and phyto-translocation of arsenic, chromium and zinc by Jatropha curcas L.: Impact of dairy sludge and biofertilizer. *Bioresource Technology, 100*(20), 4616–4622.

Yang, Y., Li, Y., Li, R., & Wang, Z. (2024). Research progress on arsenic, arsenic-containing medicinal materials, and arsenic-containing preparations: Clinical application, pharmacological effects, and toxicity. *Frontiers in Pharmacology, 15*, 1338725.

Zhang, B., Wang, M., Sun, L., Liu, J., Yin, L., Xia, M., Zhang, L., Liu, X., & Cheng, Y. (2024). Recent advances in targeted cancer therapy: Are PDCs the next generation of ADCs? *Journal of Medicinal Chemistry, 67*(14), 11469–11487.

Zheng, Y. (2020). Global solutions to a silent poison. *Science, 368*(6493), 818–819.

Chapter 9
Arsenic Biological Treatment Using Microorganisms in the Environment

Rushda Sharf, Ambrina Sardar Khan, and Yusra Sharf

Abstract The ecology and public health are seriously threatened by arsenic pollution of soil and groundwater. Natural sources such as aquifers and man-made sources account for the majority of primary releases of arsenic into the environment. Both organic (such as methylated) and inorganic (such as arsenate and arsenite) forms of arsenic (As) are prevalent in the environment. Because it is the most effective, economical and environmentally benign means of treating arsenic, the use of microorganisms in biological treatment provides an intriguing substitute for traditional techniques. Biological techniques in certain locations, methods such as arsenic microbial detoxification or phytoremediation employing aquatic plants are commonly accepted. Other techniques such as immobilization by microbial oxidation biological activity and microbiological molecular analysis were necessary, as was coprecipitation or adsorption with Fe oxyhydroxides in the solid phase. Arsenic oxygen is used by certain autotrophic and heterotrophic microorganism species to recover energy. Arsenate is a nutrient that some microorganism species can use in their respiratory processes. One prevalent type of arsenic resistance in microbes is detoxification operons. Therefore, bioremediation may be a practical and affordable means of removing this pollutant from the environment. This chapter summarizes what is now known about the various biological methods for removing arsenic from soil and groundwater.

Keywords Phytoremediation · Arsenic resistance microorganism · Microbial oxidation

R. Sharf (✉) · A. S. Khan
Department of Environmental Science, Integral University, Lucknow, India
e-mail: rsharf@iul.ac.in

Y. Sharf
Department of Zoology, Aligarh Muslim University, Aligarh, India
e-mail: rsharf@iul.ac.in

A. K. Yadav and S. Shirin (eds.), *Emerging and Innovative Arsenic Removal Technologies for a Sustainable Future*, Water Science and Technology Library 131,
https://doi.org/10.1007/978-3-032-01176-3_9

9.1 Introduction

In the periodic table, arsenic is a metallic element that belongs to the 'nitrogen' family and has characteristics in common with nitrogen, phosphorus, antimony, and bismuth. There are two main oxidation states for the element: pentavalent arsenate and trivalent arsenite. Although arsenite (As III) is more common in decrease conditions, such as in anaerobic groundwater, arsenate (As V) is typically found in soil and surface water. More poisonous, the As III form can dissolve in water and perhaps infiltrate living cells. Arsenic can enter the environment through natural processes, such as atmospheric emissions and the dissolution of minerals rich in arsenic, and anthropogenic sources, including mining, fossil fuel combustion, metal extraction, and the use of wood preservatives. Both inorganic and organic forms of As are known to cause cancer in humans; the latter is linked to cancers of the skin, lungs, bladder, kidney, and liver. Due to its toxicity, non-biodegradability, persistence, bioaccumulation, and possible carcinogenic qualities, arsenic in soil and water is a global concern (Rahaman et al., 2021; Shirin & Yadav, 2014, 2021; Simon et al., 2024). Arsenic contamination of the environment has been widespread due to rapid urbanization and industrialization. Arsenic-rich groundwater can be found in geothermal zones, mining districts, and areas where sulfide mineral oxidation has taken place. A metalloid that is frequently present in aquatic environments, arsenic negatively impacts plant growth and food production. Soil health refers to the long-term capacity of the soil to serve as a thriving ecosystem. This has a significant impact on the sustainability of agricultural ecosystems (Lehmann et al., 2020). The biological health of the soil is characterized by the presence of a diverse and abundant microbial community, the suppression of harmful pathogens and the support of crop growth (Shahane & Shivay, 2021). The biological health of the soil can be impacted by the presence or lack of hazardous agrochemicals and pollutants.

9.2 Sources of Arsenic

Exposure to arsenic can arise from natural sources, industrial sources, or accidental sources resulting from human intervention. The main causes of inhaled arsenic can be pesticide manufacturing facilities, air emissions from burning arsenic-containing fossil fuels, and tobacco smoke (Fig. 9.1).

9.3 Distribution of Arsenic in Soil

The rhizosphere ecosystem is negatively affected by arsenic contamination, which also negatively affects microbial communities, plant health, and overall soil functionality. The accumulation of arsenic in crops is significantly impacted by the levels

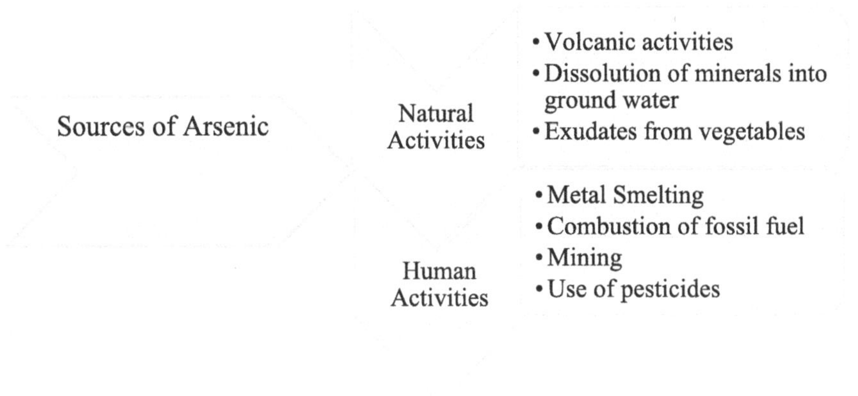

Fig. 9.1 Schematic diagram of the effects of arsenic

of arsenic in the soil and the use of contaminated water for irrigation (Sandhi et al., 2022). Numerous nations across the globe, including Vietnam, Ghana, India, China, USA, Canada, Mexico, Thailand, Brazil, and Pakistan are grappling with this issue (Al-Makishah et al., 2020). Higher arsenic can negatively affect soil fertility and properties, and harm beneficial soil microorganisms, altering plant physiology and inhibiting plant growth and development.

9.4 Distribution of Arsenic in Groundwater

Many model parameters have been used to forecast the global distribution and occurrence of As. Arsenic-contaminated groundwater has so far impacted more than 70 countries with a combined population of over 140 million. Due to their stable natural supply and high microbiological quality, groundwater resources account for over 97% of the world's freshwater resources and are frequently the primary source of drinking water. Arsenic can be found in trace amounts in natural waterways. The WHO recommends a regulation threshold of 10 μg/L for As in drinking water, while the maximum allowable content of 50 μg/L is still in effect in a few countries. Many nations' potable groundwater systems, including those in Chile, Mexico, China, Hungary, West Bengal (India), Bangladesh, and the United States, have naturally occurring as in exceptionally high quantities (Table 9.1). It has been discovered that contamination exists in groundwater of the Bihar district of India. The key sources of water supply in West Bengal, India, include ponds, open wells, and rivers. In 79 blocks spread over 8 districts in West Bengal, the limit of As is 0.05 mg/l. About 80 million people live in Bangladesh, where groundwater contains more than 50 μg/L of arsenic, and an additional 35 million people may be exposed to even greater amounts of the metal in drinking water, ranging from 50 to 300 μg/L (Rahman et al., 2015). In addition to West Bengal, the states of Assam, Bihar, Haryana, Jharkhand,

Table 9.1 Countries where arsenic concentration is found in groundwaters

Country	Water source	Arsenic source	Arsenic range µg/L
Germany	Groundwater	Mineralized sandstone	10–150
Slovakia	Groundwater	Aquifer sediments	37–39
Eastern Croatia	Shallow groundwater	Aquifer sediments	27
Spain	Groundwater	Alluvial sediments	< 1–100
Ghanaa	Groundwater	Gold mining	< 1–175
Greece	Groundwater	Mine tailings	10,000
Italy	Groundwater	Volcanic aquifers	0.1–6940
Romania and Hungary	Groundwater wells	Aquifer sediments	0.5–240
Australia	Groundwater	Pyritic sediments	7000
India	Groundwater	Alluvial sediments	1–1300
Taiwan	Groundwater	Fine-grained fluvial deposits	2.5–348

Herath et al., (2016)

Punjab, and Uttar Pradesh have been reported to have contaminated groundwater with arsenic.

9.5 Effects of Arsenic on Environment, Human Health and Social and Economic

9.5.1 Health Risk

Humans have developed several malignancies, including skin, lung, bladder, kidney, and liver cancers; skin lesions; cardiovascular and neurological illnesses; and problems with fetal development following the long-term effect of higher As in drinking water (Fig. 9.2).

9.5.2 Environmental Risk

As poisoning in water causes ecosystem disruption, soil contamination harms agriculture growth and disrupts the food chain. Because it can reach higher predator levels, arsenic can accumulate biologically in aquatic organisms as they shift up the food chain.

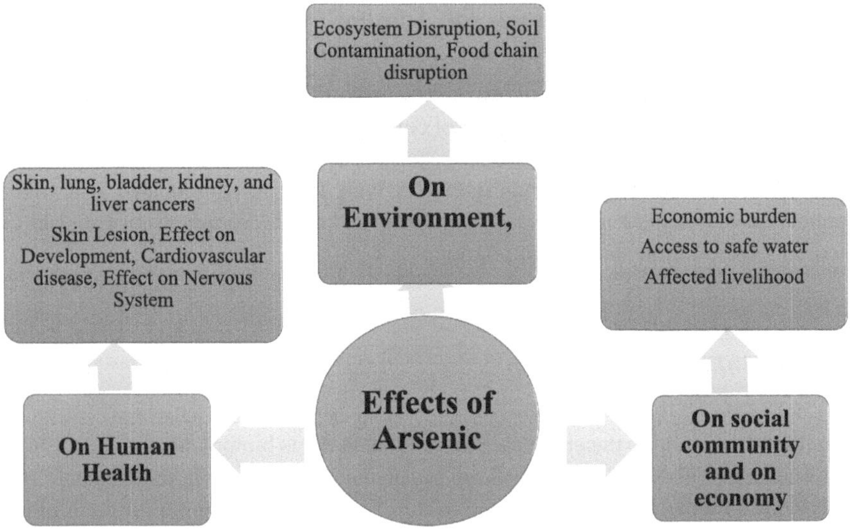

Fig. 9.2 Schematic diagram of the sources of arsenic

9.5.3 Risk in Society and Economic

Waterborne arsenic poisoning results in an economic burden for affected communities, especially in resource-poor places, as treating and mitigating the disease can be costly.

- Communities may find it difficult to obtain clean drinking water due to arsenic poisoning, forcing them to rely on tainted sources a high price for other possibilities.
- Farmers who depend on irrigation using arsenic-polluted water can experience a decrease in agricultural productivity, which could lead to monetary losses. Possible health effects on farmers could exacerbate the socioeconomic challenges faced by affected populations.

9.6 Technologies for the Removal of Arsenic from the Environment

Arsenic from As-contaminated areas can currently be cleaned using a variety of methods, such as chemical, biological, and physical ones. In addition to being costly and lacking in specificity, conventional physicochemical remediation techniques also disturb the microenvironment of the soil, rendering it unfit for agricultural use. A growing number of people are interested in environmentally friendly and sustainable methods for cleaning up As-contaminated sites as a result of these worries.

9.6.1 Biological Techniques for Arsenic Removal

Arsenic is present in a variety of industrial raw materials, finished goods, and waste. In many remediation sites, it is a pollutant of concern in groundwater and soil. Because As eagerly changes its valence state and reacts to form species with different toxicity and mobility, treatment of As exposure effectively may be complicated. The main application of arsenic is in above-ground water, where microorganisms are used to increase precipitation and coprecipitation.

9.6.2 Biological Oxidation

An alternate method to the chemical oxidation of AsIII is biological oxidation. "Iron and manganese oxidizing bacteria" are particular native bacteria that mediate the biological oxidation of arsenic. These bacteria have been functional in the biological oxidation of arsenic in groundwater treatment continuously with success. A promising technique has been identified for the elimination of arsenic from groundwater through the biological oxidation of Fe by two bacteria, *Leptothrixochracea* and *Gallionellaferruginea.* Because iron oxides and microorganisms are coated in the filter medium, arsenic can be adsorbed and eliminated from the water through this process. Even at initial concentrations of 200 mg/L of arsenic, these bacteria have been shown to oxidize trivalent arsenic under ideal experimental conditions, contributing to nearly total arsenic removal (up to 95%). Under the previously mentioned experimental circumstances, the pentavalent arsenic content can be substantially reduced, resulting in residual levels below the recently imposed limit of 10 mg/L. Furthermore, unlike other absorption processes, there is no need to monitor a penetrating point because the iron oxides are constantly produced in situ. It can be used to simultaneously remove Fe, Mn, and As from groundwater because it combines biological oxidation, filtration, and sorption processes (Katsoyiannis et al., 2004).

9.6.3 Biotransformation

The process of converting a metal from a toxic to a nontoxic form is known as biotransformation. The respiratory chains or particular enzymes are in charge of redox-transforming aspartame. These microorganisms are important components of detoxification processes because they can function as either an electron donor or an acceptor. Some bacteria can convert the toxic form of arsenite into the non-toxic form of arsenate, including *Bacillus* sp. and *Rhodobacter* sp. (Laha et al., 2021).The bacterial transformation of As, comprising As(V) reduction and As(III) oxidation for autotrophic growth, is mentioned as "arsenotrophy" (Oremland et al., 2009).

Aspartate reductase (ARR), one of the dimethyl sulfoxide reductase family, and two faintly connected arsenite oxidases, AIO and ARX, catalyze the interconversion of As(V) and As(III) in arsenotrophy (Ospino et al., 2019).

9.6.4 Bioremediation

The process of bioremediation, which uses microbes to clean soils of pollutants, bioremediation of soils contaminated by organic or inorganic pollutants. In certain regions, has become widely accepted due to the presence of pesticides and hydrocarbons. By lowering, detoxifying, breaking down, mineralizing, or changing more dangerous metals into less dangerous ones, bioremediation works. Cleaning methods are used to remove toxic waste from a contaminated area. The widespread assumption in bioremediation is that the all-encompassing activity will degrade, eradicate, and immobilize a variety of chemical pollutants and physically hazardous compounds from the surrounding environment (Sharma, 2020). Therefore, it is more beneficial to use a planned bioremediation approach in a highly polluted area.

9.7 Techniques of Bioremediation

Both ex situ and in-situ bioremediation may be done based on several variables, such as the type, concentration, and cost of the pollutants as well as the site's features. There are 2 types of bioremediation in situ: intrinsic and designed. Intrinsic bioremediation is a process that breaks down soil with low levels of pollutants using naturally occurring microbes without interacting with humans. To maximize environmental conditions and boost the activity and proliferation of local microbes, engineered bioremediation often depends on human intervention. In addition, bioremediation techniques include biostimulation, bioaugmentation, bioventing, biopiles, and bioattenuation (Fig. 9.3). The most economical, efficient, and environmentally responsible method of managing the contaminated environment is bioremediation.

9.7.1 Ex-Situ Bioremediation

It involves the removal of toxic metals from contaminated areas and moving them to another location where they will be treated. Ex situ bioremediation techniques are regularly taken into account depending on the extent, kind, and degree of pollution as well as the cost and location of the polluted site.

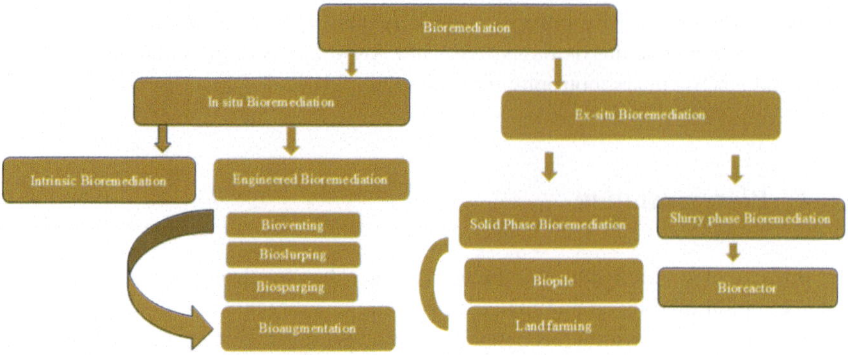

Fig. 9.3 Schematic diagram of different bioremediation methods

9.7.2 *Solid-Phase Treatment*

Solid-phase bioremediation is an ex situ method that entails excavating and piling the contaminated soil. It includes organic wastes such as leaves, animal manure and agricultural wastes in addition to garbage from homes, businesses, and municipalities. Bacterial growth is transported by pipes placed throughout the piles. Air must be drawn through pipes for microbial ventilation and respiration. Solid-phase systems are more space-intensive and require more cleanup time than slurry-phase procedures. Land farming, composting, biopiles, and other methods are used to treat solid waste (Kulshreshtha et al., 2014).

9.7.3 *Biopile*

Polluted excavated soil is piled above ground for bioremediation; aeration and nutrient addition are then added to improve bioremediation through microbial metabolism. The biopile is used to treat unstable low molecular weight pollutants; according to Gomez and Sartaj (2014), it is also an effective remediation tool for highly polluted environments.

9.7.4 *Land Farming*

Land farming frequently involves the excavation and tilling of toxic metal soils, and the type of bioremediation is mistakenly regulated at the treatment site. The term "ex situ" refers to the on-site treatment of excavated filthy soil, which is distinct from other ex situ bioremediation methods. Excavated toxic soils are generally used with caution on a fixed layer of support above the ground surface to aid in the aerobic biodegradation of contaminants by autochthonous microorganisms (Silva-Castro et al., 2012). Large amounts of hazardous soil can be treated using the land-based bioremediation process with minimal energy and environmental impact. Additionally, it is relatively simple to develop and apply.

9.7.5 Slurry-Phase Bioremediation

Slurry-phase bioremediation is a process than other treatment methods. Polluted soil is combined with nutrients, water, and oxygen in a bioreactor to provide the ideal environment for microorganisms to decompose the pollutants.

9.7.6 Bioreactor

A bioreactor is a container in which a sequence of biological reactions transforms raw materials into one or more targeted products. Bioreactors can operate in a variety of ways: batch, continuous, and multistage. The ideal growing environment for bioremediation is provided by a bioreactor.

9.8 Types of In-Situ Bioremediation

In situ bioremediation includes intrinsic and engineered bioremediation.

9.8.1 Intrinsic Bioremediation

An in situ bioremediation method called natural reduction, or intrinsic bioremediation, entails passively clearing contaminated areas without the use of external factors such as human involvement.

9.8.2 Engineered in Situ Bioremediation

The second strategy entails the introduction of precise microorganisms to the polluted area. Genetically engineered microorganisms accelerate the in-situ bioremediation process because they improve the physicochemical conditions that promote microorganism growth.

9.8.3 *Bioventing*

By providing O_2 to the unsaturated zone, bioventing techniques regulate airflow stimulation and increase the activity of microbes involved in bioremediation. Nutrients and moisture are added as amendments to bioventuring to boost bioremediation (Höhener & Ponsin, 2014).

9.8.4 *Bioslurping*

This technique combines vacuum pumping, bioventing, and soil vapor extraction to treat soil and groundwater by indirectly providing oxygen and encouraging the biodegradation of pollutants.

9.8.5 *Biosparging*

Injecting air beneath the soil to increase microbial activity, which in turn encourages the removal of harmful metals from contaminated areas, is similar to bioventing. In contrast, volatile organic molecules can be transported upward and into the unsaturated zone by infusing air into the saturated zone during bioventing, which speeds up the biodegradation process. Pollutant biodegradability and soil permeability are the two primary factors that affect the effectiveness of biosparging. Biosparging, which encourages biodegradation, and in situ air sparging, which relies on high airflow for the volatilization of harmful metals, are closely related methods in soil vapor extraction and bioventing. Biosparging has traditionally been used to cure kerosene and diesel contaminated aquifers.

9.8.6 *Bioaugmentation*

Introducing microbial strains that have been proven in laboratories into a synthetic environment is a step in the bioaugmentation process. In this process, the rates of bioremediation are first investigated in a laboratory setting, and subsequently, the most effective microbe groups are added to a larger system. It is also possible to develop and employ highly effective stains that are genetically engineered. Using recombinant DNA technology, the genetic composition can be changed to produce efficient strains with unique characteristics for the bioremediation of polluted soil (Rafeeq et al., 2023). Its advantage lies in its ability to foster the growth of microbial strains that are resistant to harsh and stressful conditions and can serve as remediators in a range of complex environmental settings. *Corynebacterium glutamicum* strains that have undergone genetic modification can accumulate AsV and AsIII more than control strains, acting as As biocontainers. Due to several environmental factors and competition from resident bacterial populations, maintaining the viability of the recombinant bacterial population in the soil is challenging. This is a significant obstacle to the use of genetically modified microorganisms in challenging field environments. Additionally, genetic engineering has produced "microbial biosensors" that allow for precise monitoring of pollution in contaminated areas (Yaashikaa et al., 2022).

9.9 Noval Approaches of Bioremediation

9.9.1 Biosorption

Passive uptake of pollutants by natural materials, such as dead biomass, microorganisms, forestry waste products, casein, and sugar-beet pulp, is known as biosorption. The innovative technology known as biosorption provides the best substitute for removing harmful metals from contaminated streams. Natural materials provide a high capacity for heavy metal decontamination. There are two categories of biosorption mechanisms. (1) Metabolism-dependent, meaning that it is transported across cell membranes and accumulates and precipitates within cells. (2) Metabolism-independent processes, such as complexation, ion exchange, precipitation, and chemical and physical adsorption. Biosorption processes are quickly reversible and independent of cell metabolism. It has been observed that the biosorption capacity of living plants occurs in two stages: a rapid 1st phase and a slow second phase. Lignin, cellulose, and other components that are kept in cell walls and vacuoles make up the majority of biomass; other materials, such as phenols, cutin, and suberin waxes, bind to cell walls and can be utilized as biosorbents. The use of *Aspergillus fumigates* as a biosorbent is an innovative approach to waste management for this common waste of the fermentation industry. Additionally, *Aspergillus fumigates* is a cost-effective and cultivable biosorbent. Aspergillus fumigate pellets for biosorption have been created. These pellets exhibit As(V) adsorption by the Freundlich isotherm model. These pellets are easily compostable, biodegradable, and disposable (Nordstrom, 2002).

9.9.2 Bioaccumulation

An energy-dependent method of transporting heavy metals is bioaccumulation. The influx of heavy metals across bacterial membranes can occur through various bioaccumulation pathways, such as ion channels, endocytosis, carrier-mediated transport, and lipid permeability. The most likely bioremediation process is the accumulation of As accumulation by bacteria. *Corynebacterium glutamicum* has been shown to accumulate As intracellularly in a mutant strain (Satyapal & Rani, 2016). The primary methods for achieving bioaccumulation are biotransformation to less toxic forms of As(V), periplasmic biosorption, and intracellular bioaccumulation. According to Singh and Sarma (2010), certain microbial strains can adsorb As (III) ions. These strains include *Lactobacillus crispatus, L. acidophilus, Acetobacter spp., Zygosaccharomyces spp., Rhizopus oryzae, Penicillium canescens, Aspergillus niger, Scytonema, and Lessonianigrescens.*

9.9.3 Methylation and Volatilization

Through the process of As methylation, microorganisms change inorganic As into organic forms. As V is changed into trimethylarsine through a sequence of successive steps in the methylation process. The reduction in toxicity and the increase in environmental volatility are the outcomes of conversion to trimethylarsine. These processes result in the formation of trimethylarsine oxide, which is subsequently reduced to trimethylarsine and involves several intermediates, such as AsIII, methyl arsenate, and dimethylarginine (Stýblo et al., 2021). Fungal organisms are primarily responsible for producing trimethylarsine, while bacterial organisms are generally responsible for synthesizing dimethylarsine and methylated mercaptans. Environmental methylation and volatilization processes are essential to the global As cycle and are critical in detoxifying As. When the thermophilic algae*Cyanidioschyzon merolaes* Cars M gene was genetically inserted into the heat-tolerant bacterium *Bacillus subtilis*, during waste composting, arsenic was methylated and volatilized at higher temperatures such as 50 °C (Huang et al., 2015).

A unique kind of As-resistant system is present in several microorganisms, including *Acinetobacter, Agrobacterium, Thiobacillus, Escherichia, Pseudomonas, Microbacterium, Ochrobactrum, Foliopsispinicola, Penicillium gladioli, Psychrobacter sp., Rhodobium, Sinorhizobium, and Clostridium.* They have the ars operon, which allows them to survive in stressful environments. As resistance operons, which bestow resistance to both As III and AsV, are found in almost all bacteria and archaea. As sensitive E. coli strains that were injected with the ars Mgene of *Rhodopseudomonas palustris* showed increased resistance to As III. The results of the study revealed that the methylated As III intermediates are formed by the Ars M enzyme, which then catalyzes the formation of trimethylarsine. Additionally, this investigation showed a significant decline in As levels in bacterial cells as well as in the growth medium (Yang et al., 2020). Plant growth-promoting microorganisms (PGPMs) resistant to arsenic have the potential to be considered long-term, low-cost, benign biological agents that can effectively reduce arsenic toxicity and manage arsenic accretion in crops.

9.10 New Technologies in the Arsenic Bioremediation

Certain novel technologies, such as the use of non-living biomass, nanoparticles and genetically engineered plants, are said to have a large and fast bioremediation capacity and can be used to eliminate metal pollution from different sources (Gaur et al., 2014).

9.10.1 Exploitation of Nanoparticle

In many different bioremediation domains, nanotechnology is widely applied to promote the development of creative, methodical, and environmentally friendly nanomaterial structures. Consequently, attempts have been made to use fungal extracellular enzymes, such as *Aspergillus foetidus*, to synthesize silver nanoparticles (Roy et al., 2013).

9.10.2 Non-living Biomass

Non-living marine algae have the capability to absorb toxic metals from diluted water solutions. The two candidates for brown algae with the highest As adsorption capacity are Cystoseira and Dictyopteris. As-specific absorption values from Dictyopteris were close to the highest ever observed, indicating excellent performance (Pennesi et al., 2012).

9.10.3 Transgenic Crops

A genetically altered *Arabidopsis thaliana* was produced with the *E. coli* arsC genes, which codes for arsenic reductase, and ECS, which code for glutamylcycteinesynthatase. Compared to normal Arabidopsis thaliana that expresses its own arsC and ECS genes, this transgenic plant was found to accumulate two to three times as much arsenic (Dhankher et al., 2002).On the other hand, research has also been done on transgenic *E. coli* that expresses the genes responsible for phytochelatin synthase (AtPC), which is derived from *Arabidopsis thaliana* (Fig. 9.4).

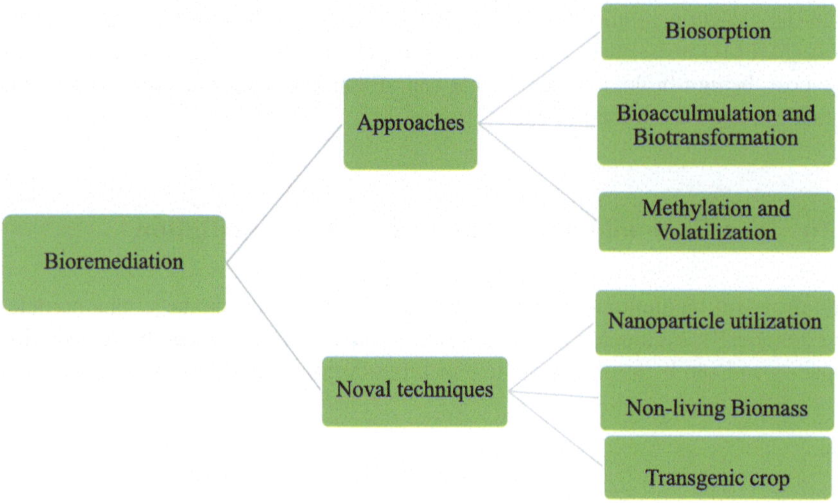

Fig. 9.4 Schematic diagram of different approaches and techniques of bioremediation

9.11 Phytoremediation

An emerging technique called phytoremediation is typically limited to a shallow enough for plant roots to reach. All chemical, physical, and biological processes that help remove contaminated materials from the environment and are impacted by plants and the rhizosphere are collectively referred to as phytoremediation. Phytoextraction, also known as phytoacceration, is the process by which pollutants are taken up by plant roots and moved into plant shoots and leaves. Phytoremediation mechanisms include phytodegradation, which breaks down pollutants inside plant tissues, and phytostabilization, which is the process by which plants produce chemical compounds that immobilize pollutants at the root-soil interface. Improved rhizosphere biodegradation occurs in the soil near the roots of the plant. Phytoremediation uses plants with large root systems, rapid growth, higher tolerance, and biomass production, and the ability to extract and absorb As from polluted soil. Osmoregulation, uptake, translocation, accretion, distribution, and elimination are some of the steps involved in the elimination of metals. Metals are taken out and directed toward the surface of the roots. Furthermore, the basis for the ability to absorb the metals is the identification of heavy metals through selective permeability, which is achieved by binding metal ions to root cells and transferring them to the aerial portion (Suman et al., 2018). Metal ion transport is carried out by the plant vascular system in conjunction with a carrierprotein. Different approaches to phytoremediation exist, such as phytoextraction, phytoexclusion, and phytostabilization. Each technique has a unique set of mechanisms for establishing a pollutant-free environment. Certain plant species, called hyperaccumulators, can absorb and concentrate pollutants at levels higher than those found in nearby soil or groundwater. Arsenic-contaminated soil, waste and water have been remedied using a hyperaccumulating fern (*Pteris vittata*). Invasive plants, such as *Parthenium hysterophorus,* are also useful in remediation strategies due to their rapid spread throughout a variety of environments and high metal accumulation (Hadi et al., 2014).

Several techniques, including phytoextraction, phyto-stabilization, phytofiltration, and phytovolatilization, fall under the umbrella of the phytoremediation method.

9.11.1 Phytoextraction

Hyperaccumulators move metal pollutants from soil to above-ground biomass through phytoextraction, phytoaccumulation, or phytomining. The biomass is then harvested to remove pollutants. Bioconcentration is the main method used in phytoextraction techniques. Through their roots, plants draw toxins from the soil, sediments, or water and move them to the biomass above the ground, where they build up in areas like shoots and other parts of the plant that can be harvested. The basic idea behind phytoextraction for contaminated locations is to cultivate suitable plant species locally, collect heavy metal-containing biomass, and then treat it to make it smaller and lighter. Biomass can be dehydrated, compressed, composted, or thermally broken to achieve this. According to Suman et al. (2018), plant species that are suitable for phytoextraction must possess the following qualities: (i) resistance to dangerous metals; (ii) high biomass output; and (iii) active heavy metals in easily harvested parts.

9.11.2 Phyto-Stabilization

By absorbing substances onto their roots, allowing them to accumulate or causing precipitation within the rhizosphere, plants are utilized in the process of phytostabilization to immobilize and stabilize pollutants in groundwater and soil. By reducing the mobility of the contaminant, this process stops erosion and migration into the groundwater and air. Transpiration and root development immobilize hazardous metals by reducing leaching, establishing aerobic conditions in the root, and incorporating organic components into to substrate that bind the contaminants (Bolan et al., 2011). Additionally, it enhances the physiology, quality and biomass output of maize. In addition, metal translocation to the shoot is improved by applying poultry manure alone or in conjunction with biogas residues.

9.11.3 Rhizofiltration

Rhizofiltration is the process by which the plant absorbs capillary water through its root xylem and absorbs any heavy metals that may be present in the water. The processes of phytoextraction and phyofiltration are comparable to the rhizofiltration technique. Plants are employed in this to extract impurities from the solution around the root zone. Using hyperaccumulators and arsenic-tolerant macrophytes such as pondweed (*Potamogetonlucens*), yellow water lily (Nupharlutea) and holly-leaved Naiad (*Najas marina*), this method cleans polluted water such as groundwater or wastewater (Mazej & Germ, 2009).

9.11.4 Phytovolatilization

Through transpiration, plants absorb toxic metals and convert them into unstable and less dangerous chemical species, a process known as phytovolatilization. Plants transpire or disperse unstable arsenic through their stems, roots, or leaves, during the phytovolatilization process. By this mechanism, pollutants, both organic and inorganic, are taken up by the plants in the transpiration stream from the soil or water, and at relatively low concentrations, they are volatilized into the atmosphere in an adapted or unadapted form. Remediating soil contamination involves the use of certain fungal species that employ the volatilization technique. Marketable vegetable crops such as broccoli, cabbage, and rice are very efficient in phytovolatilization; however, it has been noted that the volatilization process is hampered by salinity in the soil (Terry & Zayed, 1994).Research indicates that some plants—whether wild or created through genetic engineering, are capable of absorbing heavy pollutants and converting them into gaseous forms inside the plant before eliminating them into the surrounding atmosphere. Examples of these plants are Indian mustard (*Brassica juncea*), Arabidopsis thaliana, and muskgrass (*Chara canescens*) (Ghosh & Singh, 2005) (Fig. 9.5).

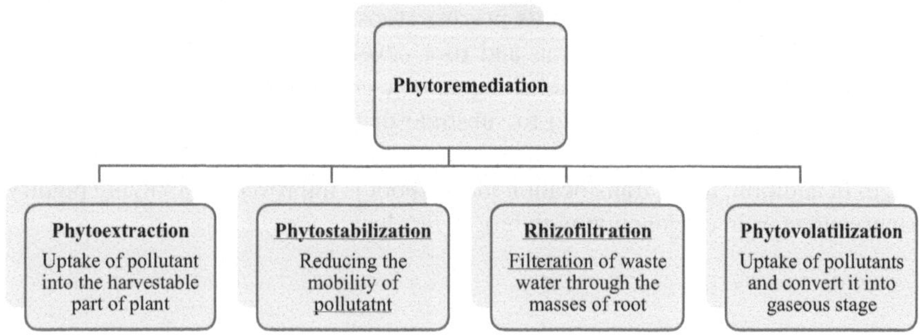

Fig. 9.5 Schematic diagram of phytoremediation techniques for arsenic removal

9.11.5 Role of Biochar in the Remediation of Arsenic-Polluted Water and Soils

Biochar is a porous, carbon rich material that is made by heating organic biomass, which can be derivative of plants or animals to temperatures between 300 and 800 °C without the presence of oxygen (such as pinewood, oak wood, perilla leaf, etc.). Biochar treatment of As-contaminated soils and water has the potential to reduce arsenic contamination and stop its harmful impact on human health. Much research has been done on the effectiveness of various forms of biochar in removing As from water. Arsenic removal from water has been achieved through the use of biochar derived from wood biomass, Pristine, and modified/activated biochar (Srivastav et al., 2021).

9.11.6 Conclusion

An alternate strategy for removing arsenic from contaminated soil and water is biological methods. Techniques such as phytoremediation and bioremediation can be applied as a reasonably inexpensive means of eliminating or stabilizing arsenic, which will lessen its negative effects on the environment and gain more support from the general public. In the future, a coordinated application of molecular biology and conventional breeding methods may help with the bioremediation of contaminated areas. These findings could lead to some encouraging discoveries when combined with in-depth evolutionary analysis across reputed genes linked to heavy pollutant tolerance.

Acknowledgements The author is grateful to Dr. Ambrina Sardar Khan, Integral University Department of Environmental Sciences, Lucknow, and to Dr. Yusra Sharf, Department of Zoology, Aligarh Muslim University, for their valuable advice and inputs.

References

Al-Makishah, N. H., Taleb, M. A., & Barakat, M. A. (2020). Arsenic bioaccumulation in arsenic contaminated soil: A review. *Chemical Paper, 74*, 2743–2757.

Bolan, N. S., Park, J. H., Robinson, B., Naidu, R., & Huh, K. Y. (2011). Phytostabilization: A green approach to contaminant containment. *Adverse Agronomy, 112*, 145–204. https://doi.org/10.1016/B978-0-12-385538-1.00004-4

Dhankher, O., Li, Y., Rosen, B., Shi, J., Salt, D., Senecoff, J., Sashti, N., & Meagher, R. (2002). Engineering tolerance and hyperaccumulation of arsenic in plants by combining arsenate reductase and γ-glutamylcysteinesynthetase expression. *Nature Biotechnology, 20*, 1140–1145.

Gaur, N., Flora, G., Yadav, M., & Tiwari, A. (2014). A review with recent advancements on bioremediation-based abolition of heavy metals. *Environmental Science Process Impacts, 16*(2), 180–193.

Ghosh, M., & Singh, S. P. (2005). A review on phytoremediation of heavy metals and utilization of it's by products. *Asian Journal of Energy Environment, 6*, 18.

Gomez, F., & Sartaj, M. (2014). Optimization of field scale biopiles for bioremediation of petroleum hydrocarbon contaminated soil at low temperature conditions by response surface methodology (RSM). *Integrated Journal of Biodeterioration Biodegradation, 89*, 103–109. https://doi.org/10.1016/j.ibiod.2014.01.010

Hadi, F., Ali, N., & Ahmad, A. (2014). Enhanced phytoremediation of cadmium contaminated soil by Parthenium hysterophorus plant: Effect of gibberelic acid (GA3) and synthetic chelator, alone and in combinations. *Bioremediation Journal, 18*, 46–55.

Herath, I., Vithanage, M., Bundschuh, J., Maity, J. P., & Bhattacharya, P. (2016). Natural arsenic in global groundwaters: Distribution and geochemical triggers for mobilization current pollution. *Reproduction, 2*, 68–89. https://doi.org/10.1007/s40726-016-0028-2

Höhener, P., & Ponsin, V. (2014). In situ vadose zone bioremediation. *Current Opinion in Biotechnology, 27*, 1–7. https://doi.org/10.1016/j.copbio.2013.08.018

Huang, K., Chen, C., Shen, Q., Rosen, B. P., & Zhao, F. J. (2015). Genetically engineering Bacillus subtilis with a heat-resistant arsenite methyltransferase for bioremediation of arsenic-contaminated organic waste. *Applied and Environmental Microbiology, 81*, 6718–6724.

Katsoyiannis, I. A., Zouboulis, A. I., & Jekel, M. (2004). Kinetics of bacterial As(III) oxidation and subsequent As(V) removal by sorption onto biogenic manganese oxides during groundwater treatment. *Industrial and Engineering Chemistry Research, 43*, 486–493.

Kulshreshtha, A., Agrawal, R., Barar, M., & Saxena, S. (2014). A review on bioremediation of heavy metals in contaminated water. *Journal of Environmental Science Toxicology and Food Technology, 8*(7), 44–50.

Laha, A., Bhattacharyya, S., Sengupta, S., Bhattacharyya, K., & GuhaRoy, S. (2021). Investigation of arsenicresistant, arsenite-oxidizing bacteria for plant growth promoting traits isolated from arsenic contaminated soils. *Archive of Microbiology, 203*(7), 4677–4692.

Lehmann, J., Bossio, D. A., Knabner, I. K., & Rillig, M. C. (2020). The concept and future prospects of soil health. *Nature Review Earth and Environment, 1*, 544–553.

Mazej, Z., & Germ, M. (2009). Trace element accumulation and distribution in four aquatic macrophytes. *Chemosphere, 74*, 642–647.

Nordstrom, D. (2002). Public health: Enhanced: Worldwide occurrences of arsenic in ground water. *Science, 296*, 2143–2145.

Oremland, R. S., Saltikov, C. W., Wolfe-Simon, F., & Stolz, J. F. (2009). Arsenic in the evolution of earth and extraterrestrial ecosystems. *Geomicrobiology Journal, 26*(7), 522–536.

Ospino, M. C., Kojima, H., & Fukui, M. (2019). Arsenite oxidation by a newly isolated betaproteobacterium possessing arx genes and diversity of the arx gene cluster in bacterial genomes. *Frontier of Microbiology, 29*(10), 1210.

Pennesi, C., Vegliò, F., Totti, C., Romagnoli, T., & Beolchini, F. (2012). Non-living biomass of marine macrophytes as arsenic (V) biosorbents. *Journal of Applied Phycology, 6*, 1495–1502.

Rafeeq, H., Afsheen, N., Rafique, S., Arshad, A., Intisar, M., Hussain, A., Bilal, M., & Iqbal, H. M. (2023). Genetically engineered microorganisms for environmental remediation. *Chemosphere, 310*, Article 136751.

Rahman, M. M., Dong, Z., & Naidu, R. (2015). Concentrations of arsenic and other elements in groundwater of Bangladesh and West Bengal, India: Potential cancer risk. *Chemosphere, 139*, 54–64.

Rahaman, M. S., Rahman, M. M., Mise, N., Sikder, M. T., Ichihara, G., Uddin, M. K., Kurasaki, M., Ichihara, S. (2021). *Environmental arsenic exposure and its contribution to human diseases, toxicity mechanism and management*. Environmental Pollution, 289, 117940. https://doi.org/10.1016/j.envpol.2021.117940

Roy, P., Mondal, N. K., Bhattacharya, S., Das, B., & Das, K. (2013). Removal of arsenic (III) and arsenic (V) on chemically modified low-cost adsorbent: Batch and column operations. *Applied Water Science, 3*(1), 293–309.

Sandhi, A., Yu, C., Rahman, M. M., & Amin, M. N. (2022). Arsenic in the water and agricultural crop production system: Bangladesh perspectives. *Environment of Science Polluion Research, 29*, 51354–51366.

Satyapal, G., & Rani, S. (2016). Potential role of arsenic resistant bacteria in bioremediation: Current status and future prospects. *Journal of Microbial & Biochemical Technology, 8*(3), 256–258.

Shahane, A. A., & Shivay, Y. S. (2021). Soil health and its improvement through novel agronomic and innovative approaches. *Frontiers in Agronomy, 3*, 1–31.

Sharma, I. (2020). Bioremediation techniques for polluted environment: concept, advantages, limitations, and prospects. In M. A. Murillo-Tovar et al. (Eds.), *Trace metals in the environment-new approaches and recent advances*. IntechOpen.

Shirin, S., & Yadav, A. K. (2014). Physico-chemical analysis of municipal wastewater discharge in Ganga river, Haridwar district of Uttarakhand, India. *Current World Environment, 9*(2), 536–543. https://doi.org/10.12944/CWE.9.2.39

Shirin, S., & Yadav, A. K. (2021). Assessment and suitability for the dumping of flyash in an abandoned mine. *Environmental Challenges, 4*, Article 100111. https://doi.org/10.1016/j.envc.2021.100111

Silva-Castro, G. A., Uad, I., Gònzalez-Lòpez, J., Fandiño, C. G., Toledo, F. L., & Calvo, C. (2012). Application of selected microbial consortia combined with inorganic and oleophilic fertilizers to recuperate oil-polluted soil using land farming technology. *Clean Technology of Environment Policy., 14*, 719–726. https://doi.org/10.1007/s10098-011-0439-0

Simon, M., Joshi, H., Yadav, A. K., & Hopke, P. K. (2024). Simultaneous nitrogen and phosphorus removal from polluted water using a novel bacterial consortium (BACON-38): Promising application for wastewater remediation. *Journal of Water Process Engineering, 66*, Article 106044. https://doi.org/10.1016/j.jwpe.2024.106044

Singh, A., & Sarma, P. (2010). Removal of Arsenic(III) from waste water using Lactobacillus acidophilus. *Bioremediation Journal, 14*, 92–97.

Srivastav, A. L., Pham, T. D., Izah, S. C., Singh, N., & Singh, P. K. (2021). Biochar adsorbents for arsenic removal from water environment: A review. *Bulletin Environment Contaminted Toxicology, 108*, 616–628.

Stýblo, M., Venkatratnam, A., Fry, R. C., & Thomas, D. J. (2021). Origins, fate, and actions of methylated trivalent metabolites of inorganic arsenic: Progress and prospects. *Archives of Toxicology, 95*, 1547–1572.

Suman, J., Uhlik, O., Viktorova, J., & Macek, T. (2018). Phytoextraction of heavy metals: A promising tool for clean-up of polluted environment? *Frontiers in Plant Science, 9*, 1476.

Terry, T., & Zayed, A. M. (1994). Sellenium volatilization in plants. In W. T. Frankenberger Jr, & S. Benson (Eds.), *Sellenium in the environment* (pp. 343–367). Marcel Dekker.

Yaashikaa, P. R., Devi, M. K., & Kumar, P. S. (2022). Engineering microbes for enhancing the degradation of environmental pollutants: A detailed review on synthetic biology. *Environmental Research, 214*(1), Article 113868.

Yang, P., Ke, C., Zhao, C., Liu, B., Xue, X., Rensing, C., & Yang, S. (2020). ArsM-mediated arsenite volatilization is limited by efflux catalyzed by as efflux transporters. *Chemosphere, 239*, Article 124822.

Chapter 10
Bioremediation of Arsenic-Poisoned Environments Using Bacteria and Fungi

Sri Keerthana Gopalakrishnan, Shiva Krithika Srinivasan, S. Rahul, and A. Arumugam

Abstract Aquifer-based geogenic processes are frequently responsible for the prevalence of arsenic in aquatic habitats. Pentavalent arsenate is less harmful to aquatic biota than trivalent arsenite, which is one of the two common oxidation states of arsenic. Conventional physical and chemical techniques such as oxidation, filtration, and adsorption have been examined for arsenic removal; however, these techniques frequently have high operating expenses, poor efficiency, and difficult disposal situations for hazardous waste. Biological treatment methods offer a cost-effective and sustainable way to minimize arsenic pollution. This chapter looks at the bioremediation of arsenic-poisoned environments using plants and microorganisms such as bacteria and fungi. Important mechanisms such as bioaccumulation, biotransformation, and biovolatilization that contribute to the immobilization of arsenic within biological matrices or its transformation into less hazardous forms are emphasized. The chapter also examines potential integrations of these biological systems with other treatment modalities to increase efficacy. It also covers the limitations and real-world problems associated with applying bioremediation strategies. Future directions in genetic engineering and management of the microbial community are also explored as means of refining biological arsenic cleaning methods.

Keywords Arsenic remediation · Biological treatment · Microbial bioremediation · Phytoremediation · Bioaccumulation

S. K. Gopalakrishnan · S. K. Srinivasan · S. Rahul · A. Arumugam (✉)
Advanced Biorefinery and Catalysis (ABC) Lab, Centre for Bioenergy, School of Chemical and Biotechnology, SASTRA Deemed University, Thirumalaisamudram, Thanjavur, Tamil Nadu, India
e-mail: aruchemxl@scbt.sastra.ac.in

A. K. Yadav and S. Shirin (eds.), *Emerging and Innovative Arsenic Removal Technologies for a Sustainable Future*, Water Science and Technology Library 131,
https://doi.org/10.1007/978-3-032-01176-3_10

10.1 Introduction

Arsenic is known to be a toxic metalloid that possesses complex biological and chemical properties that can exist in more than 200 mineral forms. It can be found in air, water, and many geographical locations. Out of these 20% are sulphosalts and sulfides, 60% are arsenates and the remaining 20% are arsenites and silicates. Inorganic and organic are the main forms of arsenic. Among these, inorganic forms are considered to be more toxic. Arsenic content is found to be higher in lakes, rivers, drinking water, etc. Factors such as arsenic extraction and overuse are the key factors for the increase in arsenic concentrations. Although it exists in many other allotropes, only the gray, metallic-looking variety is significant for industry. Arsenate species ($H_3A_sO_4$, $H_2A_sO_{4-1}$, and HA_sO_{4-2}) are the stable forms in oxidized surface waters. AsS (realgar) and As_2S_3 (orpiment) exist in stable solid form at pH values approximately below 5.5 (Gupta et al., 2022).

Activities such as overexploitation and unethical human activity toward nature are the main causes of arsenic pollution and this exposure leads to many potentially fatal problems for humans through contaminated food and tainted water bodies (Rahidul Hassan, 2023). Not only humans but also aquatic lifeforms have been severely affected (Chen & He, 2008). Arsenic pollution can lead to several health hazards such as respiratory problems, cancer, skin-related problems, neurological effects, reproductive and developmental effects, etc. Arsenic acts as an endocrine disruptor, which is one of the key factors for health hazards (Meakin et al., 2020). Notably, important hotspots for arsenic pollution are found in Central Africa, South Asia, Latin America, and West Europe (Fig. 10.1). About 38.8% groundwater area was contaminated with arsenic (Sevak & Pushkar, 2024).

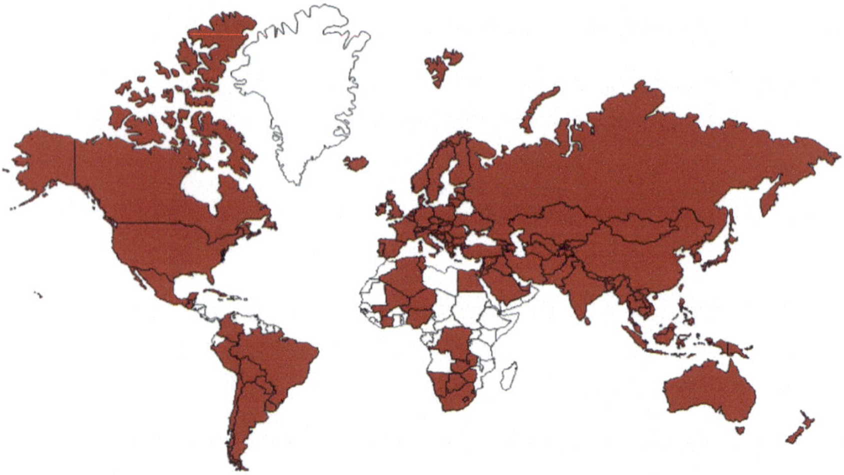

Fig. 10.1 A global map showing the 150 nations and territories where arsenic contamination has occurred from different sources

Arsenic contamination is also observed in soil which causes severe harmful effects on plant growth. Toxic metals (loids), including As, are released into the soil as a result of rapid worldwide urbanization and agricultural, and industrial growth. Plant health, microbial communities, and the general functionality of soil are all impacted by arsenic poisoning, which alters the rhizosphere ecosystem (Goren & Kobya, 2021). Volatilization, demethylation, immobilization, and arsenic mobilization are some of the mechanisms that lead to arsenic poisoning in the soil (Kaya et al., 2024). In consideration of the Monte Carlo simulation, the hazard index is greater than 1 for Cr, Zn, As, and Cu, which are considered to be non-carcinogenic from eating rice from the area of Malakand, Pakistan (Khan et al., 2024).

Vital micronutrients, including zinc, iron, magnesium and boron are present in coconut water, which aids in the formulation of suitable plans to reduce PTE (potential toxic element) pollution and protect public health (Biswas et al., 2024). A recent Latin Hypercube Sampling (LHS) model study was performed in San Luis Potosi. The location with the highest mean As concentration (1374 mg/kg) was Villa de la Paz. It has the highest levels of arsenic in both children (39.32) and rodents (11.994) according to the Hazard Quotient (HQ) calculation (Fernández-Macías et al., 2020). The wetland substrate, alkalinity, pH, dissolved oxygen present in water, temperature, chemical elements (sulfur and Phosphate), a supply of carbon, and other variables are the main determinants of arsenic speciation. One possible explanation for the occurrence of arsenic is the use of chemical weapons for defense during World War I. The soil in this region of France and Germany is extremely contaminated as a result of the inappropriate disposal of weapons, which has detrimental impacts on health (Bausinger & Preuß, 2005). Microbial resistance to arsenic is achieved by increasing the concentration of carbon sources as reported by Bausinger & Preuß, 2005.

As is a hazardous metalloid with numerous negative consequences on human health. It has been documented to occur not only in water but also in soil. Arsenic contamination is mostly caused by mining, pesticides, chemical weapons used in combat, and other related activities. Bangladesh, India, Nepal, and some regions of China are the most affected (Fig. 10.1). As may be seen from the preceding conclusion, places poisoned by arsenic need to be treated. Here are a few traditional and biological solutions to these issues:

10.2 Conventional Methods

10.2.1 Electrokinetic Treatment

By using constant voltage or direct current, the electric field is created between the electrodes; Electrokinetic remediation removes harmful metals from contaminated bodies through electromigration and electrosmosis, which causes arsenic dissolved in the matrix's aqueous phase to migrate toward the electrodes (Yao et al., 2020).

There are two processes involved in the removal of arsenate:

(i) Arsenate is deposited into interstitial water by desorption;
(ii) The dissolved arsenate migrates toward the electrode through electro-osmosis or ionic migration (Suzuki et al., 2013). Under polarity reversal, alternate current appears to function better than direct current (Acosta Hernández et al., 2023).

It is a labor-intensive procedure that frequently requires chemical additions to improve metal mobilization and stop pH shifts (Johansson et al., 2024; Shirin & Yadav, 2014). Recent studies show that the pH of electrode solutions increases, which also increases the elimination capacity of arsenic. Cyclodextrin (10% HPCD) is used as an activating agent which increases the desorption of arsenic (Hawal & Al-Sulttani, 2023). The electrokinetic method works well, but it depends on several key variables, including pH and the polarity of the current supply. While maintaining a constant pH and current, supply can be achieved in a pilot project; scaling up to an industrial scale can be challenging. As a result, switching to a more reliable and simple process is a reliable way to reduce arsenic contamination.

10.2.2 Adsorption

Compared to other technologies, absorption is a relatively inexpensive, passive, and straightforward process. The breakthrough point is reached when arsenic can no longer be adequately removed by the media. The most common ways to remove arsenic from a material are complexation with polyvalent metal species like ferric iron and co-precipitation with metal hydroxide. Other methods include adsorption on activated carbon and alumina, coagulant floc, lime softening, and sulfide precipitation (ferrous sulfide beds) (Gupta & Chen, 1978). Activated charcoal is the most widely used absorbent and is easily available, cheap, and easy to use. Recent studies show that the combined effect of hematite and manganese oxide was successful in reducing the arsenic (III) content (Zheng et al., 2020). The adsorbate matrix is constructed in such a way that the matrix hole is smaller than the arsenic's size. Either a fixed or non-fixed method of filter creation can be used. Unwanted debris clogging can cause membrane fouling, which makes filter cleaning difficult. Here, certain metal oxides or chemicals can be used to carry out the adsorption process. It must be ensured that these are inert and should be utilized exclusively in the removal of arsenic from the environment.

10.2.3 Coagulation

Coagulation is a process that leads to the accumulation of arsenic contaminants into a larger floc, which leads to easy settlement and removal of arsenic particles. It does not require a matrix, which ultimately reduces the cost. Usually, coagulation treatment is done using iron and aluminum salts when drinking or industrial wastewater is considered. The presence of organic matter decreases the efficiency of removal (Ge et al., 2020; Simon et al., 2024). According to recent research, arsenate removal can be performed through an electrocoagulation reactor using 3D AI electrodes. The anode material dissolves to yield Al^{3+} ions, which are then converted to aluminum hydroxides, such as boehmite, gibbsite and bayerite, as well as aluminum oxides.

Steps involved in the electrocoagulation process:

1. Adsorption of inorganic anions inside a precipitate of iron or aluminum hydroxide.

Direct precipitation of anions. However, this method is not reliable, since the presence of phosphate and sulfate will compete with arsenic, thus reducing the efficiency of removal (Goren & Kobya, 2021).

10.2.4 Ion-Exchange Method

Ions in a solution are converted into a solid and the release of ions of the same polarity but of a different kind is known as ion exchange. This indicates that distinct ions that were initially present in the solid are substituted for the ions in solutions. Anion exchange is considered to be a more suitable method to remove nitrate from drinking. If high concentrations of competing ions are present, then ion exchange becomes less appealing due to the expenses associated with brine preparation (Kim & Benjamin, 2004). The removal of arsenic through the IEM process is shown in Fig. 10.2. The ion exchange method has certain advantages, such as it can remove ions quantitatively, being insensitive to the pH of the solution, and long lifetime for the resins. To overcome this, a two-way ion exchanger is used. An argon-aspartate chromatographic peak is performed to fully flow into the downstream column. Each cycle of ion-exchange columns is connected in series before the upstream column is renewed. When the generation is completed, the column is re-seated downstream position within the treatment train (Dadwal & Mishra, 2017). Although different types of strategies enhance the efficacy, it is important to look at the economic prospects. Moreover, the disposal costs and the costs of brine preparation diminish the attractiveness of the ion-exchange method.

10.2.5 Electrocoagulation

Electrochemical processes involve the transfer of electrons from one end of the electrode to another. Electrons can be transferred in solid, liquid, or gaseous medium (Bharti et al., 2023). Compared to conventional flocculation-coagulation, electrocoagulation adds an advantage in eliminating the smallest colloidal particles because the electric field that propels these charged particles increases their likelihood of coagulation and destabilization. Furthermore, electrocoagulation flotation can shorten the

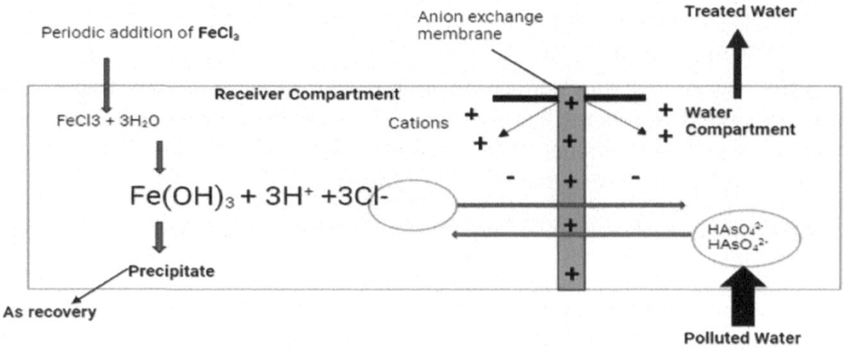

Fig. 10.2 Schematic of the removal of arsenate through the IEM process

time needed for treatment, while also minimizing the amount of trash that is produced during wastewater treatment (Nguyen et al., 2019). A study was conducted that integrated natural siderite into an electrocoagulation system to increase the efficiency of removal. Several studies have been performed on electrocoagulation for the removal of arsenic from contaminated water. The arsenic removal rate dropped from 99.30% ± 0.37% to 98.60% ± 0.45% after six testing cycles, indicating general stability in the effectiveness of arsenic removal. After treating actual groundwater with a high arsenic level, the removal rate was 96.48% ± 0.27%, and the arsenic concentration in the effluent was reduced according to WHO standards. The removal of arsenic from contaminated water samples is achieved using electrocoagulation methods that are 80.0 to 99.99% efficient. In terms of operational costs, the As removal efficiency with iron plate electrodes- along with the EC method ranged from 75 to 99.9%, or 0.0020. Arsenic removal using hybrid iron-aluminum plates costs less than using iron scrap electrodes (0.017 US $/m^3). The efficiency of electrocoagulation can be improved by increasing the applied current density, which reduces the response time for pollutant removal.

Taking this method as a whole into account, it is efficient for the removal of both VIII oxidation states of arsenic. The electrocoagulation process appears to be simple and easy to operate and it does not require any addition of chemicals, which makes downstream processes less complex. In addition, electrocoagulation does not release any waste particles like in membrane processes. Moreover, coexisting ions in solution do not have a major effect in the EC method. In contrast, EC has certain disadvantages such as the high amount of energy required, waste disposal, etc. Renewable sources of energy, such as solar energy, are equipped to neutralize the high energy demand. This technique also reduces the cost of water (Goren & Kobya, 2021).

10.2.6 Membrane Treatment

Membrane technology is one of the most effective technologies and can reduce groundwater contamination by 96%. It stops microorganisms from getting through the barrier and dissolves arsenic from contaminated water without collecting it (Li et al., 2024). The selectivity is extended, and the ion removal efficiency is increased in large part due to the surface charge present on the membrane surface. There are two main methods for creating charged membranes: one is by adding charged components to materials that form membranes, and the other is by utilizing the coating method to create charged groups on the membrane surface, grafting, and interfacial polymerization are considered to be the alternate method (Maity et al., 2021).

According to recent reports, different types of membranes are used in water filtration systems such as nanofiltration, ultrafiltration, microfiltration, and reverse osmosis (Sarango et al., 2018; Shirin et al., 2021). Arsenic-contaminated fluids can be successfully treated by membrane processes, which are thermally regulated, such as membrane distillation, which is regulated by atmospheric pressure. Membrane distillation should be considered as a way to filter high-quality water that has a high concentration of contaminants, such as As (Tomaszewska et al., 2020). Arsenic is removed to a safe level with remarkable efficiency using this large-scale processing technology. However, this method is considered to be a cost sink because it uses a high-cost membrane, as well as maintenance and energy requirements. Additionally, this process results in the disposal of large amounts of leftover water. Microfiltration and ultrafiltration are ineffective because the arsenite and arsenate molecules are tiny. Furthermore, an oxidation phase increases total removal effectiveness by converting smaller arsenite molecules into larger arsenate molecules, just like in precipitation (Jain, 2022).

10.2.7 Chemical Precipitation

This method is used for removing heavy metals like arsenic by using chemical agents like sulfides and salts of calcium, ferric, and magnesium salts. By changing the nature of dissolved As into low-solubility substances, the reagents help in the removal of arsenic from wastewater and can also be utilized to stabilize solid waste. Ferric and calcium arsenate precipitates are the techniques that are most often utilized to remove arsenic from wastewater. However, these reagents are highly unstable in the presence of carbon dioxide and water (Long et al., 2019). According to Zama et al. (2022), it is the possibility that unmodified biochar made from aspen wood and rice husk at 400 and 700 °C will help precipitate more calcium/arsenic compounds, improving the As(III) removal efficiency. On the basis of the results, it was shown that calcium and arsenates were precipitated and formed as aggregates at pH = 6.5, which is nearly neutral. Around pH 4.5, arsenite precipitation is very little, whereas in alkaline around pH 9.5, no arsenite precipitated at all. At pH 6.5, arsenite desorption from wood biochar was at its lowest. In another study by Lim et al. (2014), investigated pyrite is used as a sulfur and iron in a novel approach to remediate wastewaters with elevated arsenic levels, producing environmentally benign scorodite and As2S3 precipitates. Oxidation of sulfide ions to H_2SO_4 aided in the breakdown of pyrite. The wastewater of copper smelting eliminated up to 99.4% of arsenic, which had an initial concentration of 11,200 mg/L.

Ammonium magnesium arsenate ($MgNH_4AsO$) was the specific precipitate of arsenate that was obtained from an alkaline solution. Arsenic in copper and cobalt slag was eliminated by a unique alkaline cycle leaching method and ammonium magnesium arsenate precipitation technology. The results of the phase analysis showed that most of the arsenic species were present in the slag as a Cu–As alloy. Under ideal circumstances, the arsenic leaching achieved using the $NaOH–H_2O_2$ leaching method could reach 94.5%. The filter residue containing arsenic was present primarily as $MgNH_4AsO_4{\cdot}6H_2O$, which has good crystallinity and helps to separate arsenic and alkali more effectively. The purified liquid and residue had arsenic content of 39 mg/L and 24.15%, respectively, under ideal circumstances (Zhang et al., 2021).

Limonite can also be used to remove arsenic from wastewater. For arsenic precipitation, limonite serves as an iron donor, neutralizer, and seed. To efficiently remove arsenic from wastewater, scorodite crystallization can be promoted via heterogeneous nucleation using limonite as an in situ solid iron source. During the precipitation of arsenic and its conversion to scorodite, low iron supersaturation in the solution is made possible by the in situ iron donator and seed function of limonite. Furthermore, because fresh limonite has a high arsenic absorbability, the residual arsenic concentration from scorodite synthesis can be further reduced to 0.2 μg/L. The suggested method has great promise for removing arsenic from smelting wastewater (Gupta et al., 2022).

10.3 Biological Arsenic Removal Technologies

10.3.1 Phytoremediation

Noccaea caerulescens can accumulate up to 30,000 parts per million, suggesting that hyperaccumulators may be used in metal remediation. Phytoremediation is becoming recognized as an effective and sustainable technique that uses specific plant species to absorb and stabilize pollutants. Furthermore, a comparison of various phytoremediation techniques, such as phytodegradation, phytostabilization, phytovolatilization, hemofiltration, and phytoextraction, was compared. The study of nanophytoremediation has looked into ways to raise the molecular efficiency of phytoremediation. Numerous studies have shown that techniques such as phytostabilization, biodegradation, and phytovolatilization can effectively reduce heavy metal concentrations by up to 80% (Fig. 10.3) (Gallo et al., 2021).

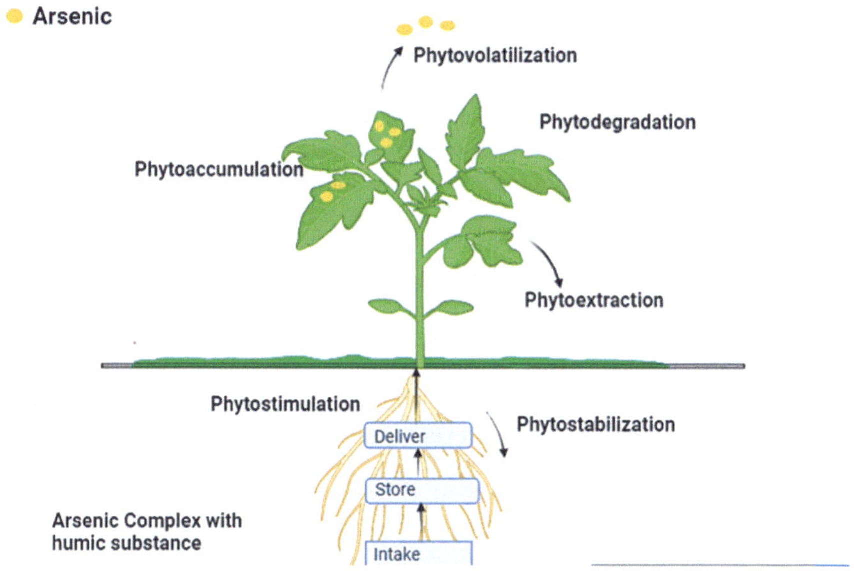

Fig. 10.3 Potential phyto-remediation (phytostabilization) methods of As in soil and water environment

Phytoremediation is an environmentally safe, plant-based technique for cleaning up contaminated areas that use bacteria and plants to purify contaminated water (Rahidul Hassan, 2023). Phytoremediation is an alternative method that reduces the amount of metal and metalloid pollutants in polluted settings by removing or detoxifying pollutants using hyperaccumulator plants and the bacteria of the rhizosphere that dwell in them. When plant species were found that could both endure and accumulate high metal concentrations in their biomass, this method was developed (Abou-Shanab et al., 2020). First identified as an arsenic hyperaccumulator, *P. vittata*, a Chinese brake fern, collects up to 22,630 mg/kg of arsenic in soil modified with 1500 mg / kg, while its fronds can accumulate 12–64 mg/kg in uncontaminated soil with 0.5–7.5 mg/kg. Certain additional fern species, including *Agrostis tenuis, Agrostis stolonifera, Pityrogrammacalomelanos, Pteriscretica, Pterislongifolia, and Pterisumbrosa,* have also been shown to have the capacity to hyperaccumulate arsenic (Wang & Zhao, 2009).

Certain arsenic-hyperaccumulating aquatic plants complete their life cycle in the presence of elevated levels of arsenic. Another strategy to lower arsenic absorption and promote tolerance is to impede phosphate transport activity, a crucial pathway for arsenate entry. Some unicellular organisms are arsenic-tolerant, which depends on aggressively removing arsenite from the cytosol while limiting the absorption of arsenate (Mosa et al., 2016).

Two tests have been carried out by Cantamessa et al. (2020). In the first, soil from the metallurgic site was used to grow colonized ferns under controlled conditions for two years. Beneficial impacts on phytoextraction and accumulation were found, as on well as plant health. After these findings, an in situ experiment was carried out at an industrial site to evaluate the As cleanup at two distinct depths. According to our findings, *P. vittata* significantly enhanced the As cleanup process at a depth of 0 to 0.2 m. The identification of superior rhizobacteria has received significant interest due to their crucial function in improving the effectiveness of phytoremediation. Rhizobacteria are essential to the phytoremediation process because they increase its efficiency. As a result, there has been a significant focus on the need to select the best rhizobacteria.

Furthermore, Rahman et al. (2024), reported that in conjunction with deep root systems and robust biomass production, *Pennisetum purpureum* has great promise as a phytoremediation technique for cleaning up polluted soil. *P. purpureum* phytotoxicity and phytoremediation were assessed at different concentrations of arsenic (V) over 28 days. The findings show that *P. purpureum* can withstand up to 40 mg As(V) kg^{-1} sand, at 80 and 60 mg As(V) kg^{-1} sand, the plant dried up and withered.

10.3.2 Microbial Remediation

Using bacteria to decrease the amount of heavy metals, such as arsenic, which are harmful to living things and can be found in soil and water streams. At the treatment location, physicochemical procedures introduce secondary pollutants. In heavy-metal remediation, microbes change metals into a less hazardous form. Microorganisms that secrete metabolic products, such as low molecular weight, can dissolve minerals and heavy metals that are present in the soil. Numerous methods can lead to microbial remediation. Several processes involve the absorption, oxidation/reduction, volatilization, methylation, and biosorption of arsenic.

10.3.3 Oxidation and Reduction

Oxidation refers to an increase in the charge of arsenic. Take the oxidation states from + 3 to + 5, for instance. Unsaturated soil columns treated with 250 μM arsenate [As(V)] or 75 μM arsenite [As(III)], the oxidation-causing microbial populations were investigated. According to Macur et al., microbial activity quickly changed As(III) into As(V), although As(V) did not appear to be diminished. By using As(V) as an electron acceptor for anaerobic respiration, the Arr4 genes are found in certain bacteria, such as *Pseudomonas aeruginosa* (Escudero et al., 2013). Bactrians and prokaryotes are found in low pH areas, which help in arsenic transformation. Gut microbes help in the bioaccumulation and biotransformation of arsenic.

10.3.4 Methylation and Volatilization

Arsenic methylation, which removes volatile arsine or low-toxic pentavalent arsenic from ambient media via volatilization, could be thought of as a detoxifying process, particularly under aerobic conditions. Increasing the volatilization of methylated As is a potential mitigation point for arsenic mobility and toxicity in the environment (Di et al., 2019). Arsenic methylation forms of arsenic are believed to be a component of a detoxification process in living things. Arsenobetaine and arsenocholine are two more quaternary organic arsenic compounds that are not harmful. Microorganisms can convert arsine, methylarsine, dimethyl arsine, and trimethylarsine to inorganic arsenic and/or volatile arsinestime (Frankenberger, 2002).

For the aim of bioremediation, the biotic and abiotic variables that affect arsenic biovolatilization can be changed (Wang et al., 2014). The combination of straw and charcoal greatly improved As mobility. The main way that straw biochar mobilized As was by increasing the pH content of the soil. The primary method of straw incorporation is done by releasing As from iron minerals and arsenate is reduced to arsenite.

10.3.5 Biosorption

Biological materials can absorb heavy metals from wastewater by physicochemical (not ATP-dependent) or metabolically mediated (ATP-dependent) routes, which is known as biosorption. Arsenic can be removed from groundwater ex situ or in situ by biosorption, which involves sorption to biomass and coprecipitation using biological compounds or sulfides. The secret to the success of this procedure is the introduction of appropriate biosorbents or microbes that produce active biosorbents (Zhang et al., 2020).

Fruit waste, algae, activated carbons, submerged plants, bacteria, and fungi are only a few of the low-cost adsorbents and biosorbents that can be used for arsenic biosorption (Mishra & Srivastava, 2020). *Penicillium purpurogenumis is* used to extract ions from aqueous solutions. The biosorption of heavy metal ions achieved equilibrium after four hours. It was evident that the binding of heavy metal ions by *Penicillium purpurogenum* was pH-dependent. Due to heavy metal speciation and H^+ competition at the same binding sites, the heavy metal loading capacity increased under acidic circumstances as pH increased. The adsorption of heavy metal ions reached a plateau at pH 5.0.

10.3.6 Constructed Wetlands

Low-energy, "green" methods called constructed wetlands are being used more and more in wastewater treatment. The usage of wetland systems has increased, mainly as a result of growing prices for energy sources produced from fossil fuels. Studies show that primary mechanisms of precipitation and sorption remove arsenic from artificial wetlands. Through their mediation of redox and precipitation processes, microorganisms can improve the elimination of As. In a man-made wetland, the primary pathways for arsenic transformation are those that include other organisms. Nguyen et al. studied the removal of arsenic and heavy metals from wastewater by adsorption with modified iron-ore drainage sludge leads to an increase in efficiency. In conducting the pilot experiment, they found that the combined system with limestone substrate eliminated an average of 80.3%, 96.9%, 79.6%, 52.9% and 38.7% of arsenic, manganese, cadmium, zinc and lead from wastewater.

The main effect of the kind of medium was its ability to change pH: zeolite cells raised pH to around 3.8, whereas limestone cells raised pH to approximately 7.1. With minimal assistance from plant absorption, As, Fe and Zn were eliminated from the environment, less than 0.02%, 0.07%, and 0.7%, respectively. For a wetland medium, limestone is preferred over zeolite primarily because of its neutralizing ability (Lizama-Allende et al., 2021). To lower the concentration of As in the horizontally flow-constructed wastes, the subsurface wetlands (CW) proposed in another (Corroto et al., 2019). The arsenic in the reeds had a translocation factor of less than 1, with the majority of the element building up in the roots. Arsenic elimination was mainly attributed to substrate buildup (19.9–30.4%), with lower plant levels (3.8–9.5%), according to the arsenic mass balance. *Aspergillus* strains that are capable of adsorbing arsenate to remove it from contaminated locations. Arsenic methyltransferase is an enzyme that changes arsenic from its inorganic form to a less dangerous and volatile organic form. This enzyme is encoded by the WaarsM gene. *Saccharomyces cerevisiae* that have undergone genetic modification express WaarsM to develop arsenic tolerance. Arsenate reductase and plasma membrane arsenite efflux transporter are two more *S. cerevisiae* genes that provide resistance to arsenic (Naseem et al., 2021).

10.3.7 Mycoremediation

A new form of innovative technology is called mycoremediation, which, without affecting any natural component of the soil, uses extracellular enzymes to break down environmental pollutants such as heavy metals, organic chemicals, polychlorinated compounds, and aromatic hydrocarbons (Mishra & Srivastava, 2020). As levels close to polluted locations can be lowered with mycoremediation techniques, which are becoming increasingly well liked for their affordability and environmental friendliness. A variety of fungi possess unique metallothionein proteins that bind to metals. These proteins are utilized to immobilize soil As concentrations, which helps eliminate accumulated As in crops. Aside from the aforementioned methods, certain fungi can also reduce As pollution by employing glutathione production, cell surface precipitation, bioaugmentation, biostimulation, biosorption, bioaccumulation, volatilization, methylation, and As chelation. Removal from polluted regions could be significantly improved using recombinant yeast and arsenic-resistant fungus (Gupta et al., 2022). The cellular mechanism of the *S. cerevisiae* cell involves the conversion of arsenate to a detoxified form. The capacity to eliminate arsenic from the soil exists in *Aspergillus nidulans*, which was separated from the soil polluted with As. *A. nidulans* of contaminated soil was found to have an arsenic adsorption potential after 11 days at pH 4 and 35 °C (Maheswari & Murugesan, 2009).

10.3.8 Phytoremediation

According to the data compared to fungi and bacteria, algae can remove 84.6% of heavy metals from the environment. This provides several benefits over others. The effectiveness of green algae *C. reinhardtii, C. vulgaris, Scenedesmus almeriensis*, and an indigenous *Chlorophyceae sp.* in removing arsenic was documented by (Chugh et al., 2022). Among these four, *S. almeriensis* removed 40.7% of arsenic in three hours. Additionally, it was found that the effectiveness of arsenic is increased by 1.7 times in multimetallic solutions. To improve *C. reinhardtii*'s ability to remove arsenic, Ramrez-Rodrguez and colleagues altered the acr3 gene. They found that, compared to the wild strain, the abatement capacity increased by a factor of 1.5 to 3 for varying concentrations of arsenic. The red algae Sarcodiasuiae were treated in three distinct temperature ranges (15, 20, and 25 °C) by Chugh et al., who found that total arsenic and arsenite decreased at lower temperatures (15 °C). While concentration has a substantial impact on the accumulation rate, light has no discernible effect on the clearance efficiency. Therefore, it is inferred that both temperature change and the initial concentration of arsenic affect the elimination of different oxidation states of arsenic. Salinity has little bearing on the effectiveness of arsenic removal in the case of living seaweed (Henriques et al., 2019). Within 72 h of working with *Ulva lactuca*, they recorded 48% of the abatement capacity.

10.3.9 Nano Remediation

A great deal of progress has been made in the field of nanobiotechnology toward the creation and improvement of nanoparticles with specific shape and composition, increased active surface area, and a notable effect on biogeochemistry, as well as decreased mobility and phytoaccumulation in soil conditions. The most efficient and economical way to remove contaminants is by combining nanotechnology and bioremediation. An increasing number of studies are using biomolecules, extracts, and components of bacterial, fungal, and algal cultures as catalysts for the environmentally friendly synthesis of nanomaterials. By immobilizing or stimulating the manufacture of remediating microbial enzymes, they can act as catalysts in the bioremediation of hazardous substances. Carbon particles, particularly carbon nanotubes, zero-valent iron, metallic, and magnetic nanoparticles, are the most frequently utilized nanomaterials for remediation. Because of their ability to absorb things, carbon nanotubes are mostly employed to eliminate organic contaminants. The primary factor behind the extensive application of zero-valent iron nanoparticles in soil and groundwater remediation is their strong reactiveness (Singh et al., 2015).

Karimi et al. (2019) created iron nanoparticles using two plant extracts, *Teucriumpolium* and *Prangosferulacea*, and applied them to remove As(III) from copper mining effluents. Iron nanoparticles produced via green synthesis showed an impressive As adsorption effectiveness (over 70% in both cases). Using potassium borohydride as a reducing agent, soluble arsenic was changed to (AsH_3) or solid arsenic to extract arsenic from wastewater. After 30 min, the reaction reached equilibrium and the rate of arsenic removal increased to more than 73% in just 0.5 min. Numerous influencing factors were examined, such as hydrogen ion (H^+) concentrations, aeration, addition technique, and arsenic valence (Feng et al., 2023). Investigated the possibility of removing inorganic arsenic species from water using composites of hydroxyapatite-alginate, nanozero valent zinc, and biochar. Enough zerovalent iron is used to remove As(V) and As(III) from the configuration. Zerovalent iron could immobilize a variety of different toxicants in addition to treating degraded groundwater caused by chlorinated solvents.

10.4 Molecular Techniques for Arsenic Removal

Expression genes linked to glutamic acid (GA) and indole acetic acid (IAA) biosynthesis, such as cyp112, cyp114, trpA, and trpB, were measured using RT-PCR. IU31 inoculation significantly improved plant growth and arsenic accumulation by up to 50%. These results show that by producing IAA and GA, which may facilitate As uptake and detoxification, IU31 improves plant growth and health. Compared to chemically based methods, plant-based genetically engineered phytoremediation techniques are less harmful to human health and the environment (Ullah et al., 2024). Genetically modified plants can be used for the phytoremediation of As through a variety of strategies, such as improved root uptake, xylem loading, reduced arsenate content, vacuole sequestration, and increased tolerance to As (Mosa et al., 2016). In recent years, the use of genetically modified microorganisms (GMOs) for the detoxification or removal has gained prominence. However, the biosafety controversy about GMOs still presents a major barrier. Xue et al. (2024) suggested incorporating charcoal and a controlled GMO in a bioremediation technique to effectively absorb arsenic from mine effluent. In mining wastewater environments, a single metagenomic gene element (ArsR) with high arsenic adsorption efficiency was recently discovered. *The Pseudomonas putida* KT2440 strain for ecological bioremediation was used. The further application of a combined biochar-microorganism recovery technique in mine wastewater treatment resulted in increased arsenic removal capabilities. This approach may now be applied in scenarios where the overall amount of arsenic contamination is up to 20 mg/L. The procedure produced 1.32×10^{-9} GMO escapees at most, which is in accordance with the US NIH GMO release restrictions ($< 10^{-8}$) (Thakur et al., 2020).

Mesophilic temperature-dependent *Escherichia coli* that expressed TtarsM heterologously detoxified arsenite. Despite missing a typical arsenite binding site, TtArsM is a pure protein that catalyzes SAM-dependent arsenite methylation, leading to the synthesis of monomethyl arsenites and dimethyl arsenites. Subsequently, the TtArsM-encoding gene was eliminated from the *T. thermophilus genome* and its role in the arsenite detoxification mechanism was confirmed using a highly efficient genome-editing method based on ThermoCas9. In the end, the *T. thermophiles* ΔTtarsM genome's efflux pump gene was replaced with a gene that produces stabilized yellow fluorescent protein (sYFP) to produce a sensitive genome-based bioreporter (Gallo et al., 2021).

10.5 Hybrid and Integrated Technologies

Arsenic remediation is done with complex chemical mechanisms. Technologies such as ultrafiltration, adsorption and coagulation, ion exchange, adsorption, and photocatalytic oxidation were used for the reduction of arsenic. There are various advantages to hybrid and integrated systems, including high removal efficiency, reduced sludge production, and effective pH regulation, which are shown in Table 10.1 (Mahamallik & Swain, 2023).

The approach proved insufficient to achieve the required purity and mainly removed high molecular weight, even though flocculation and coagulation were employed before membrane techniques. Before membrane processes, ion exchange

Table 10.1 List of hybrid technology

Technology	Material
Electrocatalytic membrane reactor (ECMR) and nanofiltration (NF)	MnO_2 nanowires (NWs) in one dimension (1D) loaded on a flat carbon membrane
Hybrid adsorbent-microfiltration	Iron content and ferric salt-based additive
Electro-assisted self-alkalization and oxidant-free system	As(III) electrooxidation in situ and As(III, V) electro-sorption
Adsorption	Materials based on metal–organic frameworks (MOFs) and mMOFs
Integrated process of filtration and electro-sorption	Rice husk biochar compound with active manganese dioxide (active BC)

can be used to prevent membrane fouling caused by organic compounds (Mahamallik & Swain, 2023).

10.6 Future Prospects and Conclusion

The present study reported that biological treatment of arsenic is more environmentally friendly and provides better decreases in the quantity of arsenic in contaminated soil and water. In chemical treatment, the use of chemicals or some oxidized metal is used, increasing the complexity of the treatment. The selection of membrane and filter pore size according to the heavy metals is a cumbersome process. Moreover, it should be a cost-reliable process. In adsorption or membrane treatment, fouling is another tough problem that needs to be resolved. When membrane and coagulation processes are considered, not only is the removal of arsenic important, but oxidization to a less toxic form is also important. Certain factors such as environmental stability, cost-reliability, and reusability should also be considered. Taking into account these factors, methods such as phytoremediation, microbial remediation, and phytoremediation are reliable processes. Certain plant and microbial species have been used to regulate the amount of arsenic (*Noccaeacaerulescens, Pseudomonas aeruginosa*). This type of remediation not only focuses on reducing the arsenic content but also oxidizes into higher oxidation states which are less toxic than the earlier form. Culturing plants and microbes in vitro is an easy process, which is less toxic than the use of harsh chemicals. In conclusion, it should be highlighted that techniques that avoid causing more environmental problems include considered sustainable recycling technologies, use less energy, and require attention before they can be used to reduce arsenic contamination. Under ideal circumstances, the clearance rate of arsenic can approach nearly 100% using the most effective methods.

Acknowledgements Arumugam A greatly acknowledges the financial support from the DST/SERB, India (CRG/2022/003877).

References

Abou-Shanab, R. A. I., Mathai, P. P., Santelli, C., & Sadowsky, M. J. (2020). Indigenous soil bacteria and the hyperaccumulator Pteris vittata mediate phytoremediation of soil contaminated with arsenic species. *Ecotoxicology and Environmental Safety, 195*. https://doi.org/10.1016/j.ecoenv.2020.110458

Acosta Hernández, I., Muñoz Morales, M., Fernández Morales, F. J., Rodríguez Romero, L., & Villaseñor Camacho, J. (2023). Removal of heavy metals from mine tailings by in-situ bioleaching coupled to electrokinetics. *Environmental Research, 238*. https://doi.org/10.1016/j.envres.2023.117183

Bausinger, T., & Preuß, J. (2005). Environmental remnants of the first world war: Soil contamination of a burning ground for arsenical ammunition. *Bulletin of Environmental Contamination and Toxicology, 74*(6), 1045–1052. https://doi.org/10.1007/s00128-005-0686-z
Bharti, M., Das, P. P., & Purkait, M. K. (2023). A review on the treatment of water and wastewater by electrocoagulation process: Advances and emerging applications. *Journal of Environmental Chemical Engineering, 11*(6). https://doi.org/10.1016/j.jece.2023.111558
Biswas, A., Choudhary, A., & Darbha, G. K. (2024). From ground to gut: Evaluating the human health risk of potentially toxic elements in soil, groundwater, and their uptake by Cocos nucifera in arsenic-contaminated environments. *Environmental Pollution*. https://doi.org/10.1016/j.envpol.2024.123342
Cantamessa, S., Massa, N., Gamalero, E., & Berta, G. (2020). Phytoremediation of a highly arsenic polluted site, using Pteris vittata L. and arbuscular mycorrhizal fungi. *Plants, 9*(9), 1211. https://doi.org/10.3390/plants9091211
Chen, P. R., & He, C. (2008). Selective recognition of metal ions by metalloregulatory proteins. *Current Opinion in Chemical Biology, 12*(2), 214–221. https://doi.org/10.1016/j.cbpa.2007.12.010
Chugh, M., Kumar, L., Shah, M. P., & Bharadvaja, N. (2022). Algal Bioremediation of heavy metals: An insight into removal mechanisms, recovery of by-products, challenges, and future opportunities. *Energy Nexus, 7*, Article 100129. https://doi.org/10.1016/j.nexus.2022.100129
Corroto, C., Iriel, A., Cirelli, A. F., & Carrera, A. L. P. (2019). Constructed wetlands as an alternative for arsenic removal from reverse osmosis effluent. *Science of the Total Environment, 691*, 1242–1250. https://doi.org/10.1016/j.scitotenv.2019.07.234
Dadwal, A., & Mishra, V. (2017). Review on biosorption of arsenic from contaminated water. *Clean—Soil, Air, Water, 45*(7). https://doi.org/10.1002/clen.201600364
Di, X., Beesley, L., Zhang, Z., Zhi, S., Jia, Y., & Ding, Y. (2019). Microbial arsenic methylation in soil and uptake and metabolism of methylated arsenic in plants: A review. *International Journal of Environmental Research and Public Health, 16*(24). https://doi.org/10.3390/ijerph16245012
Escudero, L. V., Casamayor, E. O., Chong, G., Pedrós-Alió, C., & Demergasso, C. (2013). Distribution of microbial arsenic reduction, oxidation and extrusion genes along a wide range of environmental arsenic concentrations. *PloS One, 8*(10). https://doi.org/10.1371/journal.pone.0078890
Feng, Z., Ning, Y., Yang, S., Yu, J., Ouyang, W., & Li, Y. (2023). A novel strategy for arsenic removal from acid wastewater via strong reduction processing. *Environmental Science and Pollution Research, 30*(15), 43886–43900. https://doi.org/10.1007/s11356-022-24919-0
Fernández-Macías, J. C., González-Mille, D. J., García-Arreola, M. E., Cruz-Santiago, O., Rivero-Pérez, N. E., Pérez-Vázquez, F., & Ilizaliturri-Hernández, C. A. (2020). Integrated probabilistic risk assessment in sites contaminated with arsenic and lead by long-term mining liabilities in San Luis Potosi, Mexico. *Ecotoxicology and Environmental Safety, 197*. https://doi.org/10.1016/j.ecoenv.2020.110568
Frankenberger, W. T. (2002). *Environmental chemistry of arsenic*. Marcel Dekker.
Gallo, G., Mougiakos, I., Bianco, M., Carbonaro, M., Carpentieri, A., Illiano, A., Pucci, P., Bartolucci, S., van der Oost, J., & Fiorentino, G. (2021). A hyperthermoactive-Cas9 editing tool reveals the role of a unique arsenite methyltransferase in the arsenic resistance system of Thermus thermophilus HB27. *MBio, 12*(6). https://doi.org/10.1128/mBio.02813-21
Ge, J., Guha, B., Lippincott, L., Cach, S., Wei, J., Su, T. L., & Meng, X. (2020). Challenges of arsenic removal from municipal wastewater by coagulation with ferric chloride and alum. *Science of the Total Environment*. https://doi.org/10.1016/j.scitotenv.2020.138351
Goren, A. Y., & Kobya, M. (2021). Arsenic removal from groundwater using an aerated electrocoagulation reactor with 3D Al electrodes in the presence of anions. *Chemosphere, 263*. https://doi.org/10.1016/j.chemosphere.2020.128253
Gupta, A., Dubey, P., Kumar, M., Roy, A., Sharma, D., Khan, M. M., Bajpai, A. B., Shukla, R. P., Pathak, N., & Hasanuzzaman, M. (2022). Consequences of arsenic contamination on plants and

mycoremediation-mediated arsenic stress tolerance for sustainable agriculture. *Plants, 11*(23), 3220. https://doi.org/10.3390/plants11233220

Gupta, S. K., & Chen, K. Y. (1978). Arsenic removal by adsorption. *Water Pollution Control Federation, 50*(3).

Hawal, L. H., & Al-Sulttani, A. O. (2023). The technique of arsenic elimination from contaminated soil with enhanced conditions by electro-kinetic remediation. *Environmental Monitoring and Assessment, 195*(11). https://doi.org/10.1007/s10661-023-11979-z

Henriques, B., Teixeira, A., Figueira, P., Reis, A. T., Almeida, J., Vale, C., & Pereira, E. (2019). Simultaneous removal of trace elements from contaminated waters by living Ulva lactuca. *Science of the Total Environment, 652*, 880–888. https://doi.org/10.1016/j.scitotenv.2018.10.282

Jain, R. (2022). Recent advances of magnetite nanomaterials to remove arsenic from water. *RSC Advances, 12*(50), 32197–32209. https://doi.org/10.1039/D2RA05832D

Johansson, K., Carabante, I., & Kumpiene, J. (2024). Immobilisation of arsenic in contaminated soil by electrokinetics in an outdoor experiment. *Science of the Total Environment, 918*. https://doi.org/10.1016/j.scitotenv.2024.170656

Karimi, P., Javanshir, S., Sayadi, M. H., & Arabyarmohammadi, H. (2019). Arsenic removal from mining effluents using plant-mediated, green-synthesized iron nanoparticles. *Processes, 7*(10), 759. https://doi.org/10.3390/pr7100759

Kaya, C., Uğurlar, F., Ashraf, M., Hou, D., Kirkham, M. B., & Bolan, N. (2024). Microbial consortia-mediated arsenic bioremediation in agricultural soils: Current status, challenges, and solutions. *Science of the Total Environment, 917*. https://doi.org/10.1016/j.scitotenv.2024.170297

Khan, A., Khan, M. S., Shafique, M. A., Khan, Q., & Saddiq, G. (2024). Assessment of potentially toxic and mineral elements in paddy soils and their uptake by rice (Oryza sativa L.) with associated health hazards in district Malakand, Pakistan. *Heliyon, 10*(7). https://doi.org/10.1016/j.heliyon.2024.e28043

Kim, J., & Benjamin, M. M. (2004). Modeling a novel ion exchange process for arsenic and nitrate removal. *Water Research, 38*(8), 2053–2062. https://doi.org/10.1016/j.watres.2004.01.012

Li, H., Wang, Z., Feng, B., Shi, J., Liao, M., He, K., Tian, H., Megharaj, M., & He, W. (2024). Arsenic stress on soil microbial nutrient metabolism interpreted by microbial utilization of dissolved organic carbon. *Journal of Hazardous Materials, 470*. https://doi.org/10.1016/j.jhazmat.2024.134232

Lim, K. T., Shukor, M. Y., & Wasoh, H. (2014). Physical, chemical, and biological methods for the removal of arsenic compounds. *BioMed Research International, 2014*, 1–9. https://doi.org/10.1155/2014/503784

Lizama-Allende, K., Ayala, J., Jaque, I., & Echeverría, P. (2021). The removal of arsenic and metals from highly acidic water in horizontal subsurface flow constructed wetlands with alternative supporting media. *Journal of Hazardous Materials, 408*, Article 124832. https://doi.org/10.1016/j.jhazmat.2020.124832

Long, H., Zheng, Y., Peng, Y., Jin, G., Deng, W., & Zhang, S. (2019). Comparison of arsenic(V) removal with different lead-containing substances and process optimization in aqueous chloride solution. *Hydrometallurgy, 183*, 199–206. https://doi.org/10.1016/j.hydromet.2018.12.006

Mahamallik, P., & Swain, R. (2023). A mini-review on arsenic remediation techniques from water and future trends. *Water Science & Technology, 87*(12), 3108–3123. https://doi.org/10.2166/wst.2023.190

Maheswari, S., & Murugesan, A. G. (2009). Remediation of arsenic in soil by Aspergillus nidulans isolated from an arsenic-contaminated site. *Environmental Technology, 30*(9), 921–926. https://doi.org/10.1080/09593330902971279

Maity, J. P., Chen, C. Y., Bhattacharya, P., Sharma, R. K., Ahmad, A., Patnaik, S., & Bundschuh, J. (2021). Advanced application of nano-technological and biological processes as well as mitigation options for arsenic removal. *Journal of Hazardous Materials, 405*. https://doi.org/10.1016/j.jhazmat.2020.123885

Meakin, C. J., Szilagyi, J. T., Avula, V., & Fry, R. C. (2020). Inorganic arsenic and its methylated metabolites as endocrine disruptors in the placenta: Mechanisms underpinning glucocorticoid receptor (GR) pathway perturbations. *Toxicology and Applied Pharmacology, 409*. https://doi.org/10.1016/j.taap.2020.115305

Mishra, M., & Srivastava, D. (2020). Mycoremediation: A step towards sustainability. *International Journal of Plant And Environment, 6*(04), 298–305. https://doi.org/10.18811/ijpen.v6i04.09

Mosa, K. A., Saadoun, I., Kumar, K., Helmy, M., & Dhankher, O. P. (2016). Potential biotechnological strategies for the cleanup of heavy metals and metalloids. *Frontiers in Plant Science*. https://doi.org/10.3389/fpls.2016.00303

Naseem, M., Raghuwanshi, R., Verma, P. C., & Srivastava, P. K. (2021). Mycoremediation-effective strategy to ameliorate arsenic toxicity. In *Fungi bio-prospects in sustainable agriculture, environment and nano-technology* (pp. 433–458). Elsevier. https://doi.org/10.1016/B978-0-12-821925-6.00019-8

Nguyen, H., Nguyen, B., Duong, T., Bui, A., Nguyen, H., Cao, H., Mai, N., Nguyen, K., Pham, T., & Kim, K.-W. (2019). Pilot-scale removal of arsenic and heavy metals from mining wastewater using adsorption combined with constructed wetland. *Minerals, 9*(6), 379. https://doi.org/10.3390/min9060379

Rahidul Hassan, H. (2023). A review on different arsenic removal techniques used for decontamination of drinking water. *Environmental Pollutants and Bioavailability, 35*(1). https://doi.org/10.1080/26395940.2023.2165964

Rahman, M. E., Shamsuzzaman, S. M., Mahmud, K., Uddin, M. K., Ghani, S. S. A., Shukor, M. Y. A., Chompa, S. S., Akter, A., Nabayi, A., Sadeq, B. M., & Halmi, M. I. E. B. (2024). Isolation, screening, identification, and characterization of arsenic resistant plant growth promoting rhizobacteria isolated from Pennisetum purpureum. *Biocatalysis and Agricultural Biotechnology, 59*, Article 103237. https://doi.org/10.1016/j.bcab.2024.103237

Sarango, L., Paseta, L., Navarro, M., Zornoza, B., & Coronas, J. (2018). Controlled deposition of MOFs by dip-coating in thin film nanocomposite membranes for organic solvent nanofiltration. *Journal of Industrial and Engineering Chemistry, 59*, 8–16. https://doi.org/10.1016/j.jiec.2017.09.053

Sevak, P., & Pushkar, B. (2024). Arsenic pollution cycle, toxicity and sustainable remediation technologies: A comprehensive review and bibliometric analysis. *Journal of Environmental Management, 349*. https://doi.org/10.1016/j.jenvman.2023.119504

Shirin, S., & Yadav, A. K. (2014). Physico-chemical analysis of municipal wastewater discharge in Ganga river, Haridwar district of Uttarakhand, India. *Current World Environment, 9*(2), 536–543. https://doi.org/10.12944/CWE.9.2.39

Shirin, S., Jamal, A., Emmanouil, C., & Yadav, A. K. (2021). Assessment of characteristics of acid mine drainage treated with fly ash. *Applied Sciences, 11*(9), 3910. https://doi.org/10.3390/app11093910

Simon, M., Joshi, H., Yadav, A. K., & Hopke, P. K. (2024). Simultaneous nitrogen and phosphorus removal from polluted water using a novel bacterial consortium (BACON-38): Promising application for wastewater remediation. *Journal of Water Process Engineering, 66*, Article 106044. https://doi.org/10.1016/j.jwpe.2024.106044

Singh, M., Srivastava, P. K., Verma, P. C., Kharwar, R. N., Singh, N., & Tripathi, R. D. (2015). Soil fungi for mycoremediation of arsenic pollution in agriculture soils. *Journal of Applied Microbiology, 119*(5), 1278–1290. https://doi.org/10.1111/jam.12948

Suzuki, T., Moribe, M., Okabe, Y., & Niinae, M. (2013). A mechanistic study of arsenate removal from artificially contaminated clay soils by electrokinetic remediation. *Journal of Hazardous Materials, 254–255*(1), 310–317. https://doi.org/10.1016/j.jhazmat.2013.04.013

Thakur, S., Choudhary, S., Majeed, A., Singh, A., & Bhardwaj, P. (2020). Insights into the molecular mechanism of arsenic phytoremediation. *Journal of Plant Growth Regulation, 39*(2), 532–543. https://doi.org/10.1007/s00344-019-10019-w

Tomaszewska, B., Bundschuh, J., Pająk, L., Dendys, M., Delgado Quezada, V., Bodzek, M., Armienta, M. A., Muñoz, M. O., & Kasztelewicz, A. (2020). Use of low-enthalpy and waste geothermal energy sources to solve arsenic problems in freshwater production in selected regions

of Latin America using a process membrane distillation—Research into model solutions. *Science of the Total Environment, 714*, Article 136853. https://doi.org/10.1016/j.scitotenv.2020.136853
Ullah, I., Anwar, Y., Siddiqui, M. F., Alsulami, N., & Ullah, R. (2024). Phytoremediation of Arsenic (As) in rice plants, mediated by strain IU31 through antioxidant responses and phytohormones synthesis. *Environmental Pollution, 355*, Article 124207. https://doi.org/10.1016/j.envpol.2024.124207
Wang, P., Sun, G., Jia, Y., Meharg, A. A., & Zhu, Y. (2014). A review on completing arsenic biogeochemical cycle: Microbial volatilization of arsines in environment. *Journal of Environmental Sciences (China), 26*(2), 371–381. https://doi.org/10.1016/S1001-0742(13)60432-5
Wang, S., & Zhao, X. (2009). On the potential of biological treatment for arsenic contaminated soils and groundwater. *Journal of Environmental Management, 90*(8), 2367–2376. https://doi.org/10.1016/j.jenvman.2009.02.001
Xue, Y., Li, Y., Li, X., Zheng, J., Hua, D., Jiang, C., & Yu, B. (2024). Arsenic bioremediation in mining wastewater by controllable genetically modified bacteria with biochar. *Environmental Technology & Innovation, 33*, Article 103514. https://doi.org/10.1016/j.eti.2023.103514
Yao, W., Cai, Z., Sun, S., Romantschuk, M., Sinkkonen, A., Sun, Y., & Wang, Q. (2020). Electrokinetic-enhanced remediation of actual arsenic-contaminated soils with approaching cathode and Fe0 permeable reactive barrier. *Journal of Soils and Sediments, 20*(3), 1526–1533. https://doi.org/10.1007/s11368-019-02459-4
Zama, E. F., Li, G., Tang, Y.-T., Reid, B. J., Ngwabie, N. M., & Sun, G.-X. (2022). The removal of arsenic from solution through biochar-enhanced precipitation of calcium-arsenic derivatives. *Environmental Pollution, 292*, Article 118241. https://doi.org/10.1016/j.envpol.2021.118241
Zhang, S., Manasa, P., Wang, Q., Li, D., Dang, X., Xiaoqin, N., & Ran, F. (2020). Grafting copolymer of thermo-responsive and polysaccharide chains for surface modification of high performance membrane. *Separation and Purification Technology, 240*, 116585. https://doi.org/10.1016/j.seppur.2020.116585
Zhang, Y., Xu, X., Yue, C., Song, L., Lv, Y., Liu, F., & Li, A. (2021). Insight into the efficient co-removal of Cr(VI) and Cr(III) by positively charged UiO-66-NH_2 decorated ultrafiltration membrane. *Chemical Engineering Journal, 404*, Article 126546. https://doi.org/10.1016/j.cej.2020.126546
Zheng, Q., Hou, J., Hartley, W., Ren, L., Wang, M., Tu, S., & Tan, W. (2020). As(III) adsorption on Fe–Mn binary oxides: Are Fe and Mn oxides synergistic or antagonistic for arsenic removal? *Chemical Engineering Journal, 389*. https://doi.org/10.1016/j.cej.2020.124470

Chapter 11
Ex-Situ and In-Situ Arsenic Removal from Contaminated Environments

Subhajit Dutta, Manik Chandra Kundu, Animesh Chowdhury, Biswajit Pramanik, Umalaxmi Thingujam, and Sandip Debnath

Abstract The chapter focuses on biological treatment approaches for the removal of arsenic from contaminated environments, such as soil and water, highlighting their great potential in the context of current treatment technologies. These methods depend on microbial activity to mobilize or immobilize arsenic through processes such as sorption, biomethylation–demethylation, complexation, coprecipitation, and reduction of oxidation. Ex situ bioleaching is identified as a dependable approach for removing large amounts of arsenic from polluted soils, utilizing biostimulation techniques like the addition of carbon sources and nutrients to enhance arsenic leaching rates. Biosorption can be applied ex situ and in-situ to remove arsenic from groundwater either through sorption with biomass or coprecipitation with biogenically precipitated solid products, such as sulfurides. In situ, the effectiveness of this method hinges on the addition of appropriate biosorbents or arsenic-oxidizing microbes. Phytoremediation, another key focus of the chapter, involves using hyperaccumulator plants to extract arsenic from soils and shallow groundwater. This is achieved by genetically enhancing the plants' ability to accumulate high levels of arsenic in their tissues, a method that can also be studied and applied ex situ, along with additional microorganisms. Biological treatments are capable of addressing arsenic contamination along with co-contaminants such as ammonia, iron, and manganese in groundwater. Two important groups of bacteria in this process include

S. Dutta · M. C. Kundu · U. Thingujam
Department of Soil Science and Agricultural Chemistry, Palli Siksha Bhavana (Institute of Agriculture), Visva Bharati, Sriniketan, India

A. Chowdhury
Department of Agronomy, Palli Siksha Bhavana (Institute of Agriculture), Visva Bharati, Sriniketan, India

B. Pramanik
Biotechnology Unit, Crop Improvement Division, ICAR-Central Research Institute for Jute and Allied Fibres, Barrackpore, India

S. Debnath (✉)
Department of Genetics and Plant Breeding, Palli Siksha Bhavana (Institute of Agriculture), Visva Bharati, Sriniketan, India
e-mail: sandip.debnath@visva-bharati.ac.in

A. K. Yadav and S. Shirin (eds.), *Emerging and Innovative Arsenic Removal Technologies for a Sustainable Future*, Water Science and Technology Library 131,
https://doi.org/10.1007/978-3-032-01176-3_11

ammonia oxidizers and arsenic oxidizers, which help convert arsenic(III) to more easily removable arsenic(V). In biological treatment systems, an aeration contactor is placed before a standard granular media filter, where orthophosphate is introduced as a biological nutrient. The chapter underscores that plant-based techniques are environmentally friendly and potentially cost-effective; however, additional research and enhancement are essential to fully realize their potential.

Keywords Arsenic · Arsenic removal · Biological treatment · Ex-situ bioleaching · Microbial activity · Oxidation reduction

11.1 Introduction

In the Earth's crust, arsenic is a mineral that is extensively dispersed and is mainly found in the minerals of igneous and sedimentary rocks and soils. Certain minerals that include arsenic, such as sulfarsenides and arsenides, are deemed nontoxic as a result of their high degree of insolubilities. But complications arise when these bare minerals disintegrate or combine to generate more soluble species, such as oxides. Arsenic can be released into the atmosphere by natural and man-made processes or dissolved in water. Arsenic contamination in water and soil is a persistent global environmental problem that poses significant health risks to human populations and requires effective remediation strategies. Traditionally, chemical precipitation and adsorption have been the methods most employed to remove inorganic arsenic from water (Biswas et al., 2024; Patel et al., 2022). Growing interest has been shown in recent years that biological treatment methods of arsenic removal are eco-friendly and more sustainable for mankind (Anirudhan & Unnithan, 2007). In arsenic biogeochemistry, Microbial activities have been reported to play an essential role (Gadd, 2010), which influences As speciation and mobility in paddy soil–water systems through different As biotransformation pathways (Mishra & Mishra, 2018; Ojuederie & Babalola, 2017; Simon et al., 2024). Some microorganisms have the metabolic potential to lower Arsenic concentration by some biological treatments in soil and plant (Tripathi et al., 2022). One possible biological tactic is to use arsenic resistant bacteria that can oxidize, reduce, or methylate the metal to alter its speciation and facilitate removal and can be used as a promising biological approach (Anirudhan & Unnithan, 2007). Using plants, fungi, or algae that can bioaccumulate or bioabsorb arsenic is another strategy that can be used to remove it from the environment (Ngo et al., 2002).

Zecchin et al. (2017) reported numerous microbial genes encoding As-transformation enzymes and transporters. Microbes exposed to low levels of nutrients can grow by consuming minerals containing arsenic. This process releases arsenic into the environment and creates an abiotic environment that can alter arsenic speciation (Mailloux et al., 2009). Because they can possibly attain low concentrations of arsenic and are both environmentally friendly and cost-effective, biological treatment systems have emerged as viable techniques to remove arsenic from the environment. Arsenite methylation (through arsM genes), respiration (through the arr genes), reduction (through the ars genes) and oxidation (through the aio genes) have been shown to have an impact on the microbial-mediated arsenic cycle in agronomic systems (Xiao et al., 2016). Microorganisms can methylate inorganic arsenic species; This process produces mono-, di-, tri-, and tetramethyl As species by the use of arsenite methyltransferases (Zhao et al., 2013). However, these conventional techniques can be energy intensive and may generate hazardous sludge (Ngo et al., 2002).

Because rice plants cannot methylate arsenic, human dietary exposure to arsenic increases by microbial transformation of organic arsenic and subsequent root uptake (Punshon et al., 2017; Zhao et al., 2013). Furthermore, microbes can oxidize AsIII to less toxic and less As(V) through detoxification (Zhang et al., 2015b). Both flooded and non-flooded settings are used for this process. Zhang et al. (2015a) showed that by employing nitrate and oxygen as electron acceptors, Paracoccus species can oxidize As(III) in paddy soil in aerobic and anoxic conditions, respectively. Therefore, arsenic toxicity and accumulation in rice plants can be reduced by inoculating As-polluted soils with As(III) oxidizing bacteria (Kumarathilaka et al., 2020). Phytoremediation, using plants to extract, stabilize, or detoxify environmental contaminants, has shown particular promise for the removal of arsenic (Gupta et al., 2023). Furthermore, microorganisms reduce As(V) to As(III) mainly under anaerobic conditions through detoxification and respiratory pathways. Reducing As(V) to more mobile and toxic As(III) may increase As mobility and bioavailability in paddy soils (Kumarathilaka et al., 2020). It has been shown that introducing microorganisms into growth media contaminated with arsenic can decrease the phytotoxicity and build-up of arsenic in plants. Many bacterial strains have been shown to reduce the amount of arsenic in various organs of rice plants, including *Bacillus pumilus* (Dolphen & Thiravetyan, 2019), *Rhodopseudomonas palustris*-C1, *Brevundimonas diminuta*-NBRI012, *Bacillus aryabhattai*-MCC3374, and KKU2500 strains of Cd-tolerant bacteria. In a more recent study, Moens et al. (2020) showed that when the hyperaccumulator rhizobacterial strain of *Ochrobactrum tritici* was injected into the rhizosphere of rice plants and recorded lowest concentration of arsenic in inoculated plants which was 2.6 times lower than sterile plants. Inoculation with the As5 strain also alleviated some of the toxic effects on shoot growth compared to inoculation with the unmodified strain.

For an efficient biological treatment plan for the remediation of this hazardous chemical, one must take into consideration its efficacy under different environmental conditions. In areas prone to arsenic pollution, these strategies must be clearly evaluated and periodically monitored to have a healthier and cleaner habitat for us. This chapter focuses on the principles, methodologies, and recent progress in the biological treatment strategies that secure the remedy for arsenic contamination. It also delves into the merits and demerits of ecofriendly bioremediation processes emphasizing on the current scenario and future prospects.

11.2 Mechanisms of Biological Arsenic Remediation

Arsenic, being a hazardous toxicant, must be removed from the environment, preferentially through biological processes using microorganisms and plants that possess the ability to detoxify it from contaminated environments (Mondal et al., 2024). This section focuses primarily on several efficient bio-mechanisms to be adopted using plant and microbes for the same (Fig. 11.1).

11.2.1 Biotransformation

Arsenic exists in several oxidation states viz., − 3, 0, + 3 and + 5, having different degrees of toxicity. Arsenite, As(III), is more toxic than Arsenate, As(V). Therefore, biotransformation processes generally include shifting of the arsenic molecule between its different oxidation states depending on its mobility, toxicity, and removal efficiency. In addition, two crucial in-depth mechanisms lie in As(III) oxidation and As(V) reduction.

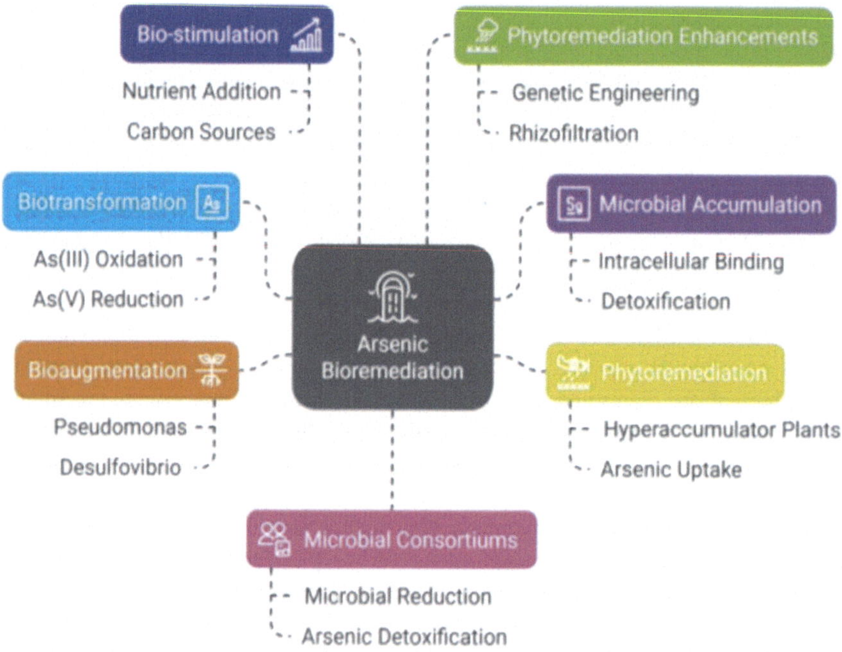

Fig. 11.1 Visual representation of arsenic bioremediation protocols (Generated with Napkin.ai; final layout edited by author.)

11.2.1.1 As(III) Oxidation

As stated above, As(III) is more toxic and mobile than As(V). Hence, As(III) oxidation can be an efficient strategy to a convert it to less toxic and less mobile form, arsenate. This oxidation is generally executed by several microorganisms. Bacteria such as *Oxalobacter* and *Thermoplasma* play vital role in this oxidation process. arsenite oxidase, the enzyme present in these bacteria, is responsible for catalyzing the oxidation of As(III) to As(V) (Biswas et al., 2019). As(III), formed in this oxidation process, can be removed from the system after adsorption and subsequent precipitation (Mandal et al., 2021). These microorganisms also mediate the dissolution of arsenic-containing minerals, producing a variety of compounds with variable ultimate arsenic oxidation states. The dissolution of minerals is promoted by prokaryotes and acidophilic bacteria that improve the oxidation of Fe. The mesophilic microbes *Acidithiobacillus ferroxidans*, *Acidithiobacillus thioxidans*, and *Leptospirillum ferroxidans* are the most efficient among them in arsenic mineral biooxidation (Drewniak & Sklodowska, 2013). The adaptability of this biotransformation process across various microbial communities can be seen in *Thermoplasma* species, which have a significant contribution in the oxidation of As(III) to As(V).

11.2.1.2 As(V) Reduction

The reduction of As(V), which could represent an essential stage in the bioremediation process, involves the conversion of As(V) to As (III). Bacteria that reduce arsenate, such as those of the genera *Geobacter* and *Shewanella*, use it as a terminal electron acceptor when they respire anaerobically. By reducing As(V) to As(III), this technique improves its amenability for subsequent processing or immobilization (Weeger et al., 1999). As(III) formed after reduction is then removed by precipitation or complexation with sulfides. Microorganisms that can promote joint reduction of As(V) and sulfate play an important role in this precipitation process. The sulfuric and alkalinity resulting from this process provide a favorable condition for sequestration of arsenic. In nature, more than 200 minerals exist in nature, mainly arsenides, sulfides, oxides, As (V) and As(III). The most abundant arsenic ore minerals are As sulfides, formed under high temperatures in the Earth's crust and occur in hydrothermal and magmatic ore deposits. After reduction, As(III) can be treated using various remediation techniques, including adsorption to solid matrices or precipitation with sulfides. Arsenic bio-mining-assisted microbial activities promoted by *Geobacter sulfurreducens* and *Shewanella oneidensis* help clean up the environment (Darma et al., 2022). These microbes also generate energy under anaerobic conditions by reducing As(V) and immobilizing As(III).

11.2.1.3 Methylation/Demethylation

The surface of methylation and demethylation is crucial pathways for arsenic biotransformation, in addition to oxidation and reduction. Several bacterial species can generate monomethylarsenic acid (MMA) and dimethylarsenic acid (DMA) from the inorganic form. It is proven that methylated forms, being less toxic, are easy to eliminate. *Bacillus* are efficient at methylating arsenic, aiding in its detoxification and elimination. Furthermore, demethylation transforms organic arsenic into an inorganic one, affecting its overall mobility as its well as toxicity.

11.2.1.4 In-Situ Oxidation

Arsenic concentrations could be reduced prior to groundwater pumping. Application methods include using permeable reactive barriers (PRBs) to treat groundwater as it passes through, bioreactors to cultivate bacteria that oxidize arsenic, and injection wells to send chemicals or microorganisms directly into the aquifer. This approach minimizes environmental effects, reduces excavation needs, and delivers site-specific treatment. The dissolved oxygen content in the water oxidizes As(III) to less mobile As(V) and the ferrous iron in the aquifer to ferric iron, which lowers the As content in the water of the tube well.

11.2.2 Microbial Accumulation

Arsenic bioremediation is highly dependent on microbial accumulation of arsenic. Arsenic is often found to bind to intracellular proteins or vacuoles inside bacterial or fungal cells for effective sequestration, which eventually reducing its toxicity as well as bioavailability (Gupta et al., 2022). *Bacillus flexus* and *Acinetobacter junii* are renowned as potent microbes for effective arsenic bioremediation due to their adaptability in arsenic detoxification (Marwa et al., 2019).

11.2.3 Phytoremediation

Phytoremediation leverages the natural ability of certain plants to absorb and concentrate arsenic from the soil, for which several hyperaccumulator plants, such as *Pteris vittata* and *Brassica juncea*, are known to be potent contributors. After acquiring arsenic from soil, these plants accumulate it at their tissues in significantly higher concentration than in the surrounding environment (Roy et al., 2015), followed by harvesting, which helps to remove arsenic from the environment.

11.2.4 Bioaugmentation

This process introduces specific microbes into an environment to break down arsenic into lesser harmful forms in those soil or water systems where innate microbial population is not strong enough. To reduce arsenic toxicity, strains of *Pseudomonas* or *Desulfovibrio* are employed to convert As(V) to As(III), which are further precipitated or eliminated.

11.2.5 Bio-stimulation

Biostimulation is basically a process through which nutrients or growth-enhancing minerals are added to the indigenous microorganisms to degrade the toxicity level of arsenic. In this process, the main deciding factor behind microbes' activity of microbes is their ability to react with arsenic biologically, including microbial reduction, and oxidation (Macy et al., 2000). Furthermore, pH, temperature, availability of nutrients, and presence of competing compounds can also impact the efficiency of this method.

The potential of biostimulation in enhancing arsenic leaching from contaminated soils and sediments is increasingly being recognized. The strategy involves the deliberate application of carbon sources and essential nutrients to stimulate microbial activities, thereby facilitating arsenic mobilization. Organic carbon compounds, including sugars and organic acids, act as substrates and energy sources for microbial communities, promoting critical biochemical processes such as the oxidation, methylation, and reduction of arsenic species. Supplementation with key nutrients, notably phosphorus and nitrogen, further supports microbial proliferation and activity, leading to improved arsenic transformation and mobilization.

In practice, the addition of inorganic and organic nutrients or the direct injection of carbon-rich substrates into contaminated matrices has shown promising results in boosting microbially mediated arsenic mobilisation. Laboratory investigations have successfully delineated optimal regimes for carbon and nutrient amendments to achieve effective arsenic removal. Field studies have further substantiated the practical viability of these biostimulation techniques, demonstrating notable improvements in arsenic bioavailability and leaching. Additionally, careful management of site-specific variables and potential environmental risks, particularly the unintended amplification of arsenic mobility, remains imperative. Future research efforts are expected to focus on integrating biostimulation with other complementary remediation technologies. Therefore, the development of advanced monitoring and predictive modeling frameworks will be crucial to optimizing these strategies for comprehensive arsenic remediation.

11.3 Phytoremediation Enhancements

11.3.1 Genetic Engineering

Genetic engineering presents a promising avenue to improve the ability of plants to absorb, accumulate, and detoxify arsenic. This enhancement is achieved by introducing specific genes that either increase arsenic metabolism or confer increased tolerance to high arsenic concentrations. In particular, arsenic exposure has been shown to reduce the expression and functionality of key DNA repair proteins, thus affecting DNA lesion repair mechanisms and contributing to genomic instability (Tam et al., 2020). To counteract these deleterious effects, genetic modifications can be employed to introduce genes encoding phytochelatins or metallothioneins, which facilitate the binding and detoxification of arsenic within plant tissues (Zhao et al., 2017). Comparative studies between transgenic and wild-type plants have demonstrated that genetically engineered lines, such as those overexpressing genes from Nicotiana and Arabidopsis species, exhibit superior capabilities in arsenic accumulation and detoxification. Furthermore, targeted manipulation of genes associated with arsenic transport and sequestration, including those coding for arsenic efflux transporters, offers an additional strategy to enhance the efficiency with which plants extract arsenic from contaminated soils and aquatic environments. Collectively, these approaches signify the potential of genetic interventions in the development of phytoremediation technologies tailored for arsenic-contaminated ecosystems.

11.3.2 Rhizofiltration

This approach has proven particularly effective for the remediation of arsenic-contaminated water bodies, where plant roots serve as natural biofilters, efficiently capturing and concentrating arsenic from the surrounding aqueous environment (Singh et al., 2015). Few plant species, especially *Pteris vittata,* exhibited considerable potential in arsenic removal through this mechanism. These plants not only uptake arsenic from water but also sequester and precipitate it within their tissues, often converting it into less toxic chemical forms. Such biological processes underscore the vital role of specific plant species in phytoremediation strategies aimed at mitigating arsenic pollution. This can be enhanced by selecting plant species with naturally high arsenic uptake capabilities or by genetically engineering plants to improve their efficiency.

11.3.3 Microbial Consortiums

A significant bioremediation method to control arsenic contamination is microbial reduction, which uses microorganisms to change As(V) into As(III). Microbes have the potential to detoxify arsenic in a sustainable manner, with arsenic concentrations in land plants generally less than 10,000 ppb. Because of its greater affinity for sorption onto solid phases, As(III) is more amenable to subsequent removal techniques, such as precipitation or adsorption, mainly carried out by bacteria such as *Shewanella* species. Arsenate is used as a terminal electron acceptor during anaerobic respiration. The method is often conducted in bioreactors or permeable reactive barriers designed to promote anaerobic conditions, where the addition of electron donors such as organic carbon sources (e.g. acetate or lactate) promotes microbial activity. Although microbial reduction can increase the efficiency of arsenic removal when utilized in anaerobic settings, it has limitations as well, including the need for precise control over environmental variables such as pH and redox potential and the potential for sulfide creation in anaerobic settings.

11.3.4 Enzyme-Based Approaches

Enzyme-based arsenic remediation strategies take advantage of certain enzymes that microorganisms create, which can break down or change arsenic molecules. These enzymes, which break down arsenic into less hazardous or more readily removed forms, are essential for the biotransformation of arsenic species. In treatment systems, As(III), although comparatively less toxic than As(V), can be effectively precipitated or further processed because of the catalytic function of arsenate reductase enzymes. Conversely, the oxidation of As(III) to As(V) by arsenite oxidase facilitates the subsequent application of conventional removal techniques, including adsorption and precipitation, as As(V) exhibits a higher affinity for such processes. These enzymatic transformations significantly improve the efficiency of arsenic remediation strategies and can be integrated into various bioreactor configurations to detoxify arsenic-laden environments. By utilizing the distinct catalytic properties of arsenate reductase and arsenite oxidase, this approach offers a targeted and potentially more efficient alternative to traditional physical or chemical remediation methods (Singh et al., 2015).

11.3.5 Ex-Situ Bioleaching for Arsenic Remediation

This process helps arsenic to be mobilized from solid matrices, such as polluted soils, by stimulating microbial activities. Numerous research studies illustrating the application and scalability of ex situ bioleaching technologies are discussed in conjunction with strategies for biostimulation, notably the supplementation of carbon sources and essential nutrients. By enhancing microbial metabolism, arsenic becomes more soluble and thus more amenable to extraction from polluted soils through this process. Various methods aimed at optimizing arsenic leaching rates, particularly those involving nutrient and carbon amendments, were previously studied.

Field studies and real-world implementations shed light on the advantages and practical challenges associated with the use of bioleaching for large-scale arsenic remediation. Bioleaching, a biotechnological approach, is based on the metabolic abilities of microorganisms to extract metals from ores and mineral substrates. In the context of arsenic contamination, specific microbial strains capable of oxidizing arsenic-containing minerals facilitate chemical reactions that convert solid-phase arsenic into soluble forms. Arsenic can be further treated or precipitated after mobilization, which significantly reduces its concentration in contaminated soil and groundwater.

Compared to conventional chemical remediation methods, bioleaching serves as a more sustainable and eco-benign alternative, operating under milder chemical conditions, and reducing reliance on aggressive reagents. Optimizing the bioleaching efficiency for arsenic removal requires careful control of key parameters, including pH, temperature, and choosing appropriate microbes.

11.4 Advantages of Biological Treatments

Arsenic decontamination biological remediation protocols provide a number of advantages over other methodologies such as eco-sustainability, cost-effectiveness, diversified adaptability, and many more (Fig. 11.2).

11.4.1 Environmental Sustainability

Biological treatment techniques support ecological balance and reduce the need for artificial substances that utilize the inherent abilities of microbes, plant absorption, or biofilms to degrade or alter arsenic. These typically prevent introducing secondary pollutants into the environment, in contrast to certain chemical treatments that may include the addition of toxic substances or the production of hazardous by-products. A cleaner and more environmentally friendly way of remediation is through biological approaches, which use live organisms to transform or immobilize toxins.

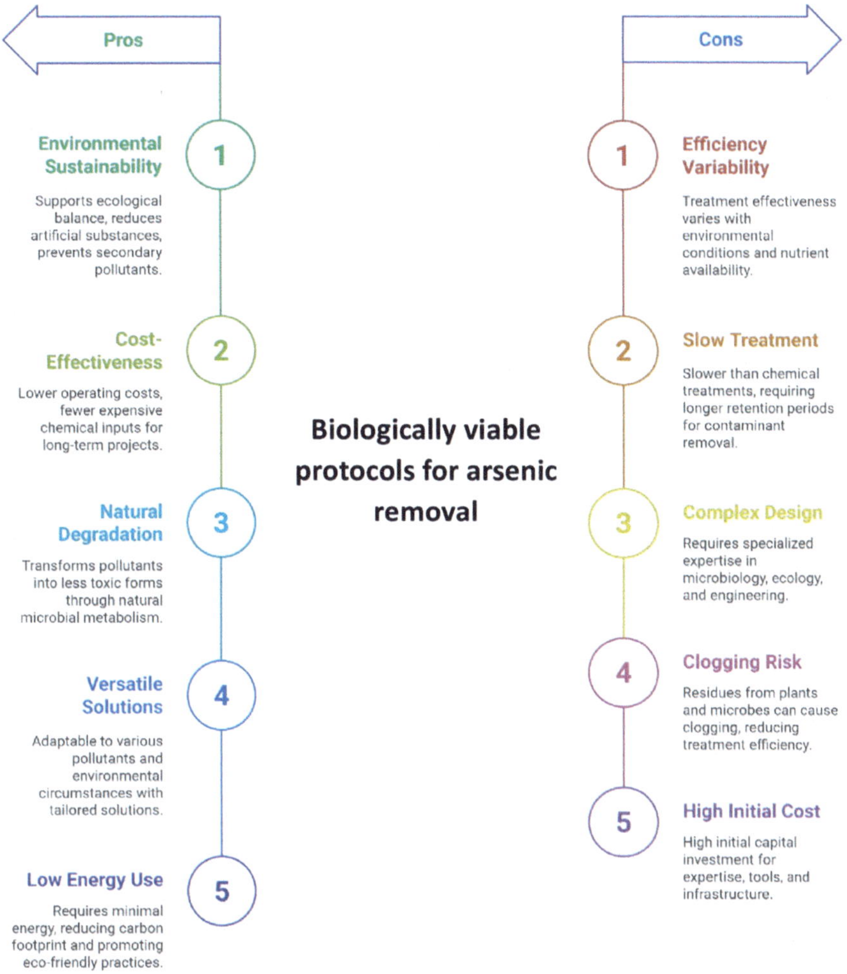

Fig. 11.2 Advantages and disadvantages of biological arsenic remediation protocols (Generated with Napkin.ai; final layout edited by author.)

11.4.2 Effectiveness in Costs

These techniques work on reduced operating expenses because biological treatments (e.g. bioreactors, phytoremediation) often have lower ongoing costs than chemical treatments. Using biological treatment with naturally occurring organisms for large-scale or long-term projects is cost effective with fewer expensive chemical inputs (Asif & Chen, 2017).

11.4.3 Natural Mechanisms in Degradation

Using naturally occurring mechanisms for degradation, biological treatment techniques transform pollutants into less toxic forms through the metabolism of microbes or plants. Bacteria can change harmful forms of arsenic into less harmful ones or immobilize them in a form that is less accessible to plants and animals. These organic methods can result in longer-lasting, more sustainable solutions without significantly changing the environmental matrix.

11.4.4 Versatility and Adaptability with Tailored Solutions

Biological therapies can be applied to various pollutants and environmental circumstances. Diverse groups of microbes are introduced, depending on the nature of the contaminants and the specificities of the contaminated sites, to transform arsenic into less hazardous form in an environmentally safer way. Several plant species can also be chosen for the phytoremediation of arsenic. Treatment procedures can be customized according to specific requirements under special circumstances in a flexible manner.

11.4.5 Less Energy Requirements

Chemical treatment procedures demand huge energy inputs for oxidation, adsorption, precipitation, filtration, etc. However, microbes are less demanding and capable of transforming arsenic to a less toxic product with minimum energy input. These biological transformation procedures take place at optimal room temperature under controlled atmospheric pressure. This less carbon-consuming nature for their survival promotes more cost-effective eco-friendly management practices.

11.4.6 Integration with Other Combined Technologies

Where more efficient remediation techniques are required to remove the pollutant within a short period of time, biological treatment techniques can be combined with conventional treatment technologies to increase efficiency and decrease the time requirement as these biotransformation processes are compatible with other physical/chemical remediation methods. For instance, biological degradation can be combined with physical filtration methods to handle organic and inorganic pollution simultaneously. In this way, we can reduce the dependence on chemicals and pharmaceuticals and increase the efficiency of the combined technologies in a cost-effective manner.

11.4.7 Public Acceptance and Perception

These biotransformation processes are mostly natural remedies and are more acceptable by people due to less or no involvement in chemical therapies. Higher public acceptance and support can result from this view, particularly in community water treatment projects where environmental and health considerations are top priorities. The application of biological techniques can improve the trust and participation in rehabilitation initiatives while being consistent with ecological principles as it is healthier.

11.4.8 Durability of Effect Sustained Performance

Biological treatment systems can provide long-term efficacy if properly cared for and observed. Established ecological systems of plants or microbial communities can adapt to shifting environmental circumstances and efficiently remove pollutants over time. This continuous performance guarantees that the therapy is reliable and effective.

11.5 Disadvantages of Biological Treatments

Although biological treatment methods for contaminants such as arsenic offer several advantages, several drawbacks and difficulties should be considered (Fig. 11.2).

11.5.1 Efficiency Variability

Several variables, including nutrient availability, the presence of competitive organisms, and environmental conditions (pH, oxygen content, and temperature), could affect how successful biological treatments. Variations in these variables can affect the effectiveness of treatment.

11.5.2 Slow Treatment Rates

Biological processes can occur more slowly than chemical treatments. Longer retention periods may be required for microbial degradation or plant absorption mechanisms to remove contaminants to the appropriate levels, especially in complicated or high-concentration contamination scenarios.

11.5.3 Complexity of Design and Operation

Microbiology, ecology, and engineering specialists are frequently required to design and operate biological treatment systems. Accomplishing regular and reliable arsenic removal is challenging due to other ions and complex microbial activity (Aredes et al., 2012). Maintaining ideal conditions for plant health or the growth in a polluted habitat can be challenging and may require constant monitoring and modification.

11.5.4 Clogging Risk

Remainders from plant or microbes and sediment depositions from adsorbents and other materials used in the remediation processes can be accumulated and deposited in treatment plants. This causes clogging in certain systems, including biofilters or other parts of the treatment plant. As time passes, this may reduce the efficiency of overall treatment, requiring maintenance and cleaning.

11.5.5 Prolonged Start-Up Times

At the initial stage, installing a biological treatment system for the first time may require a prolonged period of time. Time is also required to build up a robust microbial population and for supporting a proper plant development for efficient bioremediation. This duration may be further prolonged if the microbes require adjustment with their new environment to cope with the polluted environment. However, once installed, and the microbes and plants are adjusted to their new habitat, and once they build up a robust population, these systems run smoothly and more efficiently for cleaning up the environment in a cost-effective and less time-hungry eco-friendly manner.

11.5.6 Limited Applicability to Specific Contaminants

Specific enzymes are released by a specific group of microorganisms, and these specific groups of microbes require a particular congenial environmental condition for their proper growth and development. This is especially true for arsenic, as it can exist in several oxidation states. Some toxins are not readily biodegradable, and their removal becomes difficult with the use of plants or microbes.

11.5.7 Seasonal Variability

Plants and microbes involved in bioremediation process are very sensitive to seasonal variability in temperature, humidity, day length, wind speed, precipitation etc. Hence, there will be huge variation in the nutrient acquisition capability reflecting variation in growth and reproduction rates, which ultimately results in the diversity in the efficacy of these organisms in their pollutant cleanup rates.

11.5.8 Biomass Disposal Risk

The ultimate plant products after phytoremediation, especially in case of phytoextraction, contain a considerable amount of arsenic in the plant tissues. In case of bioremediation using microbes, microbial sludge also contains considerable amount of the pollutant. These plant products and microbial body residues are the potential source of the pollutant for further spread if not handled properly. To avoid their spread and re-entry into the environment proper disposal methods must be undertaken.

11.5.9 Regulatory Considerations

Periodical monitoring must be performed and reported to biological treatment systems to help them meet regulatory requirements and proper guidelines over time. Water quality standards must be strictly followed through regular testing and certification.

11.5.10 Initial Capital Expenses

To obtain extended benefits from biological treatment systems, the prior construction cost seems to be on the higher side, which somehow balances in the long term. Initially, it is required to invest huge capital for appointing particular expertise, installing monitoring tools, and necessary infrastructures. The first of these higher operational costs with unpredictable effectiveness and precautions in handling materials requiring materials affect the applicability and wide acceptance of biological remedial measures.

11.6 Challenges and Future Prospects

In this section, biological remedy options used in arsenic-polluted ecosystems throughout the world are discussed giving special emphasis on their advantages and disadvantages through practical applications. In Australia and South America, acidophilic bacteria are used to leach arsenic from ores and soils using a method called bioleaching, which is also promising in several other countries. However, similar to other biological remediation processes, maintaining an appropriate pH and temperature for bacterial action can be challenging, despite the fact that bioleaching can significantly reduce arsenic levels. These biological treatment systems are potentially promising strategies to control arsenic contamination in soil and water. Employing microbial activities and environmentally friendly processes offers sustainable and cost-effective solutions, demanding significantly potent environmental remediation efforts in the long run globally.

Ongoing research studies and technological advancements in biological remedies in arsenic treatment are promising, as they aim to address challenges in current scenario with their future applicability. These must focus on improving process efficiencies keeping in mind overall performance, enhancing the abilities of microbial and plant-based arsenic transformation by offering their ideal habitat, and scaling up these technologies for greater implementation with wider acceptability. Site-specific complications must be overcome by extending research in variable environmental conditions with different levels of pollution. Continual monitoring is essential for maintaining regulations for the widespread adoption of biological treatments. Future advancement may offer a worldwide perspective of widely adapted, ecologically sustainable, and more effective solutions for controlling arsenic pollution. The deployment of biological therapy to eradicate arsenic offers promising and diverse potential. The selectivity and efficiency of microbes and plants recruited in arsenic remediation can be further improved by introducing genetic engineering and synthetic biology to enhance the ability of plants and microbes to target and eliminate (Rylott & Bruce, 2020).

To increase their overall scalability and cost-effectiveness, advanced research is being done to optimize biological protocols and synchronize them with physical or chemical ones (Maity et al., 2024). Proper bioreactor design along with in situ application techniques must be anticipated to reduce the risk of contamination during digging and transport. For their wider acceptance, ecological benefits must be illustrated to common people, especially those surviving in polluted conditions. With wider acceptability and regulatory frameworks, biologically viable arsenic remediation efforts will continue for the long run. Frequent monitoring and advanced control systems are also required to optimize efficiency and flexibility under varying environmental conditions.

11.7 Summary and Conclusion

Arsenic contamination is a serious problem due to its higher toxicity, lesser biodegradability, wider distribution, and easy availability through groundwater. Several decontamination processes have been developed and practiced successfully using various physical as well as chemical means. Biological treatment methods offer a viable and durable solution for mitigating arsenic contamination in soil and water ecosystems. These technologies employ natural processes, including phyto- and microbial activity, and hence, these are eco-benign solutions in contrast to conventional remediation that rely mostly on chemicals and huge energy. However, biological remediation uses no chemicals or fewer chemicals and comparatively lower operating expenses in the long term. They recruit microbial and cost-effective inputs, such as plant metabolic processes, to overcome arsenic pollution. Furthermore, these can be combined with other physical and chemical decontamination processes to improve the efficacy of the system by reducing its overall energy utility. In situ application of these decontamination processes leaves environment less disturbed, making them more affordable, adaptable and sustainable. Hence, research must continue to enhance the progress of biological therapy technologies to achieve advancement. In this chapter, several bioremediation methodologies are presented, viz. Biotransformation in oxidative, reductive, methylation, bioaccumulation using plants and microbes, biodegradation, and bio-stimulation methodologies have been discussed in depth. Further enhancement in these techniques using genetic engineering, the development of new strains of microbes, the use of microbial consortiums, the enhancement of enzymatic activities, and the integration of bioremediation measures with other physical and chemical means is possible to increase the efficacy of the systems. The pros and cons of these processes are also emphasized critically. However, to achieve wide acceptance and sustainability, issues must be resolved, including seasonal unpredictability, operational complications, inadequate efficacy, and the implementation of stricter regulations (Singh et al., 2015). In a nutshell, these technologies can be employed at a global scale with their full potential only when strict regulatory standards can be maintained with frequent monitoring.

Acknowledgements The authors express their heartfelt gratitude to our esteemed institute and all individuals who have played a role, directly or indirectly, in contributing to this study.

References

Anirudhan, T. S., & Unnithan, M. R. (2007). Arsenic (V) removal from aqueous solutions using an anion exchanger derived from coconut coir pith and its recovery. *Chemosphere, 66*(1), 60–66. https://doi.org/10.1016/j.chemosphere.2006.05.031

Aredes, S., Klein, B., & Pawlik, M. (2012). The removal of arsenic from water using natural iron oxide minerals. *Journal of Cleaner Production, 29–30*, 208–213. https://doi.org/10.1016/j.jclepro.2012.01.029

Asif, Z., & Chen, Z. (2017). Removal of arsenic from drinking water using rice husk. *Applied Water Science, 7*, 1449–1458. https://doi.org/10.1007/s13201-015-0323-x

Biswas, R., Majhi, A. K., & Sarkar, A. (2019). The role of arsenate reducing bacteria for their prospective application in arsenic contaminated groundwater aquifer system. *Biocatalysis and Agricultural Biotechnology, 20*, Article 101218. https://doi.org/10.1016/j.bcab.2019.101218

Biswas, S., Mondal, P. C., Pramanik, B., & Debnath, S. (2024). Arsenic mitigation from contaminated ground water through chemical and biological processes: A review. In N. Kumar, M. Z. Hashmi, & S. Wang (Eds.), *Arsenic toxicity remediation. Emerging contaminants and associated treatment technologies*. Springer. https://doi.org/10.1007/978-3-031-52614-5_12

Darma, A., Yang, J., Zandi, P., Liu, J., Możdżeń, K., Xia, X., Sani, A., Wang, Y., & Schnug, E. (2022). Significance of Shewanella species for the phytoavailability and toxicity of arsenic—A review. *Biology, 11*, 472. https://doi.org/10.3390/biology11030472

Dolphen, R., & Thiravetyan, P. (2019). Reducing arsenic in rice grains by leonardite and arsenic–resistant endophytic bacteria. *Chemosphere, 223*, 448–454. https://doi.org/10.1016/j.chemosphere.2019.02.054

Drewniak, L., & Sklodowska, A. (2013). Arsenic-transforming microbes and their role in biomining processes. *Environmental Science and Pollution Research, 20–7728*, 7728–7739. https://doi.org/10.1007/s11356-012-1449-0

Gadd, G. M. (2010). Metals, minerals and microbes: Geomicrobiology and bioremediation. *Microbiology, 156*(3), 609–643. https://doi.org/10.1099/mic.0.037143-0

Gupta, A., Dubey, P., Kumar, M., Roy, A., Sharma, D., Khan, M. M., Bajpai, A. B., Shukla, R. P., Pathak, N., & Hasanuzzaman, M. (2022). Consequences of arsenic contamination on plants and mycoremediation-mediated arsenic stress tolerance for sustainable agriculture. *Plants, 11*(23), 3220. https://doi.org/10.3390/plants11233220

Gupta, R. K., Bharti, R., Pramanik, B., Duary, B., Pramanik, K., & Debnath, S. (2023). Utilizing various potentials for phytoremediation of arsenic contamination—A feasible perspective. In: N. Kumar, & S. Kumar, (Eds.), *Arsenic toxicity remediation: Biotechnological approaches*. Environmental Science and Engineering. Springer. https://doi.org/10.1007/978-3-031-37561-3_14

Kumarathilaka, P., Seneweera, S., Ok, Y. S., Meharg, A. A., & Bundschuh, J. (2020). Mitigation of arsenic accumulation in rice: An agronomical, physico-chemical, and biological approach–A critical review. *Critical Reviews in Environmental Science and Technology, 50*(1), 31–71. https://doi.org/10.1080/10643389.2019.1618691

Macy, J. M., Santini, J. M., Pauling, B. V., O'Neill, A. H., & Sly, L. I. (2000). Two new arsenate/sulfate-reducing bacteria: Mechanisms of arsenate reduction. *Archives of Microbiology, 173*(1), 49–57. https://doi.org/10.1007/s002030050007

Mailloux, B. J., Alexandrova, E., Keimowitz, A. R., Wovkulich, K., Freyer, G. A., Herron, M., Stolz, J. F., Kenna, T. C., Pichler, T., Polizzotto, M. L., & Dong, H. (2009). Microbial mineral weathering for nutrient acquisition releases arsenic. *Applied and Environmental Microbiology, 75*(8), 2558–2565. https://doi.org/10.1128/AEM.02440-07

Maity, S., Dokania, P., Goenka, M., Rahul, S., Are, R. P., & Sarkar, A. (2024). Techno-economic feasibility and life cycle assessment analysis for a developed novel biosorbent-based arsenic bio-filter system. *Environmental Geochemistry and Health, 46*(3). https://doi.org/10.1007/s10653-023-01839-7

Mandal, D., Sonar, R., Saha, I., Ahmed, S., & Basu, A. (2021). Isolation and identification of arsenic resistant bacteria: A tool for bioremediation of arsenic toxicity. *International Journal of Environmental Science and Technology, 19*(10), 9883–9900. https://doi.org/10.1007/s13762-021-03673-9

Marwa, N., Singh, N., Srivastava, S., Saxena, G., Pandey, V., & Singh, N. (2019). Characterizing the hypertolerance potential of two indigenous bacterial strains (Bacillus flexus and Acinetobacter junii) and their efficacy in arsenic bioremediation. *Journal of Applied Microbiology, 126*(4), 1117–1127. https://doi.org/10.1111/jam.14179

Mishra, P., & Mishra, M. (2018). Risk Assessment of heavy metal contamination in paddy soil, plants, and grains (Oryza sativa L.). In *Environmental pollution of paddy soils* (pp. 165–178). https://doi.org/10.1007/978-3-319-93671-0_11

Moens, M., Branco, R., & Morais, P. V. (2020). Arsenic accumulation by a rhizosphere bacterial strain Ochrobactrum tritici reduces rice plant arsenic levels. *World Journal of Microbiology and Biotechnology., 36*(2), 23. https://doi.org/10.1007/s11274-020-2800-0

Mondal, P. C., Biswas, S., Pramanik, B., & Debnath, S. (2024). Removal of heavy metals from contaminated water using hydroponics. In: N. Kumar (Eds.), *Hydroponics and environmental bioremediation*. Springer Water. Springer. https://doi.org/10.1007/978-3-031-53258-0_8

Ngo, H. H., Vigneswaran, S., Hu, J. Y., Thirunavukkarasu, O., & Viraraghavan, T. (2002). A comparison of conventional and non-conventional treatment technologies on arsenic removal from water. *Water Science and Technology: Water Supply, 2*(5–6), 119–125. https://doi.org/10.2166/ws.2002.0159

Ojuederie, O. B., & Babalola, O. O. (2017). Microbial and plant-assisted bioremediation of heavy metal polluted environments: A review. *International Journal of Environmental Research and Public Health, 14*(12), 1504. https://doi.org/10.3390/ijerph14121504

Patel, D., Pamidimukkala, P., Chakraborty, D., & Yadav, A. (2022). Bharuch district, Gujarat, India: Factor analysis and geographical distribution of water quality characteristics. *Environmental Nanotechnology, Monitoring & Management, 18*, Article 100732. https://doi.org/10.1016/j.enmm.2022.100732

Punshon, T., Jackson, B. P., Meharg, A. A., Warczack, T., Scheckel, K., & Guerinot, M. L. (2017). Understanding arsenic dynamics in agronomic systems to predict and prevent uptake by crop plants. *Science of the Total Environment, 581*, 209–220. https://doi.org/10.1016/j.scitotenv.2016.12.111

Roy, M., Giri, A. K., Dutta, S., & Mukherjee, P. (2015). Integrated phytobial remediation for sustainable management of arsenic in soil and water. *Environment International, 75*, 180–198. https://doi.org/10.1016/j.envint.2014.11.010

Rylott, E. L., & Bruce, N. C. (2020). How synthetic biology can help bioremediation. *Current Opinion in Chemical Biology, 58*, 86–95. https://doi.org/10.1016/j.cbpa.2020.07.004

Simon, M., Joshi, H., Yadav, A. K., & Hopke, P. K. (2024). Simultaneous nitrogen and phosphorus removal from polluted water using a novel bacterial consortium (BACON-38): Promising application for wastewater remediation. *Journal of Water Process Engineering, 66*, Article 106044. https://doi.org/10.1016/j.jwpe.2024.106044

Singh, R., Singh, S., & Parihar, P. (2015). Arsenic contamination, consequences and remediation techniques: A review. *Ecotoxicology and Environmental Safety*, 247–270. https://doi.org/10.1016/j.ecoenv.2014.10.009

Tam, L. M., Price, N. E., & Wang, Y. (2020). Molecular mechanisms of arsenic-induced disruption of DNA repair. *Chemical Research in Toxicology, 33*(3), 709–726. https://doi.org/10.1021/acs.chemrestox.9b00464

Tripathi, A., Ranjan, M. R., Verma, D. K., Singh, Y., Shukla, S. K., Rajput, V. D., Minkina, T., Mishra, P. K., & Garg, M. C. (2022). ANN-GA based biosorption of As (III) from water through chemo-tailored and iron impregnated fungal biofilter system. *Scientific Reports, 12*(1), 12414. https://doi.org/10.1038/s41598-022-14802-w

Weeger, W., Lièvremont, D., Perret, M., Lagarde, F., Hubert, J. C., Leroy, M., & Lett, M. C. (1999). Oxidation of arsenite to arsenate by a bacterium isolated from an aquatic environment. *BioMetals, 12*(2), 141–149. https://doi.org/10.1023/a:1009255012328

Xiao, K. Q., Li, L. G., Ma, L. P., Zhang, S. Y., Bao, P., Zhang, T., & Zhu, Y. G. (2016). Metagenomic analysis revealed highly diverse microbial arsenic metabolism genes in paddy soils with low-arsenic contents. *Environmental Pollution, 211*, 1–8. https://doi.org/10.1016/j.envpol.2015.12.023

Zecchin, S., Corsini, A., Martin, M., Romani, M., Beone, G. M., Zanchi, R., & Cavalca, L. (2017). Rhizospheric iron and arsenic bacteria affected by water regime: Implications for metalloid uptake by rice. *Soil Biology and Biochemistry., 106*, 129–137. https://doi.org/10.1016/j.soilbio.2016.12.021

Zhang, J., Zhou, W., Liu, B., He, J., Shen, Q., & Zhao, F. J. (2015a). Anaerobic arsenite oxidation by an autotrophic arsenite-oxidizing bacterium from an arsenic-contaminated paddy soil. *Environmental Science and Technology, 49*(10), 5956–5964. https://doi.org/10.1021/es506097c

Zhang, S. Y., Zhao, F. J., Sun, G. X., Su, J. Q., Yang, X. R., Li, H., & Zhu, Y. G. (2015b). Diversity and abundance of arsenic biotransformation genes in paddy soils from southern China. *Environmental Science and Technology, 49*(7), 4138–4146. https://doi.org/10.1021/acs.est.5b00028

Zhao, F. J., Zhu, Y. G., & Meharg, A. A. (2013). Methylated arsenic species in rice: Geographical variation, origin, and uptake mechanisms. *Environmental Science and Technology, 47*(9), 3957–3966. https://doi.org/10.1021/es304295n

Zhao, Z., Wang, S., & Jia, Y. (2017). Effect of sulfide on As (III) and As (V) sequestration by ferrihydrite. *Chemosphere, 185*, 321–328. https://doi.org/10.1016/j.chemosphere.2017.06.134

Chapter 12
Biological Treatment of Arsenic Contaminated Soil and Water by Using Phytoremediation and Microbial Bioremediation

Swati Sonker and Abhay B. Fulke

Abstract Arsenic contamination in water sources represents a serious environmental and public health threat. The major health threat associated with exposure to arsenic is the development of different diseases, including cancer, cardiovascular disease, and neurological disease. Environmental deterioration and long-term health problems can arise from repeated contact with arsenic-contaminated soil and water. It makes the creation of efficient treatment plans imperative. An environmentally acceptable and long-lasting option for the removal of arsenic is provided by biological treatment techniques. The use of biofilms and biosorbents, as well as phytoremediation and microbial bioremediation, are some of the biological techniques for treating arsenic that are examined in this article. By using the metabolic powers of bacteria resistant to arsenic, microbial bioremediation transforms hazardous arsenic species into less dangerous ones. Using plants, phytoremediation removes, stabilizes, and changes arsenic from polluted soils and water. Moreover, organic materials and microbial populations produce biofilms and biosorbents that effectively adsorb and accumulate arsenic ions. A viable approach to lowering arsenic levels in polluted settings is the integration of these biological approaches, which have the advantages of being inexpensive, requiring little chemical input, and perhaps having in situ uses. To provide safer water supplies and better environmental health, further study and development is required to maximize these biological therapies, increase their effectiveness, and facilitate large-scale application. This chapter aims to describe the comprehensive approach towards the implementation of biological techniques and explores the application of recent biological advancements towards arsenic treatments.

S. Sonker · A. B. Fulke (✉)
Microbiology Division, CSIR-National Institute of Oceanography (CSIR-NIO), Regional Centre, Andheri (West), Mumbai, India
e-mail: afulke.nio@csir.res.in

Academy of Scientific and Innovative Research (AcSIR), Ghaziabad, India

A. K. Yadav and S. Shirin (eds.), *Emerging and Innovative Arsenic Removal Technologies for a Sustainable Future*, Water Science and Technology Library 131,
https://doi.org/10.1007/978-3-032-01176-3_12

Keywords Arsenic · Microbial bioremediation · Phytoremediation · Biofilms · Biosorbents · Phytoextraction

12.1 Introduction

In nature, As compounds exist in three different forms: organic, inorganic, and arsenic gas. In contrast to the inorganic As form, which is a pure metalloid bound to elements, the organic As form is covalently attached to Arsenic atoms and contains an atom of carbon. Unlike inorganic arsenic compounds, organic arsenic compounds are generally thought to be less harmful to the environment. The four usual oxidation states of inorganic arsenic are As^{3-} (arsine), As^0 (elemental arsenic), As^{3+} (arsenite), and As^{5+} (arsenate). Of these, arsenate ($HAsO_4{}^{2-}$, H_2AsO^{4-}) and arsenite ($As(OH)_3$, $HAsO_3$) are most commonly encountered in both oxic and anoxic soils and aquatic environments. (Khan et al., 2024).

One very dangerous element that is becoming more and more an environmental problem is arsenic (As) contamination of soils and streams. Arsenic pollution can originate from natural sources, such as hydrothermal activity, volcanic emissions, mineral deposits, and cenozoic sediments. On the other hand, human-induced sources include mining, overextraction of near-surface groundwater, geothermal energy, and emissions from various industries. Industrial effluents, such as mining waste and the disintegration of rocks, minerals, or ores, are the major causes of water pollution. As(V) is often found in soluble form in oxygen-rich water environments like wells, whereas As(III) is found in reducing environments, such as deep lake sediments or groundwater.

In addition to being present in soils, both inorganic forms of As may be absorbed by plants and consumed by food crops, where they can have a variety of harmful consequences. The human toxicity of As(III) has been observed to be higher than that of organic As and As(V). The main channels through which humans come into contact with arsenic include the use of contaminated water for cooking and drinking, the use of tainted water for irrigation or leading to soil contamination in farming production, and the consumption of tainted seafood (Khan et al., 2024). Depending on As exposure, dose, and chemical form, once As enters the body of a human, it can produce a wide range of acute and chronic harmful consequences and disorders. (Roy et al., 2015).

There is a serious health risk to those who depend on groundwater for their drinking water, and the condition of arsenic poisoning is concerning. There is an annual identification and contamination of new sites. Repurposing groundwater for irrigation has generated additional routes for exposure to arsenic throughout the world, as it increases the bioavailability of arsenic in food crops and promotes its absorption by both people and livestock within the food chain. (Roy et al., 2015).

Some of the most efficient techniques for cleaning up As-contaminated soils and waters have been covered, including (i) microbial bioremediation, which uses microorganisms to remove or detoxify metals from the environment, and/or (ii)

phytoremediation, which uses plants to do the same. Furthermore, the most current findings on the molecular processes behind As microbial and phytoremediation, as well as the relationships and interactions between plants and microbes in the rhizosphere that are crucial to As remediation, have been thoroughly examined and presented (Roy et al., 2015). Ultimately, it is expected that sustained progress in biotechnology and materials science (such as nano- and biobased materials, genetically modified organisms) will facilitate the creation of innovative instruments, methodologies, and products to observe and simulating essential biogeochemical interactions and processes involving As, thus guaranteeing more effective As remediation.

As exposure has been shown to be a significant risk factor for many neurological and intellectual disorders in numerous experimental and clinical studies. This is especially true for the most vulnerable child population, where As can adversely affect memory and intelligence performance even at a dose of less than 10 μg/L. Lastly, exposure to this carcinogenic element has been linked to several forms of cancer, including skin, liver, kidney, bladder, lung, and breast cancer. Therefore, to reduce pollution and protect people from hazardous exposure, remediation techniques that are economical, ecologically friendly, and effective must be created and implemented (Roy et al., 2015). This chapter aims to describe the comprehensive approach toward the implementation of biological techniques and explores the application of recent biological advances towards arsenic treatments.

12.2 Arsenic Sources

Various parent rocks and minerals that contain arsenic, including silicates, carbonates, sulfides, and oxides, serve as natural geogenic sources of arsenic (Valskys et al., 2022). Sulfide minerals, such as pyrite (FeS) and arsenopyrite (FeAsS), are known to frequently possess elevated levels of arsenic (Saunders et al., 2018). Research has confirmed that arsenopyrite is a major mineral source of arsenic within the ecosystem (Flora 2015). As a result of sulfur minerals found in the crust of the Earth and the creation of bedrock in places contaminated with As, such as Shaanxi, China; many towns in Taiwan, West Bengal, India; Chattogram, Bangladesh; and Lahore, Pakistan, earlier research has identified FeAsS as a source of As. Furthermore, Shahid et al. (2018) revealed that the amounts in various parent rocks vary from 0.5 to 50 mg/kg. The levels of arsenic present in geological materials, such as rocks, sediments, and soils, vary according to specific geological conditions and rock types. Generally, the arsenic concentration in igneous rocks is low, with an average of 1.5 mg/kg. In comparison, volcanic glasses, known as obsidian and created through the rapid cooling of lava, show somewhat elevated concentrations, averaging 9.1 mg/kg (Murray et al., 2023). Among various types of rocks, pelitic metamorphic rocks, including phyllites and slate, demonstrate the highest levels of arsenic, averaging approximately 18 mg/kg. As concentrations in sedimentary rocks are usually between 5 and 10 mg/kg (Wang et al., 2023).

Environmental As pollution is mostly caused by anthropogenic activity. The concentration of arsenic in the Earth's crust is elevated as a result of extensive research across various developmental sectors and industrial and commercial activities. Herbicides, such as sodium salts of methane arsenic acid, contribute to this problem, and industrial processes related to metal, chemical, drug, and petroleum refining production are some of the primary human-caused sources of As. Burning coal can also release As into the atmosphere due to the trace levels of pyrite found in coals high in sulfur (Khan et al., 2024).

Biogenic sources are crucial in numerous natural processes that emit or alter arsenic in the environment, thereby contributing to arsenic contamination. Examples of these biogenic sources include methylation, the absorption and accumulation of arsenic by plants, and the decomposition of organic matter (Kushwaha et al., 2023). Various flora and microorganisms can absorb As and impact element transition from one oxidation state to another. Through the process of biological methylation, which is aided by microorganisms such as bacteria, fungus, algae, and phytoplankton, arsenic can also be methylated. Organo-As molecules, such as trimethylarsine [$(CH_3)_3As$], arsenobetaine [$C_5H_{11}AsO_2$], dimethylarsonous acid [$(CH_3)_2As(OH)_3$], and monomethylarsonous acid [$CH_3As(OH)_2$], can be formed as a result of this methylation process.

12.3 Arsenic Toxicity

The toxicity of arsenic is significantly influenced by its chemical forms and oxidation states. Furthermore, various factors such as the individual's preexisting health conditions, the physical state of the arsenic (whether in powder, gas, or solution form), the particle size, the absorption rate, the excretion rate, and the chemical composition of the hazardous compound all play a crucial role in determining toxicity. The relative toxicity of arsenicals decreases in the following sequence: arsines, iAsIII arsenoxides (organic AsIII), iAsV arsonium compounds, and As (Yadav et al., 2022).

Individuals who are naturally or occupationally exposed to inorganic arsenic can suffer from laryngitis, tracheae bronchitis, pharyngitis,rhinitis, and shortness of breath, nasal septal perforation, and chest noise (crepitations and/or rhonchi) (Kushwaha et al., 2023) (refer to Fig. 12.1). To some extent, toxicity may be avoided by eating a high in protein, which includes foods such as milk, fish, lentils, and meat (Mondal et al., 2013). Rich nations have the technology and means to remove As from the environment at its source, and their citizens may work to eat a diet high enough in protein. Unfortunately, those on the margins in low- and middle-income nations cannot afford a high-protein diet, which puts them at risk for As toxicity and further impedes the achievement of SDG-3. The main requirement to achieve the SDG-3 targets is the elimination of As at the commercial and residential level (Yadav et al., 2024).

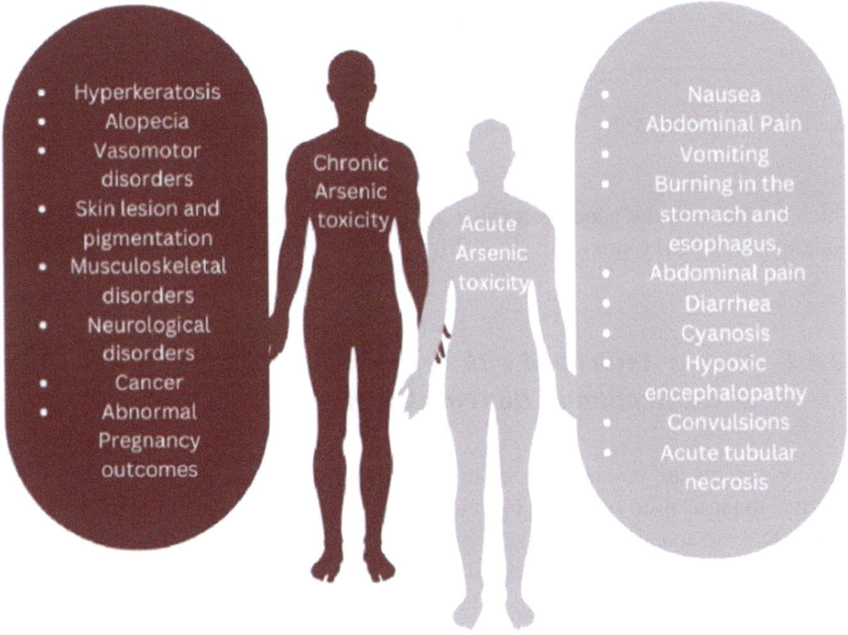

Fig. 12.1 Acute and chronic toxicity of arsenic

12.4 Microbial Bioremediation

The processes used in the bacterial bioremediation of arsenic (As) include biosorption, biotransformation, bioaccumulation, reduction of As(V)-containing calcite to arsenic sulfides, conversion of As(III) into trimethylarsine gas, phosphate solubilization, and synthesis of ACC deaminase, siderophores and indole-3-acetic acid (IAA). Bacteria such as *Achromobacter sp.* and *Aliihoeflea sp.* have demonstrated the ability to oxidize As(III), thereby reducing its toxicity compared to As(V), indicating a promising approach for groundwater remediation. Through complex metabolic pathways that involve reduction (Xiao et al., 2016), oxidation (Sharma et al., 2022), biosorption (Wang et al., 2020), methylation and volatilization (Xiao et al., 2016), these bacteria are essential in the mobilization, transformation, and bioremediation of arsenic. The processes described in Table 12.1 highlight the significance of biosorption, bioaccumulation, biotransformation, and the reduction of As(V) containing calcite to arsenic sulfides, along with phosphate solubilization and the conversion of As(III) to trimethyl arsine gas, as well as the synthesis of ACC deaminase, siderophores, and IAA. The oxidation of As(III) by bacteria such as *Achromobacter sp.* and *Aliihoeflea sp.* renders it less harmful compared to As(V), making this method a viable option for groundwater treatment (Biswas et al., 2021). Furthermore, the catalytic function of a periplasmic enzyme referred to as As(III) oxidase,

Table 12.1 Arsenic-resistant bacterial isolates

Bacteria	Mechanism	References
Bacillus cereus	As(III) efflux	Aguilar et al. (2020)
Brevibacillus sp. KUMAs2	Plasmid-mediated	Mallick et al. (2014)

which is related to the gene designated aox, plays a significant role in the bacterial oxidation of As(III) (Aguilar et al., 2020).

12.4.1 Routes Associated with Arsenic Microbial Detoxification in Bioremediation

Microorganisms are essential to the biogeochemical cycle of arsenic because they facilitate the transition from one form to another. All microorganisms encode genes for resistance to arsenic because the prevalent species of arsenic could vary depending on the biological context. In nature, As(III) and As(V) are the forms of arsenic that are most prevalent (Wang et al., 2020). The trivalent form of arsenic (As(III)) can significantly impair protein function by interacting with sulfhydryl groups, decreasing the availability of cysteine and, consequently, obstructing correct protein folding, making it the more physiologically significant of the two forms. The danger arises from the ability of arsenic in its pentavalent form (As(V)) to convert to the more toxic As(III), despite its own detrimental effects on cellular functions by obstructing phosphate anion transporters. In response to persistent environmental arsenic stress, microorganisms have developed gene expressions that are specific to their niches to improve arsenic uptake (Aguilar et al., 2020).

12.4.2 Differential Arsenic Detoxification Techniques Employed by Microbes

The mechanisms of biotransformation play a crucial role in determining the speciation and movement of arsenic within different environmental contexts. As a result, the metabolic use of arsenic as an electron donor or acceptor in respiratory redox processes serves as a significant pathway for arsenic detoxification in bacterial species. The metabolic pathways employed by these two mechanisms to use ambient arsenic differ significantly (Aguilar et al., 2020). In the presence of As(III), bacteria utilize it as an electron donor to oxidize As(III) and derive cellular energy. In contrast, during anaerobic cellular respiration, As(V) is reduced by acting as the terminal electron acceptor. Arsenic may also be converted by microorganisms into a variety of organic compounds, which has significant effects on arsenic toxicity (Wang et al., 2020) (Fig. 12.2).

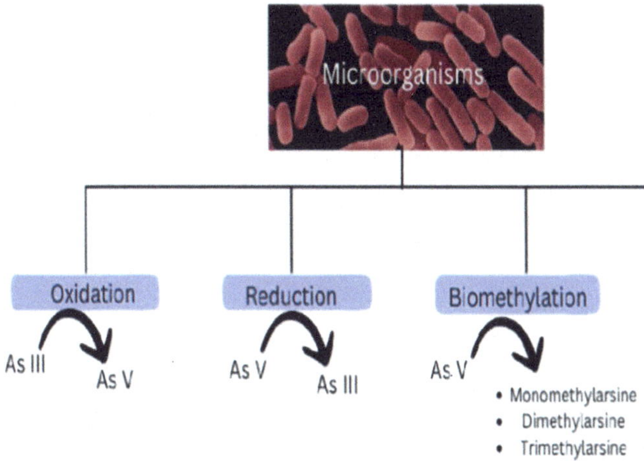

Fig. 12.2 Different techniques of arsenic detoxification by microorganisms

12.4.3 As(III) Microbial Oxidation

The main method by which microorganisms detoxify arsenic is the oxidation of the more hazardous As(III). Recent evolutionary studies have shown that the enzymes required for the oxidation of As(III) are fundamental to all arsenic degradation pathways. It is only in recent times that researchers have begun to comprehend the regulatory mechanisms and proteins involved in this process, facilitated by the development of advanced structural elucidation techniques. This has sparked an increase in interest in the use of microbial As(III) oxidation pathways for innovative bioremediation strategies (Aguilar et al., 2020). Notably, an investigation into various As(III)-oxidizing microorganisms revealed that different species exhibit variability in the cellular membranes to which Aio translocates, as well as in the unique ligands surrounding their redox-active sites. For instance, the well-known disulfide bridge that anchors the AioB Rieske cluster is present in the Aio of *Alcaligenes faecalis*, which was the first to have its structure completely characterized. This disulfide bridge was previously thought to be essential for the catalytic function of Rieske proteins (Kushwaha et al., 2023).

12.4.4 As(V) Microbial Reduction

Microorganisms are capable of reducing As(V) through two main mechanisms: the dissimilatory reduction of As(V) mediated by the arr gene system and the activation of cytoplasmic As(V) reductases that are facilitated by the ars operons. It is believed that all microbial species possess cytoplasmic As(V) reductases. Current research categorizes ArsC proteins into three separate clades based on understanding of ars genes

(Wang et al., 2020). Despite their evolutionary differences, these clades perform the same molecular function. The uptake of As(V) within cells occurs through phosphate membrane systems, triggering a metabolic cascade that leads to reduction of As(V) in the cytoplasm. Given that As(V) is a molecular analogue of phosphorus, As(V) ions can enter bacterial cells via PIT, PST, or through periplasmic binding proteins. The gene responsible for coding the As(V) reductase, arsC, is up-regulated upon its entry into the cytoplasmic environment. (Yang et al., 2017).

12.4.5 Arsenic Biomethylation

The biological transformation of metalloids into volatile as well as non-volatile methylated metabolites is sometimes called biomethylation or biovolatilization, as it occurs sometimes. Biomethylation, which was first observed in fungus, is crucial for both the biogeocycling and microbial detoxification of arsenic. The activation of the arsM gene enables microorganisms to both biomethylate and tolerate arsenic. The prokaryotic origin of the arsM gene, which explains its similarity to the ars operon genes utilized by bacteria, was identified by comparing orthologs across various microbial species (Wang et al., 2020). Currently, the biochemical pathway for arsenic biomethylation is considered the most plausible mechanism. The process involves the reduction of As(V) to As(III), followed by two additional enzyme-mediated degradation steps, which are essential for the methylation of arsenic to occur. The final product of the simultaneous methylation of each As(III) intermediate during each reduction phase is trimethyl arsine. Recent studies have revealed that the previously unidentified methyl group donor is S-adenosylmethionine. (Yang et al., 2017).

12.4.6 Bacteria

Among the most resilient bacteria used to remediate arsenic from the soil are species of bacteria found in the genus *Bacillus*. A strain of *Bacillus cereus* was identified and confirmed to As (III) exhibit resistance to concentrations that reach up to 3000 ppm in controlled laboratory experiments, as part of a research initiative aimed at identifying arsenic-tolerant bacteria from a Brazilian gold mine. The strain's ability to bioaccumulate and oxidize As(III) in vitro was assessed, taking advantage of its arsenic resistance. In particular, the mass balance of As(III) and As(V) bioaccumulated within the cell walls did not correspond to the concentrations of these ions present in the culture medium, despite the reported 94.9% efficacy in reducing As(III) (Kushwaha et al., 2023). Highly polar residues associated with the molybdopterin molecule facilitate the binding of As(III) as it approaches the funnel-shaped opening of the AioA. The center of Mo(VI) of the molybdopterin molecule is subjected to nucleophilic attack by a pair of electrons from the bound As(III), resulting in the reduction of Mo(V) to Mo(IV). The oxidation of As(III) to As(V) and the reduction of Mo(V) occur almost immediately at a reported rate exceeding 4000 s^{-1}. Following this, electron transfers from the molybdenum center to the two Rieske clusters take place, where the iron and sulfur atoms also experience rapid reduction. Electron transfer from [2Fe–2S] (Wang et al., 2020).

12.4.7 Biofilm

Microbes adhered to a surface are called biofilms. It is crucial because it aids in the bioaccumulation of arsenic in the cell of microorganisms. However, the microorganisms and biomass would have various problems, including solid–liquid separation, swelling, and unique work circumstances, if only the biofilm was implemented for metal removal. Furthermore, when biomass is present alone, accumulation can occur dynamically through a passive process that is often quick (a few minutes), although metabolism may or may not be involved (Wang et al., 2020). Therefore, immobilizing free cells on a surface makes sense realistically (Singh et al., 2015).

Finding and using arsenic-resistant mixed microbial consortiums is a largely untested method for arsenic bioremediation. The preservation of pure cultures presents significant challenges, especially in in situ applications where the effectiveness of bioremediation may be limited. Furthermore, the ability of pure cultures to address complex contamination scenarios is an important consideration. Numerous contaminants have shown successful bioremediation when using mixed microbial cultures, which is attributed to the cometabolic and symbiotic relationships among the different species within a specific consortium (Yang et al., 2017).

12.4.8 Phytoremediation

A wide range of plant-based methods, including those derived from genetically modified or naturally existing plants, are collectively referred to as phytoremediation. Their purpose is to remove contaminants from polluted areas (Nedjimi, 2021). Plant growth habits and their vulnerability to As-induced phytotoxicity have been identified as the main causes of As phytoremediation's sluggish speed and limited efficacy (Nedjimi, 2021). Further developments in phytoremediation could be facilitated by applying cutting-edge biotechnology techniques to increase plant tolerance and promote As accumulation.

Therefore, given that phytoremediation encompasses various aspects of treatment, it is essential to optimize each element to ensure that the process works with maximum efficiency (Duan & Zhu, 2018). Although research has examined the impact of incorporating phosphate, sulfur, iron, biochar and compost amendments into arsenic-contaminated soil (Yang et al., 2017), none have demonstrated effectiveness in improving arsenic stress tolerance, promoting phytoaccumulation, or preserving the biochemical integrity of the soil, nor in expediting the phytoremediation process (Kushwaha et al., 2023).

12.4.9 Microbe-Assisted Phytoremediation of Arsenic

In order to make phytoremediation more profitable and practical, it is preferable to further enhance plants' remediation potential and growth, even after choosing a suitable hyperaccumulator plant. Phytobial remediation, so named because of the synergistic interaction between plant-associated bacteria, can be a successful approach. Successful cases of increased phytoremediation efficiency for As with microbial assistance. Root colonization, survival, growth, competition with other pathogenic microorganisms, and encouragement of plant development are a few important factors to take into account. Through their mutualistic relationships, microbial communities, whether as free-living symbionts or endophytes, produce specific compounds that promote plant development, reduce stress, and aid in remediation. Plant growth hormones are produced to stimulate plant growth and siderophores enhance plant absorption of nutrients.

12.4.10 Plant Selection for Arsenic Phytoremediation

12.4.10.1 Arsenic Hyperaccumulators

The maximum amount of metal that hyperaccumulator plants may store in their shoots is 1000 mg/kg for As. In the classification of a plant as a hyperaccumulator, several additional factors are considered, such as the bioaccumulation factor, which indicates the movement of metals from the soil into the plant, and the translocation factor, which denotes the transfer of metals from the roots to the shoots, as illustrated in Fig. 12.3. With an As hyperaccumulator plant, both the bioaccumulation factor and the translocation factor should be more than one (> 1). *Pteris* fern species, such as *P. ryiunkensis. P. quadriaurita, Pityrogramma calomelanos P. longifolia, P. cretica, and P. vittata*, etc., have been the main sites of As hyperaccumulation.

About 2% of the biomass of *P. vittata* may consist of arsenic. Since *P. vittata* is a perennial, field planting does not require replanting; nevertheless, periodic harvesting and gathering of fronds are required. Research has extensively examined the application of *P. vittata* in laboratory, pot, and field experiments for the remediation of arsenic-contaminated soil. Liao et al. found that *P. vittata* removed 7.8% arsenic from soil containing 64 mg/kg of arsenic over a period of seven months. Furthermore, *P. vittata* plants grown in soil enriched with phosphate rock accumulated higher levels of arsenic compared to those cultivated in soil without this amendment, indicating that arsenic is partially mobilized when phosphorus is introduced in the form of phosphate rock.

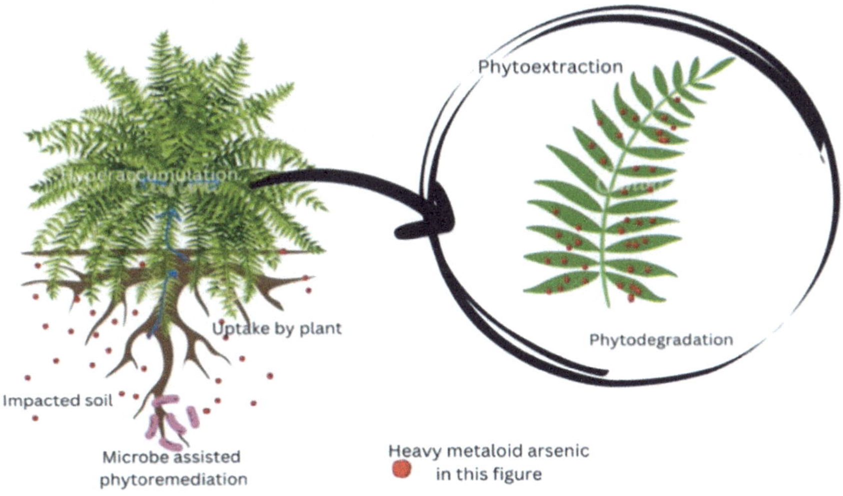

Fig. 12.3 Different modes of phytoremediation

12.4.10.2 Plants with a High Biomass for Arsenic Removal

Plants with a high biomass that have a short life cycle and moderate to high Arsenic accumulation that allow harvesting and the field's usage for successive cropping of the same or other suitable plants are necessary for the expeditious remediation of a site. This would make it possible to grow phytoremediator crops in a polluted field all year long under variable weather conditions. *Jatropha curcas* is one of the high biomass plants that has a strong chance of accumulating As. These plants can be easily harvested, grow quickly, accumulate large amounts of biomass, and self-reestablish. In addition, aquatic plants have a great tolerance to waste water and require very little input for development. The study conducted by Islam and colleagues explored the use of the water fern *Mircanthemum umbrosum*, for the remediation of arsenic and cadmium (Islam et al., 2015). Research on phytoremediation has demonstrated the effectiveness of the use of emerging aquatic plants, including *Cyperus vaginatus* and *Vetiveria zizanioides*. The effective removal of As per year can be increased by using a high biomass moderate As accumulator as opposed to a low biomass hyperaccumulator. For instance, at 500 μM As, the annual As removal by *Sesuvium* was calculated to reach as high as 1955 g As/ha/yr which is way higher than the As elimination calculated by *Pteris* (525–1470 g As/ha/yr) (Islam et al., 2015).

12.5 Algal Bioremediation

Algae have a well-known capacity to bioaccumulate heavy metals, particularly in naturally occurring environments such as lakes and seas. Several studies have revealed that different algal species, such as *Pithophora sp., H. reticulatum, and Oscillatoria sp.* Diatoms and *Phormidium sp.*, (Singh et al., 2016), *Chlorellapyre noidosa, Scenedesmus sp,* and *Pithopora sp.* (Bahar et al., 2013) carry a higher bioaccumulation capacity (760 mg kg^{-1}). Prior research revealed that the cell wall is primarily where As bioaccumulates through different algal species. Various constituents present within the cell wall, including peptidoglycan, teichuronic acid, proteins, polysaccharides, and teichoic acids, are responsible for the development of different functional groups. These include phosphoryl (–PO), sulfhydryl (–SH), hydroxyl (–OH), amino (–NH) and carboxyl (–COOH).Moreover, the cell wall absorbs and attracts more metal ions due to the negative charges these functional groups give it (Sun et al., 2015). Algal species differ in the company of different cell wall components that are responsible for different functional groupings (Singh et al., 2016).

12.6 Mycoremediation

Fungi have an advantage over bacteria in that they are considered the predominant organic matter found in soils. The close relationship between the fungus and the soil is a result of the low shear tension that the soils endure, which promotes the growth of the fungal hyphal network. Despite this, there are still limitations to the ability of fungi to bioremediate arsenic in soils. However, new research has begun to progressively take advantage of these species' diverse metabolic capacities to combat arsenic pollution. An overview of some current fungal candidates for arsenic bioremediation is provided in Table 12.2.

Research on filamentous fungi revealed a remarkable tolerance to elevated arsenic concentrations in *Penicillum, Fusarium, Trichoderma, Humicola,* and *Aspergillus.* Within screening As(V)-polluted agricultural soil in India, a strain of *Penicillium coffeae* was found to survive As(V) concentrations up to 37,461 ppm in vitro. In all cases, whether the biomass was seen to be alive, dead, or treated, because it

Table 12.2 Fungal species resistant to arsenic

Fungal species	Mechanism	References
Aspergillus oryzae FNBR_L35	Biomethylation and Biosorption	Singh et al., (2015a, 2015b)
Penicillium coffeae	Unknown	Bhargavi and Savitha (2014)
Aspergillus sp. LMC 330.04	Biomethylation	Guimarães et al. (2019)
Trichoderma sp. MG	Biomineralization	Govarthanan et al. (2019)

retained the capacity to resist the same level of As(V) under alkaline conditions. Yet, it is unclear exactly what method the fungus uses to tolerate As(V). Fungi exploit biomethylation as a means of detoxifying inorganic arsenic, as opposed to redox processes. Arsenic undergoes two types of methylation during fungal biomethylation: volatile (monomethy-larsine, dimethylarsine, and trimethylarsine) and non-volatile (methy-larsonate and dimethylarsonate). It is thought that fungi such as *Penicillium*, *Aspergillus,* and *Trichoderma* activate these genes in response to arsenic exposure.

12.7 Agricultural Waste

Constructed wetlands (CW) have become a popular ecological and sustainable solution to clean water and soil polluted by As. Because they are easily insulated and least impacted by atmospheric factors, subsurface flow (SSF) constructed wells have been proven to be successful in As cleanup (Samsó et al., 2015). As mentioned by Samsó et al. (2015), the subsurface flow CWs are thought of as composite bioreactors where interactions occur between the microbe population of the plants and the sediments and lead to several physico-chemical and biological processes that culminate in the removal of As. The substrate is an essential component of CWs in addition to plants, particularly in SSF CWs. The substrate itself serves as the growth medium, controls the wetland's hydraulic permeability, and occasionally even adsorbs As.

12.8 Biosorbents

Biosorption is a useful approach to the general practical use of bioremediation, but it is hindered by particular operating parameters related to temperature, pH, agitation, mass balance equilibrium, and remediation and waste management (Shirin et al., 2019). Furthermore, the rate and effectiveness of biosorbent are affected by the practical issues that arise during immobilization of As, biosorbent reprocessing, additional chemical costs, and most notably mass transfer limits.

12.9 Genetically Engineered Microbes

Genetically modified organism (GMO) refers to any biological creature whose genetic material has been altered by the use of recombinant DNA technology. The legal phrase "living modified organism," which is defined in the technical. As part of bioremediation, biotechnology, in the form of GMO (bacteria and plants), has also been crucial. It has been used by bacteria and fungus to increase As volatilization. Over the past few decades, the bioremediation of As using bacteria has evolved through effective accumulation of As and resistance through gene manipulation. For instance, As molecules influence the expression of the arsM gene in *Rhodopseudomonas palustris*, which codes for Cyt19 As(III) S-adenosylmethionine methyltransferase. *Sphingomonas desiccabilis* and *Bacillus idriensis* both showed significant expression of the arsM gene. Furthermore, compared to wild-type strains, it was demonstrated that transgenic bacteria expressing the arsM gene could biovolatilize and eliminate As. When arsenic is present in oxidised forms, it is extremely hazardous. Since arsenic(III) is more hazardous than arsenic(V), reducing arsenic (V) to arsenic (III) is not a viable remediation route. (Gupta & Singh, 2017).

Consequently, the conversion of these soluble forms into volatile arsenic compounds such as mono, dimethylarsine, and trimethylarsine plays a significant role in the bioremediation of arsenic. In 30 days under soil conditions, native microflora has been observed to volatilize around 2% of arsenic, despite the fact that several microorganisms have been documented to possess this ability. A growing area of study is the creation of genetically altered microbes that can volatilize arsenic compounds, as microbial biomethylation has become a key mechanism for cleaning up polluted areas (Bentley and Chasteen (2002). Volatilization of arsenic compounds initiates with the reduction of As(V) to As(III), subsequently followed by a series of methylation reactions involving As(III), ultimately resulting in the formation of trimethylarsine (TMA) (Gupta & Singh, 2017). Furthermore, it has been revealed that *Thermus thermophilus* HB8 possesses the genes TTHB128 and TTHB127, which are linked to the synthesis of arsenite oxidase, an enzyme that can convert dangerous forms of arsenic into harmless ones. More than 80% of various transgenic microorganisms have demonstrated the ability to oxidize arsenite, attributed to the cloning of these genes into the versatile broad host-range vector pBBR1MCS-5 and their subsequent introduction into suitable strains. A complete set of genes required for arsenite oxidation is found in the first known naturally occurring, self-transferable plasmid, pSinA, found in *Sinorhizobium sp.* and M14 (Alphaproteobacteria). When cells of the host strain were stripped of this plasmid, their resistance to arsenic was lost (Gupta & Singh, 2017).

12.10 Perspectives and Opportunities for Biological Treatment Approaches for Arsenic Treatment

Given the obvious drawbacks of traditional As cleanup techniques, bioremediation is currently a viable, affordable, and reliable means of treating As toxicity. More developments and scientific research will make it feasible to apply it sustainably.

It is still necessary to investigate the processes and functions of each gene and the catalytic activities connected to the phytobiological cleanup of As, despite the fact that the description of microorganisms linked to plants is gradually growing. For the purpose of determining whether the plant benefits from the increased microbiota of As stress, further attention to functional tests. The variety of life and the behavior of the plant microbiome are important. Throughout the process, the application of polygenetic markers, such as 16S rRNA fragments, underscores numerous advantageous characteristics of colonizing strains; however, it simultaneously obscures unresolved inquiries concerning the involvement of all genetic components within the genome. A greater understanding of these mechanisms is crucial since its practical consequences involve oxidation. New understanding of the usage of As in microbial fuel cell applications and bioenergy generation can result from the application of genetic engineering.

Numerous real-world implications support the phytoextraction (PE) of As by hyperaccumulators, which has shown to be one of the most promising approaches. Yet what happens to harvested biomass is still very important. In this context, composting, pyrolysis, or biogasification may be beneficial. The availability and movement of metals in wetlands are significantly influenced by microbes, particularly bacteria; however, the underlying processes are still unknown and require further investigation. There is no proof that wetlands have been injected with desirable microorganisms such as iron-oxidizing or sulfate-reducing bacteria, which are useful for seeding wetlands initially and preserving the conditions needed for improved rehabilitation. Cost–benefit analysis is the most restrictive prerequisite for the use of bioremediation in actual applications.

12.11 Conclusion

Human consumption of arsenic can cause several illnesses, such as skin lesions, neurological and/or respiratory problems, and various malignancies. It can be a crucial contributor to environmental protection by eliminating chromium toxicity in aquatic ecosystems and advocating for eco-friendly industrial activities. This approach is consistent with global efforts to secure pure water, environmental safety, and public health security across the world. A multitude of environmental variables previously discussed could also affect the methylation of arsenic within the body. Our detailed investigation of potential sources of arsenic exposure is intended to help reduce the health risks associated with arsenic exposure for people. Furthermore, different microorganisms are employed alone or in conjunction with particular plants for the bioremediation of arsenic pollution. Biological remediation is an economical, ecologically benign and on-site techniques that minimise the need to transport and dispose of contaminated material. Consequently, research aimed at clarifying the genetic and enzymatic processes and mechanisms underlying the absorption and transport of As species in the vascular systems of plants must be prioritised. In addition, physical techniques like separation, deposition, and mixing, as well as chemical techniques such as immobilization, solidification, and adsorption, can be used with bioremediation or phytoremediation.

Acknowledgements The authors are grateful to the Director, CSIR-National Institute of Oceanography (CSIR-NIO), Goa, India and Scientist-in-Charge, CSIR-NIO, Regional Center, Mumbai and

Academy of Scientific and Innovative Research (AcSIR) for their encouragement and support. Ms. Swati Sonker also like to thanks UGC for Senior Research Fellowship.

References

Aguilar, N. C., Faria, M. C., Pedron, T., Batista, B. L., Mesquita, J. P., Bomfeti, C. A., & Rodrigues, J. L. (2020). Isolation and characterization of bacteria from a brazilian gold mining area with a capacity of arsenic bioaccumulation. *Chemosphere, 240*, Article 124871. https://doi.org/10.1016/j.chemosphere.2019.124871

Bahar, M. M., Megharaj, M., & Naidu, R. (2013). Bioremediation of arsenic-contaminated water: Recent advances and future prospects. *Water, Air & Soil Pollution, 224*(12). https://doi.org/10.1007/s11270-013-1722-y

Bentley, R., & Chasteen, T. G. (2002). Microbial methylation of metalloids: Arsenic, antimony, and bismuth. *Microbiology and Molecular Biology Reviews, 66*(2), 250–271. https://doi.org/10.1128/mmbr.66.2.250-271.2002

Bhargavi, S. D., & Savitha, J. (2014). Arsenate resistant Penicillium coffeae: A potential fungus for soil bioremediation. *Bulletin of Environmental Contamination and Toxicology, 92*(3), 369–373. https://doi.org/10.1007/s00128-014-1212-y

Biosorption of Arsenic by Living and Dried Biomass of Fresh Water Microalgae—Potentials and Equilibrium Studies. (n.d.). https://www.omicsonline.org/open-access/biosorption-of-arsenic-by-living-and-dried-biomass-of-fresh-water-microalgae-potentials-and-equilibrium-studies-2155-6199.1000249.pdf

Biswas, B., Chakraborty, A., Chatterjee, D., Pramanik, S., Ganguli, B., Majumdar, K. K., Nriagu, J., Kulkarni, K. Y., Bansiwal, A., Labhasetwar, P., & Bhowmick, S. (2021). Arsenic exposure from drinking water and staple food (rice): A field scale study in rural Bengal for assessment of human health risk. *Ecotoxicology and Environmental Safety, 228*, Article 113012. https://doi.org/10.1016/j.ecoenv.2021.113012

Duan, G., & Zhu, Y. (2018). Arsenic in soil-plant system: A synthesis. In *Springer eBooks* (pp. 453–464). https://doi.org/10.1007/978-981-10-6029-8_26

Flora, S. J. (2015). Arsenic. In *Elsevier eBooks* (pp. 1–49). https://doi.org/10.1016/b978-0-12-418688-0.00001-0

Govarthanan, M., Mythili, R., Kamala-Kannan, S., Selvankumar, T., Srinivasan, P., & Kim, H. (2019). In-vitro bio-mineralization of arsenic and lead from aqueous solution and soil by wood rot fungus, Trichoderma sp. *Ecotoxicology and Environmental Safety, 174*, 699–705. https://doi.org/10.1016/j.ecoenv.2019.03.034

Guimarães, L. H. S., Segura, F. R., Tonani, L., Von-Zeska-Kress, M. R., Rodrigues, J. L., Calixto, L. A., Silva, F. F., & Batista, B. L. (2019). Arsenic volatilization by Aspergillus sp. and Penicillium sp. isolated from rice rhizosphere as a promising eco-safe tool for arsenic mitigation. *Journal of Environmental Management, 237*, 170–179. https://doi.org/10.1016/j.jenvman.2019.02.060

Gupta, S., & Singh, D. (2017). Role of genetically modified microorganisms in heavy metal bioremediation. In *Springer eBooks* (pp. 197–214). https://doi.org/10.1007/978-981-10-4041-2_12

Islam, M. S., Saito, T., & Kurasaki, M. (2015). Phytofiltration of arsenic and cadmium by using an aquatic plant, Micranthemum umbrosum: Phytotoxicity, uptake kinetics, and mechanism. *Ecotoxicology and Environmental Safety, 112*, 193–200.

Khan, A., Asif, I., Abid, R., Ghazanfar, S., Ajmal, W., Shehata, A. M., & Naiel, M. A. E. (2024). The sustainable approach of microbial bioremediation of arsenic: an updated overview. *International Journal of Environmental Science and Technology*. https://doi.org/10.1007/s13762-024-05594-9

Kushwaha, R., Singh, R. S., & Mohan, D. (2023). Arsenic resistance and accumulation by two bacteria isolated from Ratanpur Village of Varanasi District, Uttar Pradesh. *Bioresource Technology Reports, 24*, Article 101639. https://doi.org/10.1016/j.biteb.2023.101639

Mallick, I., Hossain, S. T., Sinha, S., & Mukherjee, S. K. (2014). Brevibacillus sp. KUMAs2, a bacterial isolate for possible bioremediation of arsenic in rhizosphere. *Ecotoxicology and Environmental Safety, 107*, 236–244. https://doi.org/10.1016/j.ecoenv.2014.06.007

Mondal, P., Bhowmick, S., Chatterjee, D., Figoli, A., & Van Der Bruggen, B. (2013). Remediation of inorganic arsenic in groundwater for safe water supply: A critical assessment of technological solutions. *Chemosphere, 92*(2), 157–170. https://doi.org/10.1016/j.chemosphere.2013.01.097

Murray, J., Guzmán, S., Tapia, J., & Nordstrom, D. K. (2023). Silicic volcanic rocks, a main regional source of geogenic arsenic in waters: Insights from the Altiplano-Puna plateau, Central Andes. *Chemical Geology, 629*, Article 121473. https://doi.org/10.1016/j.chemgeo.2023.121473

Nedjimi, B. (2021). Phytoremediation: a sustainable environmental technology for heavy metals decontamination. *SN Applied Sciences, 3*(3)https://doi.org/10.1007/s42452-021-04301-4

Roy, M., Giri, A. K., Dutta, S., & Mukherjee, P. (2015). Integrated phytobial remediation for sustainable management of arsenic in soil and water. *Environment International, 75*, 180–198. https://doi.org/10.1016/j.envint.2014.11.010

Samsó, R., Meyer, D., & García, J. (2015). Subsurface flow constructed wetlandmodels: review and prospects. In: *The role of natural andconstructed wetlands in nutrient cycling and retention on the landscape* (pp. 149–174). Springer.

Saunders, J. A., Lee, M. K., Dhakal, P., Ghandehari, S. S., Wilson, T., Billor, M. Z., & Uddin, A. (2018). Bioremediation of arsenic-contaminated groundwater by sequestration of arsenic in biogenic pyrite. *Applied Geochemistry, 96*, 233–243. https://doi.org/10.1016/j.apgeochem.2018.07.007

Shahid, M., Niazi, N. K., Dumat, C., Naidu, R., Khalid, S., Rahman, M. M., & Bibi, I. (2018). A meta-analysis of the distribution, sources and health risks of arsenic-contaminated groundwater in Pakistan. *Environmental Pollution, 242*, 307–319. https://doi.org/10.1016/j.envpol.2018.06.083

Sharma, R., Nanda, R., & Bhagat, N. (2022). Background level, occurrence, speciation, bioavailability, uptake, detoxification mechanisms and management of arsenic polluted soil. In *Elsevier eBooks* (pp. 221–254). https://doi.org/10.1016/b978-0-323-85621-8.00013-3

Shirin, S., Jamal, A., Ranjan, P., & Yadav, A. K. (2019). Study on assessment of slope stability and mixed disposal of overburden in voids of Singrauli Coalfield. *Environmental Quality Management, 28*(3), 131–139. https://doi.org/10.1002/tqem.21616

Singh, A. K., Majumder, C. B., & Mishra, S. (2015). Removal of Arsenic from contaminated wastewater using Eichhornia crassipes. *Integrated Research Advances, 2*(1), 1–4.

Singh, M., Srivastava, P., Verma, P., Kharwar, R., Singh, N., & Tripathi, R. (2015b). Soil fungi for mycoremediation of arsenic pollution in agriculture soils. *Journal of Applied Microbiology, 119*(5), 1278–1290. https://doi.org/10.1111/jam.12948

Singh, N., Raghubanshi, A., Upadhyay, A., & Rai, U. (2016). Arsenic and other heavy metal accumulation in plants and algae growing naturally in contaminated area of West Bengal, India. *Ecotoxicology and Environmental Safety, 130*, 224–233. https://doi.org/10.1016/j.ecoenv.2016.04.024

Sun, Y. Y., Liu, R. L., Zeng, X. B., Lin, Q. M., Bai, L. Y., Li, L. F., Su, S. M., & Wang, Y. N. (2015). Reduction of arsenic bioavailability by amending seven inorganic materials in arsenic contaminated soil. *Journal of Integrative Agriculture, 14*(7), 1414–1422. https://doi.org/10.1016/s2095-3119(14)60894-7

Valskys, V., Hassan, H. R., Wołkowicz, S., Satkūnas, J., Kibirkštis, G., & Ignatavičius, G. (2022). A review on detection techniques, health hazards and human health risk assessment of arsenic pollution in soil and groundwater. *Minerals, 12*(10), 1326. https://doi.org/10.3390/min12101326

Wang, J., Kerl, C. F., Hu, P., Martin, M., Mu, T., Brüggenwirth, L., Wu, G., Said-Pullicino, D., Romani, M., Wu, L., & Planer-Friedrich, B. (2020). Thiolated arsenic species observed in rice paddy pore waters. *Nature Geoscience, 13*(4), 282–287. https://doi.org/10.1038/s41561-020-0533-1

Wang, J., Li, Z., Zhu, Q., Wang, C., & Tang, X. (2023). Review on arsenic environment behaviors in aqueous solution and soil. *Chemosphere, 333*, Article 138869. https://doi.org/10.1016/j.chemosphere.2023.138869

Xiao, K. Q., Li, L. G., Ma, L. P., Zhang, S. Y., Bao, P., Zhang, T., & Zhu, Y. G. (2016). Metagenomic analysis revealed highly diverse microbial arsenic metabolism genes in paddy soils with low-arsenic contents. *Environmental Pollution, 211*, 1–8. https://doi.org/10.1016/j.envpol.2015.12.023

Yadav, A. K., Yadav, H. K., Naz, A., Koul, M., Chowdhury, A., & Shekhar, S. (2022). Arsenic removal technologies for middle- and low-income countries to achieve the SDG-3 and SDG-6 targets: A review. *Environmental Advances, 9*, Article 100262. https://doi.org/10.1016/j.envadv.2022.100262

Yadav, P., Ansari, M. W., Gill, R., Tuteja, N., & Gill, S. S. (2024). Arsenic transport, detoxification, and recent technologies for mitigation: A systemic review. *Plant Physiology and Biochemistry, 213*, Article 108848. https://doi.org/10.1016/j.plaphy.2024.108848

Yang, X., Li, Q., Tang, Z., Zhang, W., Yu, G., Shen, Q., & Zhao, F. J. (2017). Heavy metal concentrations and arsenic speciation in animal manure composts in China. *Waste Management, 64*, 333–339. https://doi.org/10.1016/j.wasman.2017.03.015

Chapter 13
Integration of Nanotechnologies for Sustainable Remediation of Arsenic Removal

Shivanshu Kushwaha and Zeenat Arif

Abstract Arsenic-contaminated water poses a serious global health threat, requiring the development of effective and economical remediation solutions. Nanotechnology-based technologies have emerged as a viable way to overcome the constraints of older approaches. This chapter opens with an overview of arsenic as a pollutant, including its health effects and ubiquity in wastewater. It then discusses nanotechnology and nanoparticles, highlighting their importance and benefits in Arsenic removal. Metal oxides, carbon-based nanostructures, and polymeric nanoparticles are among the nanomaterials that are being studied for arsenic adsorption. The synthesis procedures for these nanomaterials are discussed with a focus on strategies that maximize their efficiency. The chapter highlights the improved performance and economic viability of nanotechnology-based solutions through comparative examinations of various types and conventional Arsenic removal technologies. These nanomaterials' photocatalytic performance is assessed, showing that they can simultaneously remove Arsenic and break down organic pollutants. Future research directions are highlighted, with an emphasis on improving the affordability, sustainability, and scalability of Arsenic removal techniques based on nanomaterials. This chapter demonstrates the potential to greatly reduce Arsenic contamination and enhance public health outcomes worldwide through creative nanotechnological treatments.

Keywords Arsenic removal · Nanotechnology · Nanoparticles · Synthesis methods · Photocatalysis · Environmental remediation

S. Kushwaha · Z. Arif (✉)
Department of Chemical Engineering, Harcourt Butler Technical University, Nawabganj, Kanpur, India
e-mail: dariflt@hbtu.ac.in

A. K. Yadav and S. Shirin (eds.), *Emerging and Innovative Arsenic Removal Technologies for a Sustainable Future*, Water Science and Technology Library 131,
https://doi.org/10.1007/978-3-032-01176-3_13

13.1 Introduction

The availability of water in sufficient quantities and quality for major human activities, such as farming, industry, energy production, and domestic tasks, is essential to maintaining ecosystems, maintaining human livelihoods, and obtaining clean water. Conventional water supplies, including precipitation, melted snow, and river runoff, are frequently insufficient in places with a limited water supply. Although improving water use efficiency might be beneficial, methods to enhance supply and quality must also be employed in conjunction with these actions. Conventional approaches to improving supply, such as constructing reservoirs and rerouting surface water, frequently face opposition from the general public and lack evaluations of water quality.

Due to necessity, the usage of nontraditional water supplies has increased dramatically. The efficacy and cost-effectiveness of saltwater desalination and wastewater treatment have increased due to advances in technology. The water demand–supply gap can be significantly closed by utilizing unconventional water resources. Wastewater, which is "used" water that has been polluted by human activity, is becoming increasingly recognized as a dependable and reasonably priced supply of freshwater, particularly for agricultural purposes. Farm runoff is rarely collected or treated, while wastewater from residential and commercial properties can be captured in structures such as septic tanks or sewers. Recycled wastewater can be used for agroforestry, aquaculture, and groundwater recharge, in addition to sustaining river flows, particularly in dry periods. This promotes ecosystem health and reduces competition for conventional water sources (Simon et al., 2024).

Most applications of wastewater, both treated and untreated, are found in nonpotable industries such as agricultural and landscape irrigation. It is expected that dependence on alternate water sources will increase, since agriculture has the highest water demands worldwide. Better treatment facilities are clearly needed, as a large portion of recycled wastewater is either untreated or not adequately treated. Reuse safety must be guaranteed; wastewater must be adequately treated and managed safely in order to be converted into energy, nutrients, and water. Water scarcity and the harmful impacts of wastewater and contaminated water are major global challenges. Unlike home wastewater, industrial wastewater is made up of many different industrial operations. This category includes pollutants that are biological, nutritional, thermal, sensory, organic, solid, poisonous, oil-based, and acid–base. Environmental management requires an understanding of the distribution of hazardous chemicals in industrial wastewater. This chapter looks at ways to remove arsenic with a particular emphasis on using nanoparticles. The large surface areas and distinct chemical characteristics improve their capacity to absorb arsenic from tainted water. Adsorption procedures utilizing nanoparticles present a workable plan to reduce arsenic levels, enhancing public health, and water quality.

13.2 Urban Wastewater: A Growing Global Challenge

With 1.4 billion people, India produces a large amount of wastewater each day (39,604 MLD in rural regions and 72,368 MLD in urban areas) in 2020–2021. Due to the increased demand for sanitation and higher living standards, which have spurred population expansion and migration to cities, urban areas generate roughly twice as much wastewater as rural areas. More than 50% of the generation currently reside in cities, a percentage that is predicted to increase to 70% by 2050. Although it presents prospects for economic expansion, rapid urbanization puts a strain on freshwater resources in areas with limited water supplies. The gap between the supply and demand for water can be closed using more unconventional water sources. Wastewater is still "untapped" and "undervalued" in India, even though it is acknowledged as a reliable and reasonably priced supply of fresh water, particularly for agriculture (Qadir et al., 2020). Reducing pollutants for reuse or lessening environmental effects are the goals of wastewater treatment. According to Luther et al. (2015), clean wastewater helps keep rivers flowing during dry seasons. Unfortunately, many recovered wastewaters are not adequately treated due to infrastructure and technological constraints. To safeguard human and environmental well-being and encourage reuse, wastewater treatment and management must be done safely (Qadir et al., 2020). To recognize wastewater as an invaluable asset rather than trash, a paradigm shift is required.

Approximately 380 billion cubic meters of wastewater are produced annually, according to recent global studies. Due to population growth and urbanization, this number is expected to increase by 24% by 2030 and 51% by 2050. (Fig. 13.1). By 2030, Asia is predicted to produce 44% of the world's wastewater, up from its current 42% (159 billion cubic meters) (Qadir et al., 2020). A total of 359.4 billion cubic meters of wastewater will be produced annually, of which 63% (225.6 billion cubic meters) will be collected and 52% (188.1 billion cubic meters) will be treated, according to Jones et al. (2021). Each year, 40.7 billion cubic meters are recycled, or 22% of treated wastewater and 11% of all wastewater generated. This indicates that approximately 40% of wastewater is discharged without treatment (Fig. 13.2). The amount of wastewater produced worldwide by high-income countries is 42%, almost double that of low & lower-middle-income countries. Low-income nations only treat 4% of their wastewater, but high-income nations collect and treat 82% and 74% of it, respectively (Jones et al., 2021).

As a region, East Asia along with the Pacific produces the most wastewater. The rates of wastewater and treatment are highest in Western European countries and the poorest in Southern Asia, where there is also little to no reuse of treated wastewater. These findings show that even though developed nations produce more wastewater, they are equipped to reuse it. The underperformance of the southern Asian region in the treatment of wastewater and reuse was brought to light by both studies. The nation with the largest population in South Asia, India, produces the most wastewater.

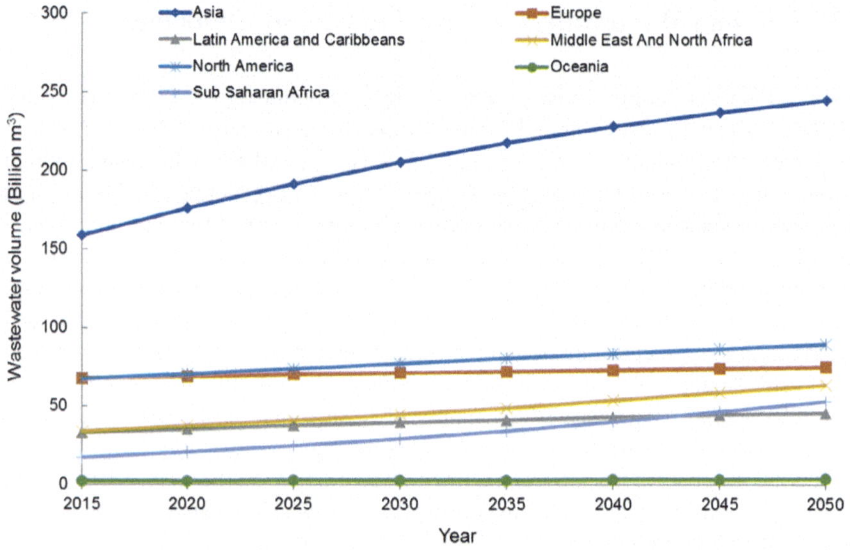

Fig. 13.1 Global wastewater production by region from 2015 to 2050 (Qadir et al., 2020)

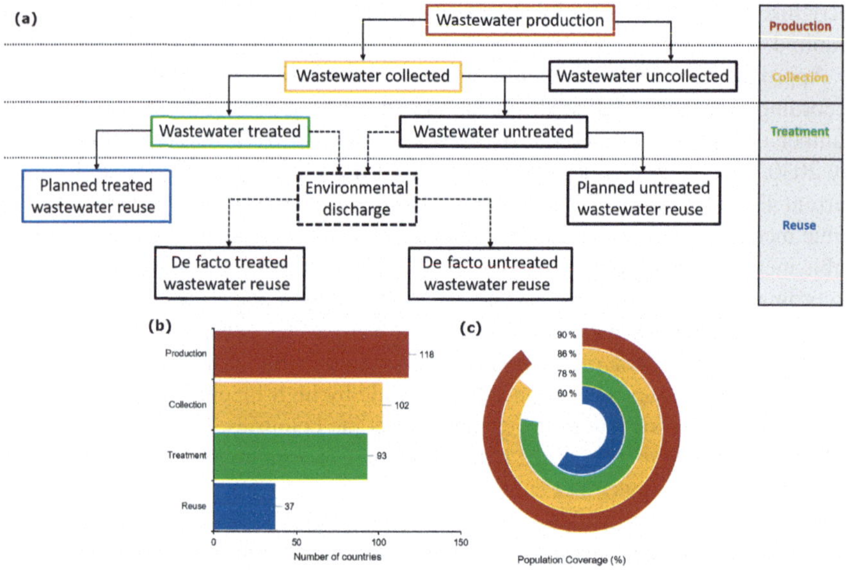

Fig. 13.2 **a**, **b** Wastewater data availability with number of countries for which wastewater data are available (**c**) and the percentage of population coverage (Jones et al., 2021)

13.3 Wastewater Scenario in India

India's urban areas are expanding rapidly as a result of both economic reforms and greater migration. With 483 million people (35%) living in cities, the population reached 1.38 billion by 2020. There were 53 cities with more than a million residents according to the 2011 census. The urban population is expected to reach 607 million by 2030 and 877 million (50%) by 2050. City planners are burdened by rapid urban growth, especially when it comes to providing basic utilities such as clean water. Increased sector-specific water needs, particularly in agriculture (74% by 2050), put pressure on shared water resources and increase competitiveness. The demand for drinking water in cities often outweighs that in rural areas, leading to water transfers. This causes problems for both rural and urban planners by redirecting freshwater and releasing effluent into river systems and contaminating irrigation water (Amerasinghe et al., 2013).

13.4 Different Types of Pollutants

Industrial wastewater contains a wide variety of dangerous chemicals, and it poses serious risks to the environment. These include radioactive materials, acids, volatile chemical compounds, and heavy metals. Aquatic environments can be harmed by alkalis such as caustic soda, which are present in effluent from industries such as electroplating, paper, leather products, oil refineries, and synthetic textiles. However, acidic pollution from the production of steel, mineral extraction, and nonferrous metals can drop pH levels and damage infrastructure and aquatic life. Smelting, mining, and manufacturing batteries release heavy metals including cadmium and mercury, which can lead to the damage to kidneys and bioaccumulation problems. Used in chemicals, medications, and insecticides, arsenic can cause serious health problems such as cardiovascular disease and cancer. The manufacturing of glass and metals releases hexavalent chromium, which is a strong carcinogen, and the extraction of gold and coal gasification releases cyanide, which is highly hazardous. As a by-product of oil refining, coal gasification, and dye manufacturing, phenol is toxic to aquatic environments and human health. Marine life is negatively affected by sulfides from gas production and leather processing, as well as oil pollution from the food and oil processing industries. Children's brain development may be hampered by lead smelting, and the textile and chemical industries may create hazardous by-products as a result of free chlorine. Long-term environmental and health harm can result from radioactive contamination of nuclear power plants and persistent polychlorinated biphenyls (PCBs) in plastics and electrical devices. Table 13.1 enlist different types of primary pollutant and their sources contaminating water.

Table 13.1 Primary pollutants and likely sources in industrial wastewater

Harmful contaminants	Sources of contaminants
Alkali contaminants	Primarily from industries such as caustic soda production, paper production, chemical fiber manufacturing, printing/dyeing, leather processing, and oil refineries
Acidic pollutants	Sources include mining, steel manufacturing, machinery operations, electroplating industries, chemical production, and nonferrous metallurgy
Cadmium compounds	Smelting, mining, electroplating, battery manufacturing industries, metal processing industries, etc.
Mercury compounds	Industries such as explosives manufacturing, mercury refining, chlor-alkali production, chemical manufacturing, mercury-based pesticides, and instrument production
Arsenic and its compounds	It enters industrial wastewater from pharmaceutical manufacturing, chemical production, metallurgical processes, ore processing, pesticide production, paint manufacturing, and fertilizer industries
Hexavalent chromium and its compounds	Significant pollutants from the smelting, mining, electroplating, specialty glass industry, chemical production, and metal processing
Cyanide	Coal gasification, coke production, the electroplating process, the gold industry, acryliconitrile refining, metal cleaning, and acrylic production
Phenol	The manufacture of synthetic resins, coal gasification, coking, oil refinement, chemicals, dyes, and pharmaceuticals
Oil pollutants	The oil sector, food processing, oil refining, mechanical operations, and natural gas processing
Sulfide	Processing leather goods, producing chemicals, producing gas, dyeing, coking, refining oil, and processing gas and oil
Lead and its compounds	Production of gasoline, smelting, enamel industry, pesticides, lead paint, and chemical industries
Organic chlorine and organic phosphorus	Chemical Manufacturing Industries and Pesticides Production
Free chlorine	Chemical, paper, and textile bleaching industries
Radioactive substances	Nuclear power plants, hospitals, radioisotope labs, and weapons manufacturing
Polychlorinated biphenyls (PCBs)	Manufacturing of lubricants, electrical equipment, polymers, and various other processes in industry

13.5 Arsenic: A Comprehensive Overview

13.5.1 Arsenic Chemistry

In group VA of the periodic chart, arsenic (atomic number 33) is located just below phosphorus. Because arsenic and phosphorus are chemically similar, arsenic can replace phosphorus in the ATP cycle, disrupting the natural processes and altering metabolism. It can exist in many oxidation states (− 3, 0, + 3, + 5) and forms both inorganic and organic compounds. Arsenic is mostly present in wastewater as inorganic oxyanions that are in pentavalent (+ 5) or trivalent (+ 3) forms of arsenite or arsenite. Arsenic hydrides, sulfides, acids, oxides and halides, such as As_2O_3, As_2O_5, and arsenic sulfides such as As_2S_3, are examples of inorganic arsenic compounds. Arsenate (V) and arsenite (III), which are components of the acid systems arsenic (H_3AsO_4) and arseneous (H_3AsO_3), respectively, are the most common forms found in wastewater. The redox potential and the pH level of the water affect the stability of these forms. Arsenite species, such as H_3AsO_3, H2AsO_3- and $HAsO_3{}^{2-}$, are stable in slightly reducing conditions, whereas arsenate species are stable in oxygenated environments. Redox potential (Eh) and pH play an important role in controlling arsenic speciation in wastewater, with pentavalent and trivalent arsenic species predominating in oxygen-rich environments and decreasing under reducing conditions, respectively.

13.5.2 Toxicity and Health Hazards

Because of its toxicity and potential for cancer, arsenic is considered a high-priority contaminant in water. It is difficult to destroy and is frequently employed in both industry and agriculture. It is challenging to remove contaminants such as mercury, lead, iron, and selenium that are frequently present in arsenic pollution. The oxidation states and chemical structures of arsenic determine its carcinogenicity and toxicity. Trivalent arsenic is more dangerous than pentavalent arsenic, and inorganic As is more harmful than organic arsenic. Trivalent arsenic affects the action of enzymes that have thiol groups, which disrupts biological activities. Pentavalent arsenic can interfere with the synthesis of ATP in human cells, resulting in energy loss. A serious environmental risk is arsenic poisoning, particularly in areas like Bangladesh and India where thousands of people are exposed to polluted groundwater. Long-term exposure to arsenic can cause several health issues, such as:

- **Cancer**: Cancers of the skin, lungs, liver, kidneys, and bladder.
- **Heart-related conditions**: Heart-related and brain-related illnesses.
- **Neurological problems** include tremors, sensory and motor polyneuritis, and neurological dysfunctions.
- **Endocrine disorders**: Diabetes and other endocrine disrupting illnesses.

- **Skin disorders** include gangrene, keratosis, and melanosis, which cause the skin to darken and stiffen.
- **Gastrointestinal problems**: recurrent diarrhea, stomach pain from colic.
- **Damage to the liver and kidneys** includes hepatitis, chronic liver disease, and hepatic hypertrophy.
- **Breathing problems**: bronchitis, chronic lung disease.
- **Additional problems** include toxic optic neuropathy (atrophy), hyperostosis, portal hypertension, aplastic anemia, and pinkeye conjunctivitis.

Hematuria, vomiting, diarrhea, burning and dry mouth and throat sensations, and difficulty swallowing are all signs of severe arsenic poisoning. Prolonged exposure can have more serious long-term consequences, such as cancer and organ malfunction. Determining efficient countermeasures and purifying water are just two of the major tasks involved in prevention and therapy.

13.6 Regulations and Contaminant Levels of Arsenic

Reports of drinking water contamination have been made ranging from 100 to over 2000 μg/L have been made for nations such as Argentina, Bangladesh, Mexico, India, Taiwan and Thailand. It is commonly known that arsenic has negative consequences even at low concentrations. In 1942, the U.S. Public Health Service established the first drinking water arsenic standard of 0.05 mg/L. Research connecting skin cancer and arsenic in the 1960s led to a reassessment. In response to fresh proof of arsenic's carcinogenicity, the World Health Organization (WHO) and the U.S. Environmental Protection Agency (USEPA) revised the Maximum Contaminant Level (MCL) from 50 to 10 ppb by the 1980s and 1990s. In India, MCL for As is 50 ppb.

13.7 Different Technologies Available

In recent years, many techniques, such as ion exchange, phyto- and phtyobial remediation, adsorption, precipitation, electrokinetic processes, and electrocoagulation, have been investigated for the removal of arsenic from environmental sources. Although each approach has unique advantages, there are major issues with cost and efficacy as well (Arif et al., 2022).

Plants are used in phytoremediation to remove arsenic from the soil, whereas ion exchange replaces harmful arsenic ions with less harmful ones. While phytobial treatment mixes microorganisms and plants to improve cleanup, adsorption uses materials to collect arsenic. Arsenic is mobilized by electrokinetic processes, which employ electrical energy, while chemical precipitation turns arsenic into a solid. Electrical currents are used in electrocoagulation to break down and remove contaminants. The best approach will depend on several variables, including local customs, legal requirements, and technology limitations. The development of these techniques is intended to improve efficiency and sustainability, which could reduce expenses and lessen the need for artificial materials. Table 13.2 provides an overview of the benefits, drawbacks, and effectiveness of various methods in removing arsenic.

13.8 Introduction to Nanotechnology

Altering materials on the nanoscale (1–100 nm) for use in a variety of fields, such as photocatalysis, electronics, and healthcare, is the rapidly developing discipline of nanotechnology. Due to their specific quantum effects and high surface area to volume ratio, nanoparticles have different properties from their bulk counterparts (Arif et al., 2023). Important nanomaterials with distinct properties and uses are fullerenes, nanotubes, and nanorods.

13.8.1 Types of Nanotechnology

Wet nanotechnology: This field examines tissues, enzymes, and membranes to better understand biological systems and how they interact with water.

Dry nanotechnology: The topic of this paper is focused on the physical and chemical properties of nonbiological materials, such as silicon and carbon, without the need for liquid media.

Computational nanotechnology: This field provides insights into atomic-level properties and interactions by modeling and predicting behavior at the nanoscale through simulations.

Each of the three categories—computational, dry, and wet—contributes to a comprehensive knowledge of nanotechnology.

13.8.2 Techniques for Nanoparticle Synthesis

There are two primary methods for synthesizing nanoparticles: top-down and bottom-up (Sethy et al., 2022).

13.8.2.1 The Top-Down Method

To use this process, bulk materials must be physically broken down into nanoparticles by lithography, the process of etching and milling. Scalability and exact regulation of size and shape are two benefits. But there are drawbacks, such as possible homogeneity and purity problems.

Table 13.2 Various methods used to remove arsenic

Removal technique	Mechanism	Pros	Cons	Removal efficiency	References
Adsorption	Utilizing a solid to remove materials from liquid or gaseous solutions	Affordability, lack of sludge, and high removal efficiency	Low-arsenic wastewater is appropriate for sorbents because they do not self-monitor and must be replaced when saturated and eventually lose their ability to separate	95%	Hao et al. (2018), Najib and Christodoulatos (2019), Wang et al., (2018a, 2018b)
Ion exchange	Solid-phase ions are exchanged with equal ion numbers from contaminated water	Complete removal and recovery; limited production of toxic sludge	Needs to be rejuvenated frequently; expensive; Specific exchanger for a specific pollutant	97.9% in the pH range from 3.5 to 7	Karakurt (2019), Shakoor et al. (2017)
Phyto-remediation	Utilises plants to address soil contamination	Ecologically benign and commercially beneficial; naturally transpires through plants absorbing arsenic from soil; stops contaminants from spreading; helps restore land	Time-consuming; microbes might release more poisonous compounds; not widely applicable;	99.9%	Yan et al. (2019)
Nanophytoremediation	Utilizing nanoparticles to reduce heavy metals and increasing increase the activity of microbes using nano-encapsul-ated enzymes	Enhances phytoremediation efficiency; increases in situ remediation; accelerates the breakdown of contaminants into less harmful forms; cost-effective	–	99.9%	Nwadinigwe and Ugwu (2018)
Phytobial remediation	–	Economical accelerates phytoremediation, increases plant resistance to stress, enhances soil structure	–	99.9%	Shukla and Srivastava (2019)

(continued)

Table 13.2 (continued)

Removal technique	Mechanism	Pros	Cons	Removal efficiency	References
Chemical precipitation	Removes dissolved metals	Enhance soil fertility; specific component removed; simple and effective	Expensive to process; invariably forms silt	95%	Vega-Hernandez et al. (2019)
Electro-kinetic processes	Effective on fine-grained soil	Economical for various heavy metals	Effective for small portion of solids	90%	Li et al. (2019), Xu et al. (2019)
Electro-coagulation	Ferric chloride; Ferric sulfate; Aluminium sulfate	Production of metallic cations by passing current between iron electrodes to dissolve soluble anodes in the treated wastes	A brand new and highly effective method uses materials that are readily available in the area, is quite affordable, requires little maintenance, and removes arsenic from drinking water	99.9%	Mohora et al. (2018), Nidheesh and Singh (2017), Rosales et al. (2018)
Membrane		Semi-permeable layer that allows specific molecules to pass through it	High prices and costlier with and water rejection is preferably high	96%	Zakhar et al. (2018)

Benefits:

- Scalability in industrial settings.
- Exact control over form and size.
- Generally accepted techniques in the business.

13.8.2.2 The Bottom-Up Method

This strategy uses chemical or biological techniques, such as sol–gel synthesis, chemical vapor deposition, and biological templating, to create nanoparticles at the molecular or atomic level. Although it can be more difficult, it delivers excellent purity, consistency, and atomic-level precision.

Benefits:

- Superior homogeneity and purity.
- Atomic accuracy.
- Various alternatives for materials.

Chemical Synthesis: Uses reactions involving precursors such as metal salts or organic molecules. It offers precise control over NPs properties but can be complex and resource-intensive.

Green Synthesis: Employs materials from nature, such as plants and microorganisms, making the process ecofriendly and cheaper. Green synthesis reduces hazardous waste and uses natural resources to minimize environmental harm. Because it requires less equipment and natural resources, it is more affordable and improves safety by reducing exposure to dangerous chemicals. This process minimizes negative responses and frequently yields medically useful biocompatible nanoparticles. It encourages scalability and sustainability and supports a wide range of applications in many domains (Kushwaha et al., 2023, Khan et al., 2022).

13.9 Nanoparticles Used for Removal of Arsenic

In water treatment, adsorption is a technique frequently employed to eliminate both organic and inorganic pollutants. The selectivity, adsorption kinetics, and surface area of conventional adsorbents are frequently constrained. Larger specific surface areas along with additional sorption sites are features of nanoadsorbents that improve performance. The adsorbate is transported to the surface of the adsorbent, diffuses into the pores, and binds to active sites in the adsorption process. When exposed to light, the widely used nanoadsorbent TiO_2 can oxidize As(III) to As(V), which leads to the formation of complexes with metal oxides. High capacity for sorption and simple separation without side effects are desirable properties of effective nanoparticles.

On the other hand, acidic environments can cause considerable nanoparticle leaching, and certain ions can either accelerate or hinder arsenic leaching. TiO_2 is the recommended photocatalyst for the treatment of water due to its low toxic effects, chemical resistance, and affordability. Table 13.3 list different NPs used for elimination of As from water.

The efficiency of nanoparticles in the adsorption of arsenic varies according to their specific surface area, pH, and particle size. For example, TiO_2 nanoparticles have shown great promise. The adsorption capabilities of anatase nanoparticles, pure TiO_2, and iron-doped TiO_2 for As(V) were 19–20.4 mg/g. Whereas amorphous TiO_2 obtained 66.8 mg/g for As(III) and 19.0 mg/g for As(V), hydrous TiO_2 showed capabilities of 33.4 mg/g for As(V) and 31.8 mg/g for As(III) (Jegadeesan et al., 2010; Pirilä et al., 2011). According to Luther et al. (2012), iron oxide nanoparticles, such as Fe_2O_3 and Fe_3O_4, also showed promise. Their capacities ranged from 4.78 to 20 mg/g for different species of arsenic. The effectiveness of nanoparticle-based adsorbents is further demonstrated by CuO and ZrO_2 nanoparticles, which have capacities of 1.09 mg/g and 83 mg/g for total arsenic and As(III), respectively (Cui et al., 2012; Goswami et al., 2012). A brief overview of the capabilities of different nanoparticles for adsorbing arsenic can be found below.

Table 13.3 Ability of various NPs to adsorb As from wastewater

Nanoparticle	Particle size (nm)	Specific surface area (m^2 g^{-1})	pH_{pzc}	Arsenic state	Adsorption capacity (mg g^{-1})	References
GPR TiO_2 pure	325		7	As(V)	19 20.4 19.9	
Hydrous TiO_2	10.8	280	4.8	As(V) As(III)	33.4 31.8	Pirilä et al. (2011)
CrystallineTiO_2/ Amorphous TiO_2		98/409	4.8/ 4.5	As(III)	38.4/66.8	Jegadeesan et al. (2010)
Fe_2O_3/Fe_3O_4	12.3/ 16.5			As(III) As(V)	20/5.68 4.9/4.78	Luther et al. (2012)
Magnetite-γ-Fe_2O_3 mixture (EWE method)	34	12		As(III) As(V)	2.9 3.1	Song et al. (2013)
CaO_2	15–25			As(III)		Olyaie et al. (2012)
ZrO_2		327.1		As(III)	83	Cui et al. (2012)

13.10 Integrated Treatment Technologies Using Nanoparticles for Arsenic Removal

When using TiO_2 nanoparticles in particular, nanoparticle adsorption offers significant benefits over conventional arsenic removal techniques. Up to 95% removal efficiency is achieved by conventional procedures such as adsorption using materials like iron-coated sand and activated granular carbon (Ghosal et al., 2018). These techniques do, however, frequently have drawbacks, such as saturation and maintenance problems. Ion exchange is more efficient, with removal rates as high as 97.9%, but it is expensive and needs to be regenerated frequently.

With an efficiency of up to 99.9%, phytoremediation is a climate-specific process that requires a lot of time and effort, but is also environmentally beneficial (Karakurt, 2019; Shukla & Srivastava, 2019). By adding nanoparticles, nano phytoremediation improves this technique and increases efficiency to 99.9% (Jesitha & Harikumar, 2018). Electrocoagulation, another effective method at 99.9% (Gilhotra et al., 2018), requires complex setups.

In contrast, TiO_2 nanoparticles, particularly those synthesized using green methods, offer high specific surface areas and adsorption capacities, making them cost-effective and sustainable. TiO_2 nanoparticles can achieve adsorption capacities of up to 38.4 mg g^{-1} for As (III) (Jegadeesan et al., 2010) and improve the removal efficiency due to their reactivity under UV light. Their green synthesis reduces the environmental impact, making them a superior choice for efficient and ecofriendly arsenic removal.

Nwadinigwe and Ugwu (2018), reported a study on the enhancement of microbe activity by incorporating nano encapsulated enzyme. This method of removing As (99.9% removal) from water and soil is referred to as nano phytoremediation; also this technique results in increased efficiency with increased degradation rate.

Zakhar et al., (2018) reported membrane-based process for As removal as they show effective removal capacity. The nanofiltration membrane showed excellent rejections of As (V) ranging from 85 to 99% and rejections of As (III) rejections between 61 and 87%.

13.11 Conclusion and Future Scope

Compared to conventional approaches, nanoparticle-based arsenic removal strategies, specifically those utilizing TiO_2 nanoparticles—offer greater efficiency, affordability, and environmental benefits. TiO_2 nanoparticles are the best option for long-term arsenic removal due to their large surface area, high reactivity, and exceptional adsorption capabilities and stability. The development of green synthesis strengthens their suitability for the environment and confirms their position as the best arsenic cleanup method. There is much hope and possibility for the future of employing nanoparticles to remove arsenic from water sources. It is anticipated that developments in nanotechnology will improve the selectivity, sustainability, and efficiency of arsenic removal methods. The improvement of green synthesis techniques for nanoparticles is a crucial area of development since it not only reduces the impact on the environment but also enhances the functioning and biocompatibility of the particles. Scalable green synthesis methods that yield high-adsorption capacity and consistently high-quality nanoparticles should be the main focus of research. The durability and reusability of nanoparticles in a range of water circumstances, such as different pH levels, temperatures, and the presence of competing ions, require further research. Furthermore, by understanding how other pollutants in water interact with nanoparticles, multifunctional treatment strategies can be developed that simultaneously target several pollutants. Including nanoparticles into hybrid systems can improve the overall efficiency of arsenic removal and offer comprehensive water treatment solutions by combining adsorption with additional treatment techniques such as membrane filtration or electrocoagulation. It is essential to create energy- and money-efficient techniques for the large-scale manufacturing and implementation of these hybrid systems. Validating laboratory results and determining the actual application of arsenic removal using nanoparticles in various environmental contexts requires field studies and pilot projects. Partnerships among government, business, and academic institutions can help turn research into practical applications. Global water safety and public health will ultimately benefit from sustainable, effective, and scalable methods for removing arsenic from water sources, which will be made possible by ongoing innovation and multidisciplinary research in nanotechnology.

Acknowledgements The authors express my gratitude to the HBTU Chemical Engineering Department for producing the available Web of Science journal access functionality.

References

Amerasinghe, P., Bhardwaj, R. M., Scott, C., Jella, K., & Marshall, F. (2013). *Urban wastewater and agricultural reuse challenges in India* (Vol. 147). IWMI.

Arif, Z., Sethy, N. K., Mishra, P. K., & Kumar, P. (2022). Adsorptive remediation of pollutants from wastewater. In S. Madhav, P. Singh, V. Mishra, S. Ahmed, & P. K. Mishra (Eds.), *Recent trends in wastewater treatment*. Springer. https://doi.org/10.1007/978-3-030-99858-5_2

Arif, Z., Sethy, N. K., Mishra, P. K., & Kumar, P. (2023). Green synthesis of nanomaterials from biomass waste for biodiesel production. In M. Srivastava, & P. K. Mishra (Eds.), *NanoBioenergy: Application and sustainability assessment. Clean energy production technologies*. Springer. https://doi.org/10.1007/978-981-19-6234-9_8

Cui, H., Li, Q., Gao, S., & Shang, J. K. (2012). Strong adsorption of arsenic species by amorphous zirconium oxide nanoparticles. *Journal of Industrial and Engineering Chemistry, 18*(4), 1418–1427.

Ghosal, P. S., Kattil, K. V., Yadav, M. K., & Gupta, A. K. (2018). Adsorptive removal of arsenic by novel iron/olivine composite: Insights into preparation and adsorption process by response surface methodology and artificial neural network. *Journal of Environmental Management, 209*, 176–187.

Gilhotra, V., Das, L., Sharma, A., Kang, T. S., Singh, P., Dhuria, R. S., & Bhatti, M. S. (2018). Electrocoagulation technology for high strength arsenic wastewater: Process optimization and mechanistic study. *Journal of Cleaner Production, 198*, 693–703.

Goswami, A., Raul, P. K., & Purkait, M. K. (2012). Arsenic adsorption using copper (II) oxide nanoparticles. *Chemical Engineering Research and Design, 90*(9), 1387–1396.

Hao, L., Wang, N., Wang, C., & Li, G. (2018). Arsenic removal from water and river water by the combined adsorption-UF membrane process. *Chemosphere, 202*, 768–776.

Jegadeesan, G., Al-Abed, S. R., Sundaram, V., Choi, H., Scheckel, K. G., & Dionysiou, D. D. (2010). Arsenic sorption on TiO_2 nanoparticles: Size and crystallinity effects. *Water Research, 44*(3), 965–973.

Jesitha, K., & Harikumar, P. S. (2018). Application of nano-phytoremediation technology for soil polluted with pesticide residues and heavy metals. In *Phytoremediation: Management of environmental contaminants* (Vol. 6, pp. 415–439).

Jones, E. R., Van Vliet, M. T., Qadir, M., & Bierkens, M. F. (2021). Country-level and gridded estimates of wastewater production, collection, treatment and reuse. *Earth System Science Data, 13*(2), 237–254.

Karakurt, S. (2019). Removal of carcinogenic arsenic from drinking water by the application of ion exchange resins. *Oncogen Journal, 2*(1), 5.

Khan, F., Shariq, M., Asif, M., Siddiqui, M. A., Malan, P., & Ahmad, F. (2022). Green nanotechnology: Plant-mediated nanoparticle synthesis and application. *Nanomaterials, 12*(4), 673.

Kushwaha, A., Arif, Z., & Singh, B. (2023). Adsorptive removal of fluoride from water using iron oxide-hydrogen nanoparticles. In A. K. Yadav, S. Shirin, & V. P. Singh (Eds.), *Advanced treatment technologies for fluoride removal in water. Water science and technology library* (vol. 125). Springer. https://doi.org/10.1007/978-3-031-38845-3_8

Li, C., Hou, H., Yang, J., Liang, S., Shi, Y., Guan, R., Hu, Y., Wu, X., Hu, J., & Wang, L. (2019). Comparison of electrokinetic remediation on lead-contaminated kaolinite and natural soils. *CLEAN–Soil, Air, Water, 47*(4), 1800337.

Luther, S., Borgfeld, N., Kim, J., & Parsons, J. G. (2012). Removal of arsenic from aqueous solution: A study of the effects of pH and interfering ions using iron oxide nanomaterials. *Microchemical Journal, 101*, 30–36.

Mohora, E., Rončević, S., Agbaba, J., Zrnić, K., Tubić, A., & Dalmacija, B. (2018). Arsenic removal from groundwater by horizontal-flow continuous electrocoagulation (EC) as a standalone process. *Journal of Environmental Chemical Engineering, 6*(1), 512–519.

Najib, N., & Christodoulatos, C. (2019). Removal of arsenic using functionalized cellulose nanofibrils from aqueous solutions. *Journal of Hazardous Materials, 367*, 256–266.

Nidheesh, P. V., & Singh, T. A. (2017). Arsenic removal by electrocoagulation process: Recent trends and removal mechanism. *Chemosphere, 181*, 418–432.

Nwadinigwe, A. O., & Ugwu, E. C. (2018). Overview of nano-phytoremediation applications. In *Phytoremediation: Management of environmental contaminants* (Vol. 6, pp. 377–382).

Olyaie, E., Banejad, H., Afkhami, A., Rahmani, A., & Khodaveisi, J. (2012). Development of a cost-effective technique to remove the arsenic contamination from aqueous solutions by calcium peroxide nanoparticles. *Separation and Purification Technology, 95*, 10–15.

Pirilä, M., Martikainen, M., Ainassaari, K., Kuokkanen, T., & Keiski, R. L. (2011). Removal of aqueous As (III) and As (V) by hydrous titanium dioxide. *Journal of Colloid and Interface Science, 353*(1), 257–262.

Qadir, M., Drechsel, P., Jiménez Cisneros, B., Kim, Y., Pramanik, A., Mehta, P., & Olaniyan, O. (2020). Global and regional potential of wastewater as a water, nutrient and energy source. *Natural Resources Forum, 44*(1), 40–51.

Rosales, M., Coreño, O., & Nava, J. L. (2018). Removal of hydrated silica, fluoride and arsenic from groundwater by electrocoagulation using a continuous reactor with a twelve-cell stack. *Chemosphere, 211*, 149–155.

Sethy, N. K., Arif, Z., Mishra, P. K., Kumar, P., & Saha, R. (2022) Nanocatalyst synthesis by the green route: Mechanism and application. In *Innovative bio-based technologies for environmental remediation* (pp. 321–341). CRC Press.

Shakoor, M. B., Nawaz, R., Hussain, F., Raza, M., Ali, S., Rizwan, M., Oh, S. E., & Ahmad, S. (2017). Human health implications, risk assessment and remediation of As-contaminated water: A critical review. *Science of the Total Environment, 601*, 756–769.

Simon, M., Joshi, H., Yadav, A. K., & Hopke, P. K. (2024). Simultaneous nitrogen and phosphorus removal from polluted water using a novel bacterial consortium (BACON-38): Promising application for wastewater remediation. *Journal of Water Process Engineering, 66*, Article 106044. https://doi.org/10.1016/j.jwpe.2024.106044

Shukla, A., & Srivastava, S. (2019). A review of phytoremediation prospects for arsenic contaminated water and soil. In *Phytomanagement of polluted sites* (pp. 243–254).

Song, K., Kim, W., Suh, C. Y., Shin, D., Ko, K. S., & Ha, K. (2013). Magnetic iron oxide nanoparticles prepared by electrical wire explosion for arsenic removal. *Powder Technology., 1*(246), 572–574.

Vega-Hernandez, S., Weijma, J., & Buisman, C. J. (2019). Immobilization of arsenic as scorodite by a thermoacidophilic mixed culture via As (III)-catalyzed oxidation with activated carbon. *Journal of Hazardous Materials, 368*, 221–227.

Wang, C., Liu, H., Zhang, Y., Zou, C., & Anthony, E. J. (2018a). Review of arsenic behavior during coal combustion: Volatilization, transformation, emission and removal technologies. *Progress in Energy and Combustion Science, 68*, 1–28.

Wang, J., Zhang, T., Li, M., Yang, Y., Lu, P., Ning, P., & Wang, Q. (2018b). Arsenic removal from water/wastewater using layered double hydroxide derived adsorbents, a critical review. *RSC Advances, 8*(40), 22694–22709.

Xu, Y., Li, J., Xia, W., Sun, Y., Qian, G., & Zhang, J. (2019). Enhanced remediation of arsenic and chromium co-contaminated soil by eletrokinetic-permeable reactive barriers with different reagents. *Environmental Science and Pollution Research, 26*, 3392–3403.

Yan, H., Gao, Y., Wu, L., Wang, L., Zhang, T., Dai, C., Xu, W., Feng, L., Ma, M., Zhu, Y. G., & He, Z. (2019). Potential use of the Pteris vittata arsenic hyperaccumulation-regulation network for phytoremediation. *Journal of Hazardous Materials, 368*, 386–396.

Zakhar, R., Derco, J., & Čacho, F. (2018). An overview of main arsenic removal technologies. *Acta Chimica Slovaca, 11*(2), 107–113.

Chapter 14
Harnessing Nanoparticles for Sustainable Arsenic Remediation in Food Crops: A Comprehensive Study

Debadrita Das, Archita Dey, Swetanjana Ghosh, Urvashi Lama, Sharmistha Majumder, Monojit Mondal, and Tarit Roychowdhury

Abstract Arsenic (As), known for its toxicity and carcinogenic effects, poses significant environmental and health risks, particularly affecting several millions of people around the world. Through contaminated drinking water. Traditional methods such as oxidation, coagulation, and adsorption have been used for arsenic removal; however, are constrained by cost and maintenance challenges. Consequently, there is a burgeoning interest in using nanotechnology and nanoparticles (NPs) to address these issues in water, soil, and food crops. The sustainable approach of green synthesis in nanotechnology stands out, employing environmentally benign methods to produce NPs for effectively mitigating arsenic contamination in staple food crops. The distinctive qualities of NPs, such as increased surface area and reactivity, enable efficient binding and elimination of arsenic in water and soil, and the goal is to mitigate health hazards linked to arsenic exposure through dietary sources. This chapter offers a detailed analysis of recent advancements and future prospects regarding the implementation of nanoparticles (NPs) for arsenic remediation in food crops. It scrutinises various mitigation strategies, evaluating synthesis techniques, material properties, arsenic removal efficiencies, and mechanisms of action of contemporary nanomaterials. By examining a diverse array of promising nanomaterials, this study aims to propel future research toward efficient, eco-friendly, and cost-effective arsenic removal solutions. Ultimately, it seeks to foster sustainable agricultural practices in food production through innovative nanotechnological approaches.

Keywords Arsenic · Nanoparticles · Green synthesis · Food crops · Mitigation strategies

D. Das · A. Dey · S. Ghosh · U. Lama · S. Majumder · M. Mondal · T. Roychowdhury (✉)
School of Environmental Studies, Jadavpur University, Kolkata, India
e-mail: tarit.roychowdhury@jadavpuruniversity.in

A. K. Yadav and S. Shirin (eds.), *Emerging and Innovative Arsenic Removal Technologies for a Sustainable Future*, Water Science and Technology Library 131,
https://doi.org/10.1007/978-3-032-01176-3_14

14.1 Introduction

Arsenic, a common element found in the environment, has been listed as one of the most toxic and carcinogenic substances (Wu et al., 2019). Arsenic usually occurs in natural groundwater as arsenite As-(III) or arsenate As-(V). Given its ubiquitous presence throughout the environment coupled with its intrinsic poisonous nature, the potential for arsenic contamination across water, air, and soil originating from both natural geological processes and human-induced activities presents a considerable and pressing environmental health challenge (Habuda-Stanić & Nujić, 2015). The World Health Organization stipulates a guideline of no more than 10 μg/L as the permissible upper limit for potable water. However, in various countries, arsenic levels in irrigation water are not strictly regulated. This noxious substance poses considerable threats to ecological balance and human well-being, insidiously diminishing the productivity of agricultural land via its infiltration into the food web. Following contact, arsenic is quickly assimilated by the root systems of the flora and subsequently transported from its initial point of entry to the aerial portions of the plant, thus negatively impacting its growth and maturation. Furthermore, it permeates fruits and seeds through the phloem's transport system, consequently leading to the contamination of the entire trophic cascade (Joardar et al., 2023). This observation is particularly true for *Oryza sativa* (paddy), where fields are often supplied with irrigation water containing elevated levels of arsenic from subterranean sources, particularly in countries of South and Southeast Asian countries. This practice poses a health risk, as arsenic is absorbed and accumulated mainly in the grains of rice, which is utilised for consumption within the region and is also distributed internationally through export (Maity et al., 2021). Extended exposure to arsenic causes reductions in photosynthetic efficiency, suppression of respiration, disruption of the homeostasis of reactive oxygen species (ROS) such as singlet oxygen ($^1O_2$), superoxide ($O_2^{\bullet -}$), and hydrogen peroxide (H_2O_2), as well as interference with protein dynamics through direct interaction with thiol groups (-SH) (Chandrakar et al., 2016). Superfluous reactive oxygen species are produced both during the enzymatic transformation of trivalent arsenic to pentavalent arsenic, a process mediated by arsenate reductase, and also during the metabolic pathway of aerobic phosphorylation. This process hinders plant growth and yield, posing a threat through food chain contamination (Banerjee et al., 2024). Considering the aforementioned challenges, it is imperative to explore innovative and environmentally sustainable remediation strategies that not only mitigate soil contamination but also restore the vitality of food crops. The integration of nanotechnology into agriculture represents a major advance in addressing the intricate issues affecting global food production. It presents unparalleled opportunities for the adoption of sustainable and efficient farming practices.

The field of nanotechnology represents the foremost and most sophisticated industrial sector of the twenty-first century. The distinctive and compelling attributes of nanoparticles (NPs), coupled with their extensive array of potential uses, have elicited considerable interest and focus within the scientific and technological communities. As a promising and multidisciplinary field of research, nanotechnology offers numerous opportunities across various sectors, including agriculture, electronics, pharmaceuticals, and medicine (Sundarajan et al., 2024) . Recently, nanotechnology has gained significant recognition as a revolutionary and effective strategy for identifying the complexities related to soil contamination in agricultural settings. The distinctive qualities of nanoparticles, include their huge surface area, reactive nature, and ability to alter physicochemical properties and make them exceptionally well-suited for soil-plant remediation. The field of nanotechnology has the capacity to fundamentally uplift traditional soil mitigation practices by offering more efficient, precise, and environmentally sustainable techniques to mitigate the effects of agricultural contamination. Various technological methodologies have been employed for the extraction of arsenic from aqueous solutions, with their effectiveness determined by the specific physicochemical characteristics of the arsenic-containing substances present. Arsenic sequestration or immobilisation is most effectively achieved when it is present in its pentavalent oxidation state, as arsenate. Methods used for arsenic elimination typically encompass coagulation, filtration, and adsorption. Of these approaches, adsorption presents several favourable attributes, such as the operational mode is uncomplicated and reliable, waste management is facilitated, no additional chemical agents are required, the infrastructure is space efficient and the overall financial outlay for operation is typically reduced (Habuda-Stani & Nujic, 2015; Shirin & Yadav, 2014; Simon et al., 2024). This chapter provides a thorough examination of recent progress and future potential in the use of NPs for arsenic remediation in food crops. Critical evaluation of various mitigation strategies, including synthesis methods, material properties, arsenic removal efficiencies, and the mechanisms of action of modern nanomaterials in food crops. Through exploration of a broad range of promising nanomaterials, this study aims to advance research toward efficient, environmentally friendly, and cost-effective arsenic removal solutions. Ultimately, it seeks to promote sustainable agricultural practices in food production through innovative nanotechnological methods.

14.2 Characteristics of NPs

Novel particles exhibit distinctive characteristics compared to their bulk counterparts, such as extraordinarily high surface areas, superior reactivities, remarkable mobilities, and remarkable mechanical resilience, chemical inertness, and electrical conductivity. However, the characteristics exhibited by specific nanoscale particles are significantly contingent upon their dimensional scale and the material from which they are derived. As a result, certain nanoparticles exhibit exceptional optical, electrical, or magnetic qualities, whereas others lack these distinctive features. This section will explore both the general properties of nanoparticles, and the specific attributes associated with particular types, focussing on the origins of their unique characteristics.

14.2.1 Size of the NPs

The dimensional scale of nanoparticles is a key factor in dictating their inherent characteristics and potential applications (Altammar, 2023). This size is modulated by the synthetic methodology and the parameters of the reaction parameters, which consequently impacts its efficacy across diverse uses (Menichetti et al., 2023). To illustrate, iron oxide nanoparticles fabricated via a hydrothermal route showed a high degree of crystalline order, presenting a composite of magnetite and maghemite crystalline structures (Ge et al., 2009). As the size increased, the magnetite-to-maghemite ratio also increased, ultimately yielding pure magnetite for particles sized at 123 ± 44 nm. Elevating the temperature of the reaction from 100 to 180 °C over a 12 h duration caused nanoparticle size to grow from 14.5 ± 4 nm to 29.9 ± 9 nm. Similarly, extending reaction time at 180 °C from 1 to 48 h led to an increase in nanoparticle size from 20.6 ± 6 nm to 123 ± 44 nm. Gold nanoparticles under 10 nm were evaluated for their optical properties, which are highly size-dependent, influencing their applications. Furthermore, the antibacterial potency of silver nanoparticles exhibits variability on their dimensions and morphology when tested against diverse bacterial species (Sajid & Płotka-Wasylka, 2020).

The dimensions of nanoparticles are a critical determinant of their suitability for various implementation. For example, a singular platinum nanoparticle was precisely formed and enlarged onto the tip of a tunneling ultramicroelectrode (TUME) to create a specialised sensor, with sizes ranging from 1 to 40 nm or even larger. This approach enhanced the mass transfer rate, making it particularly effective for measuring electron transfer parameters in rapid electron transfer (ET) reactions. Analogously, alterations in the dimensions of iron oxide nanoparticles exert an effect on their electronic band gap and the degree of internal stress within the material's structure (Sajid & Płotka-Wasylka, 2020).

14.2.2 Shape of NPs

Beyond size, the shape and structure of nanoparticles are crucial for various technological applications. Advances in controlling nanoparticle shape through synthesis parameters have been largely achieved. Consequently, nanoparticles of the same substance can be synthesized in several forms for specific uses. Shape-guided synthesis has become a well-established area of research with extensive discussion on methods of nucleation, growth, and condition control in the literature. It is well established that the electrocatalytic activities of several electrochemical processes are influenced by both the catalyst's surface area and structure, and the nanoparticle-based catalysts are no exception. For instance, cobalt disulphide with an octahedral-formed nanoparticles based on graphene/carbon nanotube composites doped with sulphur and nitrogen have been utilized for oxygen reduction in acidic electrolytes. Additionally, spherical, cubic, elliptical TiO_2 nanoparticles and nanorods were synthesised by varying surfactants during preparation, with their shape affecting their photocatalytic activity. Selenium nanoparticles (Se NPs) have been produced in various forms, including various nanostructures, including nanowires, nanoribbons, nanoplates, nanotubes, and nanospheres with biological activities. Recent studies have synthesised cubic-like Se NPs via self-assembly and examined how their shape affects antitumor activity. The size and morphology of nanoparticles are closely linked to their fabrication methods, as summarised in a recent review focusing on solar-powered photocatalytic materials based on zinc oxide nanoparticles (ZnO; NP), which details the impact of fabrication techniques on particle size and morphology (Wu et al., 2021).

14.2.3 Surface Area Properties of NPs

Nanoscale particles are held in high regard for their utility across diverse fields because of their substantial surface areas, which enhance their roles in catalysis, adsorption, and electrochemical reactions. For instance, superparamagnetic Fe_3O_4 nanoparticles coated with ascorbic acid, with a defined surface area of 179 m^2/g and a diameter below 10 nm, were synthesized hydrothermally for heavy metal adsorption, demonstrating notable adsorption capacity. The synthesis method and experimental conditions directly impact the surface area and effectiveness of nanoparticles. For example, $LaFeO_3$ nanoparticles produced using the SBA-16 template technique exhibited a wide higher surface area in contrast to those synthesised via the traditional way of using citric acid, leading to superior photocatalytic activity. Similarly, SnO_2 nanoparticles produced through a homogenous ethanol-thermal precipitation technique showed a high surface area of 200 m^2/g and excellent gas- sensing properties, in contrast to those produced by a single homogeneous precipitation method. Furthermore, the ecotoxicity of silicon dioxide (SiO_2) nanoparticles towards the green algae *Pseudokirchneriella subcapitata* was correlated with their surface area (Feng et al., 2012).

14.3 Several Synthesis Techniques of NPs

Amid global environmental concerns, chemistry looking for greener methods. Natural reagents such as microorganisms, sugars, plant extracts, vitamins, and biodegradable plastics facilitate nanoparticle synthesis (Fig. 14.1). Metal nanoparticles, oxides, and salts are produced. Plant-based materials, favoured for large-scale biosynthesis, exemplify these ecofriendly approaches, such as using green tea extract for silver nanoparticles.

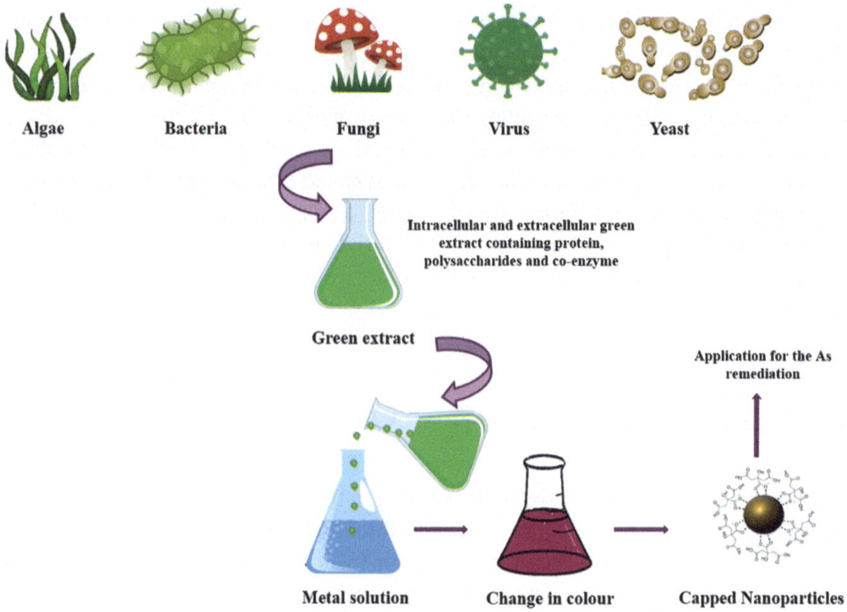

Fig. 14.1 Green synthesis of nanoparticles from different sources

14.3.1 Nanoparticle (NP) Production Aided by Fungi

Fungi are valuable reservoirs of active biomolecules and secondary metabolites essential for the synthesising of NPs. Certain fungal species, such as *Fusarium oxysporum*, secrete proteins, polymers, and enzymes that facilitate the production of metal NPs. These components significantly enhance both the yield and stability of NPs. Additionally, studies have identified several fungal species capable of synthesis of NP using residues of extracellular amino acids. For instance, when exposed to enough light, yeast surfaces rich in glutamic and aspartic acid can photochemically convert silver ions to elemental silver.

Fungal species such as *F. oxysporum* possess cytosolic reductase enzymes that catalyse the reduction of silver ions to silver metal in the presence of $NADH^{+}$, a crucial reducing agent. Furthermore, fungi belonging to the phytochelatin group produce compounds with a remarkable ability to reduce silver ions to metallic silver. The culture supernatant of fungi like *Coriolus versicolor* has been utilised effectively for synthesising silver nanoparticles (Ag-NPs). Analysis using Fourier transform infrared spectroscopy (FTIR) has verified the hydroxyl groups found in fungal mycelium, which give silver ions their electrons, thereby reducing them to Ag-NPs. The spectra also indicate that both aromatic and aliphatic amines are present, along with proteins in fungal extracts that act as capping agents, stabilising the formed Ag-NPs. In addition, studies have highlighted the role of sulphur-containing proteins from fungal extracts in capping and stabilising Ag-NPs.

Rhizopus oryzae mycelia have been employed to produce Au nanoconjugates by in-situ reduction of chloroauric acid ($HAuCl_4$) under acidic conditions (pH 3). The *Verticillium* fungus has also demonstrated proficiency in mediating the synthesis of Ag-NPs. Recent research has revealed that fungal biomass can intracellularly synthesise NPs upon exposure to $AgNO_3$ in mildly acidic environments (pH 5.5–6) (Zhao et al., 2018).

14.3.2 NPs Synthesis Aided by Bacteria

The synthesis of NPs facilitated by bacteria occurs through two distinct approaches: extracellular and intracellular mechanisms. Extracellular synthesis offers advantages over the intracellular method because of its efficiency; it circumvents the need for downstream processes to extract NPs from organisms. Within bacterial cells, reductase enzymes play a pivotal role in catalysing metal ion reduction into metallic nanoparticles. In particular, bacterial species such as *Deinococcus radiodurans* demonstrate strong resistance to oxidative damage and radiation, as well as outstanding antioxidant activity. These characteristics render it particularly suitable for green synthesis of Au-NPs from their ionic form. The Au-NPs produced in this manner demonstrate prolonged stability and enhanced antimicrobial efficacy. In a recent study, researchers utilised *Leptothrix* bacteria to synthesise Au-NP by reducing gold salt in an aqueous medium. The process revealed that residues of guanine in RNA molecules and 2-deoxyguanosine were instrumental in the reduction of gold salt, underscoring the biochemical pathways involved in this bacterial-mediated NP synthesis (Singh et al., 2020).

14.3.3 NPs Synthesis Aided by Plants

Plant-assisted nanoparticle synthesis surpasses microbial methods in terms of yield efficiency. Plants offer a rich repertoire of metabolites and biochemicals, such as polyphenols, which serve dual roles as stabilising and reducing agents in the biogenic nanoparticle (NP) synthesis process. This strategy is friendly to the environment, reducing the usage of harmful chemicals while maintaining a profitability. NPs made from substances derived from plants are notably more stable compared to those produced by microbes and fungi.

Plant-mediated NP synthesis encompasses three primary methods: extracellular, intracellular, and phytochemical-mediated. Extracellular synthesis involves using plant extracts directly as raw materials for NP production. Intracellular synthesis occurs within plant cells through the action of cellular enzymes, with NPs recovered after cell wall disruption. The use of plant extracts for the synthesis of NPs is cost-effective as a result of the abundance of phytochemicals that effectively make metal ions more stable.

NP synthesis mediated by phytochemicals, though less common, requires specific knowledge of the phytochemicals essential for stabilising NP formation. As an illustration, globular silver nanoparticles were effectively produced using extracts derived from the foliage of the *Azadirachta indica* plant. Fourier transform infrared spectroscopy (FTIR) studies revealed during NP production, flavonoids and other phytochemical components of the plant extract functioned as both stabilising and reducing agents imparting potent antimicrobial properties to the resulting NPs.

In another example, gold NPs were synthesised from root extracts of *Morinda citrifolia*, which also exhibited significant antimicrobial activity. This underscores the diverse applications of plant-mediated NP synthesis in development of antimicrobial agents and other advanced materials.

In general, leveraging plant biology for NP synthesis not only enhances yield and stability, but also aligns with sustainable practices, making it a promising avenue for future nanotechnological advancements (Singh et al., 2020).

14.4 Variables Influencing Interplay Between NPs and Food Crops Amidst Oxidative Stress

The efficiency of nanoparticles in reducing arsenic depends on a number of variables, including size, microbial contact, and surface climate. These factors have an impact on adsorption capacity and reactivity, which are essential for effective removal from soils and sources of food crops. Enhancing arsenic remediation systems based on nanoparticles requires an understanding of and focus on these aspects.

14.4.1 Dependence on Dosage and Size

The varying effects of NPs, depending on their dosage and size, necessitate thorough consideration to establish standardised applications in agriculture, crucial to achieving sustainability by managing oxidative stress. In particular, plants synthesised sulphur nanoparticles (phyto-synthesized sulfur NPs) demonstrated enhanced tolerance to oxidative stress in *Lactuca sativa* (lettuce) when administered at a lower concentration of 1 mg/mL. However, adverse phytotoxic effects were observed when the concentration was increased to 10 mg/mL (Najafi et al., 2020). Likewise, the effects of zinc oxide nanoparticles (ZnO-NP) on *Hordeum vulgare* were found to be depend both on dosage and particle size, affecting several biochemical antioxidant responses, including those of glutathione (GSH), glutathione reductase (GR), catalase (CAT), superoxide dismutase (SOD) and glutathione S-transferase (GST). Beyond their other effects, these nanoparticles also influenced genetic mechanisms, demonstrated through changes in the expression of Cat 1 along with Cat 2 genes (Azarin et al., 2022). The desired choice of an optimal (lower) amount of NPs to enhance crop yield is underpinned by hormesis, a phenomenon in which low doses generate beneficial responses. Interestingly, even minimal amounts of cadmium (Cd) can positively influence the development and yield of food crops by altering the biochemical process in plants (Carvalho et al., 2020). On the contrary, higher dosages can cause nanotoxicity in food crops, resulting in lower yields. Furthermore, exposure to elevated doses of NP can contribute to secondary environmental pollution, which poses significant health risks for humans. Therefore, NP-induced hormesis in food crops manifests as a biphasic dose-response phenomenon, which encompasses both low-level stimulatory and high-level inhibitory effects (Wei et al., 2022).

14.4.2 Interactions Between NP-Microbes and Metabolites

The interplay between microbial organisms, and metabolites constitutes a pivotal axis in strategies aimed at mitigating arsenic contamination. NPs, engineered with precise functionalities tailored for arsenic sequestration and transformation, engage dynamically with microbial communities that inhabit soil and aquatic environments. These interactions often amplify microbial activity, facilitating crucial processes such as the reduction, oxidation, or methylation of arsenic species. Consequently, such transformations can either immobilise arsenic or render it less toxic (Yan et al., 2023). A magnetic mixed oxide, neodymium orthoferrite ($NdFeO_3$), was synthesised and assessed for its efficacy in the removal of arsenic was assessed, both independently and when functionalised with a microbial biofilm. The synthesised magnetic $NdFeO_3$ nanoparticles (mNFO-NP) demonstrated the ability to adsorb both As(III) and As(V) species onto its surface (Lascano et al., 2023).

14.4.3 Change in Climatic Condition

Crop production and agricultural sustainability are negatively impacted by climate change and increased atmospheric CO_2 concentrations. Under stress caused by As_2O_3NPs, barley and maize crops have shown different reactions to increased CO_2. Maize demonstrated greater accumulation of total phenolics and flavonoids, while barley showed enhanced bioaccumulation of anthocyanins. Furthermore, it has been documented that higher levels of CO_2 in barley and maize were accompanied by elevated amounts of osmo-protectants, for instance proline, soluble sugars, and antioxidants such as anthocyanins and tocopherols. Despite these similarities, compared to maize, barley showed a higher susceptibility to oxidative stress, which was explained by increased levels of H_2O_2, lipid peroxidation, and bioaccumulation of As_2O_3NP (Selim et al., 2021). In barley, proline buildup involved mainly the glutamate pathway, while the ornithine pathway in maize was found to contain antioxidants. This distinction underscores the diverse metabolic responses of these crops to elevated CO_2 levels. Significantly, there is little doubt that rising CO_2 emissions are linked to environmental catastrophes caused by climate change, this underscores the immediate imperative for control measures to ensure continued crop production and environmental sustainability in agriculture (Rehman et al., 2022).

14.5 Mechanisms of Reducing the Impact of Arsenic on Food Crops Through NPs

Nanoparticles commonly traverse plant tissues through plasmodesmata and ion channels, as well as through endocytosis, the process by which food crops assimilate substances from the soil (Rajput et al., 2020). Mirroring the uptake systems for essential nutrients, analogous transport mechanisms, including phosphate transporters (PHTs), superfamilies of C-type ATP-binding cassette (ABC) superfamilies, and aquaporins, facilitate arsenic absorption into food crops (Farooq et al., 2022). Subsequent to entry, arsenic induces the generation of reactive oxygen species (ROS) and reactive nitrogen species (RNS), thus activating plant tolerance mechanisms in both the molecular and biochemical strata. Modulation of As(III)/As(V) uptake, translocation and accumulation within plant parts, including roots to shoots, and consumable grains, is orchestrated by PHTs and nodulin 26-like intrinsic proteins similar to nodulin 26 (NIP) and PHT (Shamshir et al., 2022). The "avoidance" mechanism, which involves the efflux of As(III) from plant roots, is controlled by aquaporin channels, phytochelatin (PC)- driven vacuolar sequestration, and ABC transmembrane protein transporters. Genes contingent on silicon (Si) and phosphate transporters (e.g., OsLsi1/OsLsi2; *OsLsi3*, *OsLsi6*, OsPHT1.1/OsPHT1.8 in rice) govern the translocation of arsenic to the stem and grains (Rai et al., 2023; Pan et al., 2020). Upon assimilation into plant organs, arsenic compromises the structural and functional integrity of chloroplast membranes, the photosystem II (PSII) electron transport chain, chlorophyll, and other photosynthetic pigments.

This has a negative impact on photosynthesis (Ulhassan et al., 2022). Despite advances, our comprehension remains incomplete concerning the impact of nanoparticles on defence-related genes, transcription factors, phytohormonal signaling pathways, mitogen-activated protein kinase (MAPK) signaling pathways, and the intricate interplay among these elements. Regarding food crops, various methods are believed to contribute to reducing the level of stress mediated by NP, as depicted in (Table 14.1). For instance, nanoparticles exhibit the capacity to adsorb both As(III) and As(V) onto their surfaces and modulate the pH of the rhizosphere. This alteration, in turn, influences the redox potential, arsenic speciation, and the biosynthesis of organic acids and root exudates within the soil environment immediately surrounding the plant roots. These physiochemical alterations are a major factor in decreased mobility bioavailability and arsenic mobility (Wu et al., 2021). Furthermore, NPs facilitate the output of phytochelatins (PC) (Fig. 14.2), which helps mitigate oxidative stress caused by arsenic in edible crops and different plants (Yan et al., 2021).

Table 14.1 Investigations into NP-mediated modulation of arsenic-derived oxidative stress in food crops, along with elucidation of the attendant mechanistic pathways

Metalloid	Food crops	NPs (size/ concentration)	Oxidative stress modulation/ mechanism	References
Arsenic	Rice	FeO (25–50 μM)	Improved photosynthetic function boosts plant growth, coupled with decreased levels of ROS, H_2O_2, and superoxide ions under arsenic stress conditions	Aqeel et al. (2022)
Arsenic	Rice	SiO_2 (0.1, 1, and 10 mM)	Silicon-mediated enhancements include enhanced cell wall thickness, elevated pectin biosynthesis, increased cation exchange capacity (CEC), and facilitated nutrient assimilation, in conjugation with mitigated oxidative stress generated by arsenic	Cui et al. (2020)
Arsenic	Rice	Graphene oxides; hydroxyapatite (20 and 40 nm)	Reduced levels of SOD and POD enzymes, with an increase observed in hydroxyapatite levels, led to decreased arsenic transfer to the shoots, resulting in enhanced biomass yield in rice crops	Ulhassan et al. (2022)
As(V)	Rice	ZnO (30–50 nm)	Elevated zinc assimilation and enhanced rice seedling development, concurrent with diminished arsenic bioaccumulation within plant tissues	Ma et al. (2020)

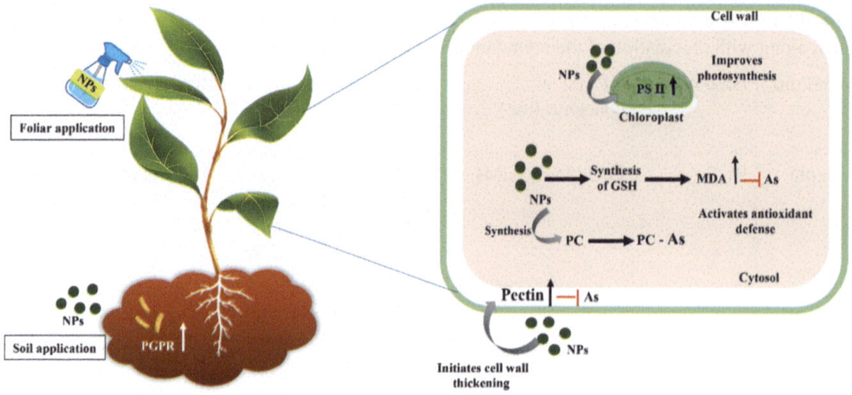

Fig. 14.2 Mechanism of nanoparticles entry and remediation of arsenic in food crop system

The application of nanoparticles to food crops experiencing arsenic stress also engenders an increase of the photochemical energy within photosystems, elevates the concentrations of proline and other osmolytes, stimulates the biosynthesis of glutathione (GSH), enhances the assimilation of minerals and other essential nutrients, and regulates malondialdehyde (MDA) levels and electrolytic leakage, among a suite of mechanisms that potentiate antioxidative responses. Furthermore, NPs affect cellular responses in food crops by increasing the thickness of the cell wall and the amount of pectin to combat arsenic stress. Despite these advances, the precise molecular mechanisms underlying the reduction of arsenic stress in food crops remain incompletely understood. However, ongoing research efforts focus on elucidating the interactions between NPs and crops in arsenic-contaminated soils to comprehend the molecular mechanisms underlying stress reduction. NPs can cooperate with phytohormones, Phytostimulants for plant development, Beneficial root-colonizing bacteria that enhance plant development and organic acids to influence the interactive signaling between arsenic and molecules such as nitric oxide (NO) and calcium ions (Ca^{2+}), Mitogen activating protein kinase (MAPKs), H_2O_2, and plant hormones (Rai et al., 2023).

14.6 Impact of Various NPs on Mitigating Stress Induced by Arsenic

Among food crops, rice is particularly impacted by arsenic contamination in the soil (Ulhassan et al., 2022). The environmentally friendly rice production of rice faces increasing challenges due to pervasive arsenic pollution in soil used for agriculture worldwide (Rai et al., 2023). This issue demands urgent attention, particularly in Asian countries, which contribute approximately 75% of the world's rice production (Ulhassan et al., 2022). Rice shows significant arsenic bioaccumulation in its root system, aerial parts, and grains, even when subjected to concentrations close to 2 mg/L. This accumulation of arsenic exerts a deleterious influence on agricultural characteristics, photosynthetic pigment concentrations, and the functional capacity of antioxidant enzymes (Yan et al., 2021). It has been observed that a variety of metal oxide NPs, including CuONPs, ZnONPs, FeONPs, MnO_2NPs and TiO_2NPs, are essential in reducing arsenic stress in food crops (Yan et al., 2021; Ulhassan et al., 2022). The implementation of selenium nanoparticles (SeNPs) under arsenic stress effectively decreases the overall arsenic concentration, particularly As(III) and As(V) species, in the aerial tissues of crops by facilitating arsenic sequestration within root organs. Furthermore, selenium nanoparticles (SeNPs) mitigate arsenic-induced stress in rice through the enhancement of photosynthesis, preservation of cellular membrane integrity, abatement of lipid peroxidation, and attenuation of reactive oxygen species (ROS) accumulation through alteration of antioxidant enzyme function (Rai et al., 2023).

14.6.1 The Influence of Varied Metallic Nanoparticles

The diverse applicability of zinc oxide nanoparticles (ZnONPs), within a concentration spectrum of 10–100 mg/l, has demonstrated efficacy in alleviating arsenic-induced oxidative stress and simultaneously improving photosynthetic activity. This effect is achieved through up-regulation of PC and the facilitation of As(III)-PC complexation primarily within the roots. This process, involving ZnONPs, effectively blocks or reduces the movement of arsenic into rice grains (Yan et al., 2021). Furthermore, ZnONPs have been observed to exert favourable influences on photosynthetic functions, the absorption of nutrients, key agricultural characteristics, and antioxidant enzymes that participate in the ascorbate-glutathione pathway (Ulhassan et al., 2022). In a different context, the conjugated application of ZnONPs and zinc (Zn) has been shown to drastically abate total arsenic levels in both shoots and roots of arsenic-cadmium-stressed rice seedlings. Specifically, this approach decreased arsenic levels by 60% in shoots and 40% in roots (Ma et al., 2020). The mechanism underlying this reduction involves a negative correlation between zinc and arsenic concentrations, which affects the bioaccumulation of arsenic in roots. Furthermore, investigations reveal a notable efficacy of zinc oxide nanoparticles (ZnONPs) in mitigating arsenic-induced oxidative stress within rice roots, which exhibits a more pronounced effect against the trivalent form of arsenic [As(III)] compared to its pentavalent counterpart [As(V)]. Similarly, copper oxide nanoparticles (CuONPs) applied at a concentration of 50 mg/L have demonstrated efficacy in reducing arsenic bioaccumulation in rice. This decrease coincided with improved crop development and yield, a phenomenon ascribed to the influence of copper oxide nanoparticles (CuONPs) on rhizospheric processes and modifications in the physiobiochemical profile of root exudates. Specifically, the application of CuONPs led to a 23% decrease in arsenic bioaccumulation in shoots and a 45% decrease in roots (Rai et al., 2023).

Imposition of stress can impede the uptake and hinder the absorption of crucial elements crucial for enzymes and photosynthetic pigments essential in the physiological and biochemical mechanisms of food crops (Bidi et al., 2021). Furthermore, by increasing levels of malondialdehyde (MDA) and POD, which damage membrane lipids and cause electrolyte leakage, arsenic stress causes oxidative bursts. Nevertheless, in response to these adverse effects, the application of iron nanoparticles (FeNPs) has demonstrated a capacity to mitigate such impacts by enhancing the glyoxalase system, enhancing antioxidant responses, protecting photosynthetic pigments, and facilitating arsenic chelation mechanisms (Bidi et al., 2021; Shah et al., 2022). Furthermore, it has been observed that, the application of FeNP in rice leads to a downregulation of gene expressions such as *Lsi1* and *Lsi2* that are linked to arsenic absorption (Bidi et al., 2021). Furthermore, FeNPs promote the upregulation of crucial genes for the transport of advantageous iron, which is a key element in cellular metabolism. The genes including *DMAS1, IRT1, IRT2, YSL2, YSL13, NAS2, NAS,* and *FRDL1*. The effects on a wider range of genes implicated and their respective mechanisms in the downregulation of arsenic are presented in Table 14.2, which also outlines the application of different nanoparticles in a variety of food crops.

Certain nanoparticle materials, notably magnesium oxide nanoparticles (MgONP), are gaining prominence in agricultural fertilisers for their capacity to augment antioxidant enzyme activity and promote increased plant biomass under conditions of arsenic exposure (Faizan et al., 2022). Manganese dioxide nanoparticles (MnO_2NPs) have demonstrated potential in mitigating arsenic bioaccumulation in rice, specifically restricting its migration to consumable grains (Rai et al., 2023). The transformation of As(III) to As(V) through oxidation, which is then eliminated by adsorption of MnO_2NP, is the cause of this reduction. Anatase, rutile, and bulk forms of titanium dioxide nanoparticles (TiO_2NPs) have been shown to significantly reduce arsenic stress in rice. In terms of arsenic bioaccumulation, rutile TiO_2NPs showed notable decreases of 88% in roots and 90% in shoots (Wu et al., 2021). The main method of reducing arsenic stress in food crops is known to be the adsorption of arsenic in TiO_2NP and iron plaques (Ulhassan et al., 2022).

The application of silicon nanoparticles (SiNPs) and titanium dioxide nanoparticles (TiO_2NP) together can reduce the phytotoxicity of rice arsenic by boosting antioxidant enzymes such as those that are involved in the photosynthetic and glyoxalase cycles. Together with genes linked to PC and GSH, this method also increases the expression of silicon/arsenic transporters, such as *Lsi1*, *Lsi2*, and *Lsi6* (Kiany et al., 2022). GSH and PCs, thereby, aid in the sequestration of arsenic inside vacuoles, reducing the phytotoxic effects of the metal on cellular metabolism. Furthermore, the use of SiNPs and TiO_2NPs promotes normal growth of arsenic-stressed rice plants by lowering levels of methylglyoxal (MG), MDA, and electrolyte leakage (Kiany et al., 2022).

Table 14.2 Effect of different NPs in various food crops and genes involved in the mechanism

Serial No.	Name	Food crop	Genes involved	Mechanism	References
1	SiNPs	Rice	*MAPK*	Stored the majority of the arsenic in the shoot and root cell walls to reduce oxidative damage	Liu et al. (2014)
2	SiO_2NPs	Rice	*Lsi1* and *Lsi2* *OsNIP1;1* *OsNIP3;3*	Reduced the gene expression profile of the *Lsi1*, *Lsi2* genes and upregulation of *OsNIP1;1 and OsNIP3;3. This helps* to limit the absorption of arsenic by strengthening the mechanical strength and pectin production	Cui et al. (2020)
3	Fe_2O_3NPs	Rice	*OsSULTR1;2*, *OsSULTR2;2*, *OsGS*, *OsGST*, *OsECS*, *OsCS* and *OsSiR*	Increased expression of genes related to sulphur alleviate arsenic stress	Meselhy et al. (2021)
4	Fe_3O_4NPs	*Brassica juncea*	genes responsible for ATP sulfurylase and APS reductase production	Upregulation of the genes remediate arsenic	Praveen et al. (2018)
5	ZnONPs	Soybean	*GmSOD*, *GmPOD*, *GmCAT*, *GmAPX*, *GmGR*, *GmGST* and *GmP5CS*	Ascribed to the primary stress-relieving the mechanism of arsenic-induced toxicity	Zeeshan et al. (2021)

14.6.2 Concurrent Use of Nanoparticles with Other Substances

Synergistic application of zinc oxide nanoparticles (ZnONPs) and salicylic acid (SA) has been found to be efficient in reducing the effects of arsenic in tomatoes by increasing the role of antioxidant metabolites (Faizan et al., 2021). Tomatoes treated with this combination, called SAZnONPs, had much higher amounts of proline, catalase (CAT), peroxidase (POD), superoxide dismutase (SOD), and total soluble proteins. Furthermore, it triggers transcriptional mechanisms that increase plants' resistance to oxidative damage caused by arsenic.

The combined application of ZnONPs, oxalic acid (OA), and PGPR such as *Provibencia. vermicola* showed significant mitigation of arsenic-derived oxidative stress in a different study, including *Luffa acutangula.* The improvement in relative water content, higher concentrations of total sugars, proteins, proline, and indole-3-acetic acid (IAA), higher levels of photosynthetic pigments, increased soil phosphatase and invertase activity, and improved nutrient uptake were all results of this treatment program. On the other hand, MDA levels, lipid peroxidation, and electrolyte leakage were significantly reduced. Interestingly, the findings indicated that this combined application resulted in a 58% reduction in arsenic bioaccumulation within the roots of *Luffa acutangula* and a 40% reduction in its shoots (Tanveer et al., 2022).

By lowering MDA and H_2O_2 levels, the administration of iron oxide nanoparticles (Fe_2O_3NPs) and citric acid (CA) separately reduced arsenic stress in soybeans (*Glycine max*). Fe_2O_3NPs and CA showed synergistic effects when applied simultaneously, improving their capacity to reduce arsenic phytotoxicity through metalloid chelation and enhanced arsenic dissolution.

Furthermore, arsenic-induced oxidative stress was successfully managed by depositing silicon dioxide nanoparticles (SiO_2NPs) on the cell walls and leaves of maize plants. Combining SeNPs with biostimulators that promote plant development has also demonstrated potential to reduce arsenic oxidative stress. For example, in *Brassica napus*, melatonin (MT) and SeNP combined improved biomass accumulation, antioxidant enzyme activities, leaf chlorophyll fluorescence, and agronomic characteristics (Farooq et al., 2022). In oil-producing crops such as *B. napus*, this combined treatment increased food security and safety in a synergistic way.

Furthermore, increasing the content of GSH, the levels of antioxidant enzymes and ascorbic acid levels while simultaneously lowering the concentrations of ROS and MDA. The combination treatment of ZnONPs and SeNPs in soybeans showed improved efficacy in mitigating arsenic stress. Similarly, ZnONPs and MT worked in concert to mitigate arsenic phytotoxicity in soybeans, resulting in significant drops in MDA and H_2O_2 concentrations (Table 14.3).

In summary, these findings underscore the potential for a synergistic effect between biostimulants and nanoparticle-based strategies to effectively mitigating arsenic-induced oxidative stress within food crops, improving both food safety and

Table 14.3 Impact of nanoparticles conjugated with other substances on the remediation of arsenic in food crops

Serial no.	Nanoparticles	Conjugated material	Food crops	Mechanism	References
1	ZnONPs	Melatonin and salicylic acid	Soybean and rice	Cooperative benefits on reducing arsenic stress	Bhat et al. (2022)
2	SeNPs	Melatonin	*Brassica napus*	Enhanced reactions to antioxidants and photosynthetic processes	Farooq et al. (2022)

agricultural production. To investigate and improve these promising approaches, more research is recommended at the soil crop system (Cao et al., 2025).

14.7 Obstacles and Future Prospective

Ensuring future food security faces considerable challenges exacerbated by the increasing impact of abiotic environmental stresses from arsenic toxicity on the soil and food crop system, a situation further compounded by global climate change. Although advances have been made in alleviating arsenic stress through nanoparticle applications, significant gaps in understanding remain. These knowledge gaps underscore the necessity for more systematic and meticulously designed future studies to comprehensively address the complexities and optimise strategies for sustainable agricultural practices.

14.7.1 Obstacles

1. **Properties and Behavior of Nanoparticles**

Size and Surface Properties: The size, shape, and surface properties of NPs can affect how well they interact with soil constituents and are absorbed by plants, which in turn can show how effective they are at removing arsenic.

Agglomeration and Dispersion: NPs have the potential to aggregate in soil, which would limit their mobility and effectiveness in removing arsenic. This would make their application and efficacy more difficult.

Fate and transport: Assessing the prolonged effects of NPs on soil vitality and ecosystem dynamics requires an understanding of their fate, transport, and persistence in soil.

2. **Environment and Ecological Concerns**

Ecotoxicological Effects: NPs may interact with beneficial insects, plant roots, and soil microorganisms, potentially leading to unforeseen ecological effects that need careful evaluation.

Bioaccumulation: The possibility of nanoparticles (NPs) bioaccumulating in food crops and then entering the food chain raises concerns about potential health effects on humans, which calls for thorough safety assessments.

Risk of Nanoparticle Release: When NPs are accidentally released into the environment during stages of production, application, or disposal, threats to ecosystems and human health arise, necessitating strict containment protocols and regulatory monitoring.

3. **Technological and Practical Challenges**

Cost-Effectiveness: The high production costs of NPs suitable for agricultural applications may hinder their widespread adoption, particularly in resource-constrained agricultural settings.

Scale-Up and Application: Scaling up NP production and ensuring uniform application in diverse agricultural landscapes remain technical challenges that need to be addressed for practical implementation.

Effectiveness under Field Conditions: Variability in soil properties, climatic conditions, and crop-specific interactions can influence the efficacy of NPs in arsenic remediation under real-world field conditions, necessitating robust field trials and validation studies.

4. **Regulatory and Societal Acceptance**

Regulatory Frameworks: The absence of standardized regulations and guidelines specific to NP use in agriculture complicates is regulatory approval and commercialization, requiring collaborative efforts among stakeholders.

Public Perception: Addressing public concerns and promoting awareness of the benefits and risks associated with the use of NP use in food production is crucial to fostering societal acceptance and trust.

14.7.2 *Future Perspectives*

1. **Advanced Nanoparticle Design and DevelopmentTailored Nanomaterials**: Continued research on novel NP formulations with improved stability, controlled release mechanisms, and biocompatibility to optimize their efficacy in arsenic remediation while minimizing environmental impacts.

   Functionalization Strategies: Surface functionalization of NPs with organic ligands or polymers to improve their dispersibility, soil mobility, and targeted arsenic uptake by plants.

2. **Integration with Sustainable Agricultural PracticesSynergistic Approaches**: To mitigate the harmful effects of arsenic in food crops, more research is needed to fully understand the importance of signalling pathways and defensive interactions among NPs, biostimulators, hormones, and chelators. Investigating the possible synergistic interactions between NPs and chelators or biostimulators such as MT and salicylic acid (SA) to boost their effectiveness in the reduction of arsenic stress is crucial. Furthermore, it is essential to clarify the ways in which NPs interact with other signalling molecules and MAPKs when exposed to arsenic stress. These kinds of findings are essential to create successful nanoformulations that support sustainable agricultural practices.

   Biofortification: Exploring the potential of NP-mediated arsenic remediation to simultaneously improve nutrient uptake and biofortify crops with essential minerals, contributing to improved food security and nutrition.
3. **Multidisciplinary Collaboration and Knowledge SharingCross-Sector Collaboration**: Encourage partnerships between nanotechnologists, agronomists, environmental scientists, policymakers, and industry stakeholders to address knowledge gaps, harmonize research efforts, and facilitate technology transfer.

   Data Sharing and Standardization: Establishing databases, protocols, and standardized methodologies to assess the safety, efficacy, and environmental impact of NP-based arsenic remediation technologies.
4. **Field-Scale Validation and Demonstration ProjectsLong-Term Field Trials**: Conduct extensive field trials across diverse agroecosystems to evaluate the performance, durability, and economic feasibility of NP applications in arsenic-contaminated soils.

   Demonstration Projects: Implementing pilot-scale NP remediation projects in collaboration with farmers and local communities to showcase practical benefits, build confidence, and promote adoption.

14.8 Conclusion

Understanding how NPs affect molecular and metabolic processes could improve oxidative stress mitigation in agricultural crops. Furthermore, tolerance to arsenic stress in food crops can be considerably impacted by understanding how NPs interact with interactions between plant microbes, bioregulators of plant development, phytohormones, signal molecules, and the dynamics of climate change. Novel approaches to improving the efficiency of NPs in creating resilient cultivars appropriate for arsenic-stressed agricultural situations, such as CRISPR gene editing and omics technology, could lead to more sustainable farming methods. Enhancing public awareness and stakeholder engagement through education campaigns, workshops, and transparent communication on the benefits, risks, and regulatory aspects of the use of NPs in agriculture. Advocating for evidence-based policy frameworks that promote safe and sustainable use of NPs in arsenic remediation, in accordance with national and international agricultural and environmental standards. In conclusion, while nanoparticles offer promising solutions for mitigating arsenic contamination in food crops and soil, addressing current obstacles and advancing future perspectives requires concerted efforts across scientific, regulatory, and societal domains. Through the adoption of novel approaches, cooperative endeavours, and a conscientious management, nanoparticles-based technologies possess the potential to foster long-lasting agricultural practices, ensure global food availability, and promote responsible environmental custodianship.

Acknowledgement The present research work is supported by the research fellowship program of the first author under University Grant Commission (UGC), Govt. of India (NTA Ref. No. 231610030618).

References

Altammar, K. A. (2023). A review on nanoparticles: Characteristics, synthesis, applications, and challenges. *Frontiers in Microbiology, 14*, 1155622. https://doi.org/10.3389/fmicb.2023.1155622

Aqeel, U., Aftab, T., Khan, M. M. A., Naeem, M., & Khan, M. N. (2022). A comprehensive review of impacts of diverse nanoparticles on growth, development and physiological adjustments in plants under changing environment. *Chemosphere, 291*, 132672. https://doi.org/10.1016/j.chemosphere.2021.132672

Azarin, K., Usatov, A., Minkina, T., Plotnikov, A., Kasyanova, A., Fedorenko, A., Fedorenko, A., Duplii, N., Vechkanov, E., Rajput, V. D., Mandzhieva, S., & Alamri, S. (2022). Effects of ZnO nanoparticles and its bulk form on growth, antioxidant defense system and expression of oxidative stress related genes in *Hordeum vulgare* L. *Chemosphere, 287*, 132167. https://doi.org/10.1016/j.chemosphere.2021.132167

Banerjee, S., Mondal, S., Islam, J., Sarkar, R., Saha, B., & Sen, A. (2024). Rhizospheric nano-remediation salvages arsenic genotoxicity: Zinc-oxide nanoparticles articulate better oxidative stress management, reduce arsenic uptake, and increase yield in *Pisum sativum* (L.). *Science of the Total Environment, 913*, 169493. https://doi.org/10.1016/j.scitotenv.2023.169493

Bhat, J. A., Faizan, M., Bhat, M. A., Huang, F., Yu, D., Ahmad, A., Bajguz, A., & Ahmad, P. (2022). Defense interplay of the zinc-oxide nanoparticles and melatonin in alleviating the arsenic stress in soybean (*Glycine max* L.). *Chemosphere, 288*, 132471. https://doi.org/10.1016/j.chemosphere.2021.132471

Bidi, H., Fallah, H., Niknejad, Y., & Tari, D. B. (2021). Iron oxide nanoparticles alleviate arsenic phytotoxicity in rice by improving iron uptake, oxidative stress tolerance and diminishing arsenic accumulation. *Plant Physiology and Biochemistry, 163*, 348–357. https://doi.org/10.1016/j.plaphy.2021.04.020

Cao, Y., Turk, K., Bibi, N., Ghafoor, A., Ahmed, N., Azmat, M., Ghani, M. I., & Ahanger, M. A. (2025). Nanoparticles as catalysts of agricultural revolution: Enhancing crop tolerance to abiotic stress: A review. *Frontiers in Plant Science, 15*, 1510482. https://doi.org/10.3389/fpls.2024.1510482

Carvalho, M. E., Castro, P. R., & Azevedo, R. A. (2020). Hormesis in plants under Cd exposure: From toxic to beneficial element? *Journal of Hazardous Materials, 384*, 121434. https://doi.org/10.1016/j.jhazmat.2019.121434

Chandrakar, V., Dubey, A., & Keshavkant, S. (2016). Modulation of antioxidant enzymes by salicylic acid in arsenic exposed *Glycine max* L. *Journal of Soil Science and Plant Nutrition, 16*(3), 662–676. https://doi.org/10.4067/S0718-95162016005000048

Cui, J., Li, Y., Jin, Q., & Li, F. (2020). Silica nanoparticles inhibit arsenic uptake into rice suspension cells via improving pectin synthesis and the mechanical force of the cell wall. *Environmental Science: Nano, 7*(1), 162–171. https://doi.org/10.1039/C9EN01035A

Faizan, M., Bhat, J. A., El-Serehy, H. A., Moustakas, M., & Ahmad, P. (2022). Magnesium oxide nanoparticles (MgO-NPs) alleviate arsenic toxicity in soybean by modulating photosynthetic function, nutrient uptake and antioxidant potential. *Metals, 12*(12), 2030. https://doi.org/10.3390/met12122030

Faizan, M., Sehar, S., Rajput, V. D., Faraz, A., Afzal, S., Minkina, T., Sushkova, S., Adil, M. F., Yu, F., Alatar, A. A., Akhter, F., & Faisal, M. (2021). Modulation of cellular redox status and antioxidant defense system after synergistic application of zinc oxide nanoparticles and salicylic acid in rice (*Oryza sativa*) plant under arsenic stress. *Plants, 10*(11), 2254. https://doi.org/10.3390/plants10112254

Farooq, M. A., Islam, F., Ayyaz, A., Chen, W., Noor, Y., Hu, W., Hannan, F., & Zhou, W. (2022). Mitigation effects of exogenous melatonin-selenium nanoparticles on arsenic-induced stress in *Brassica napus*. *Environmental Pollution, 292*, 118473. https://doi.org/10.1016/j.envpol.2021.118473

Feng, L., Cao, M., Ma, X., Zhu, Y., & Hu, C. (2012). Superparamagnetic high-surface-area Fe_3O_4 nanoparticles as adsorbents for arsenic removal. *Journal of Hazardous Materials, 217*, 439–446. https://doi.org/10.1016/j.jhazmat.2012.03.073

Meselhy, G. A., Sharma, S., Guo, Z., Singh, G., Yuan, H., Tripathi, R. D., Xing, B., Musante, C., White, J. C., & Dhankher, O. P. (2021). Nanoscale sulfur improves plant growth and reduces arsenic toxicity and accumulation in rice (*Oryza sativa* L.). *Environmental Science and Technology, 55*(20), 13490–13503. https://doi.org/10.1021/acs.est.1c05495

Ge, S., Shi, X., Sun, K., Li, C., Uher, C., Baker, J. R., Jr., Banaszak Holl, M. M., & Orr, B. G. (2009). Facile hydrothermal synthesis of iron oxide nanoparticles with tunable magnetic properties. *The Journal of Physical Chemistry C, 113*(31), 13593–13599. https://doi.org/10.1021/jp902953t

Habuda-Stanić, M., & Nujić, M. (2015). Arsenic removal by nanoparticles: A review. *Environmental Science and Pollution Research, 22*, 8094–8123. https://doi.org/10.1007/s11356-015-4307-z

Joardar, M., Mukherjee, P., Das, A., Mridha, D., De, A., Chowdhury, N. R., Majumder, S., Ghosh, S., Das, J., Alam, M. R., Rahman, M. M., & Roychowdhury, T. (2023). Different levels of arsenic exposure through cooked rice and its associated benefit-risk assessment from rural and urban populations of West Bengal, India: A probabilistic approach with sensitivity analysis. *Environmental Science and Pollution Research, 30*(27), 70950–70973. https://doi.org/10.1007/s11356-023-27249-x

Kiany, T., Pishkar, L., Sartipnia, N., Iranbakhsh, A., & Barzin, G. (2022). Effects of silicon and titanium dioxide nanoparticles on arsenic accumulation, phytochelatin metabolism, and antioxidant system by rice under arsenic toxicity. *Environmental Science and Pollution Research, 29*(23), 34725–34737. https://doi.org/10.1007/s11356-021-17927-z

Lascano, G. A., Spuches, C. F., Gómez, M. I., Ferrero, M. A., Sales, A., Romero, C. M., & Navarro, M. C. (2023). Bioremediation of arsenic using magnetic $NdFeO_3$ nanoparticles functionalized with microbial biofilm. *Journal of Environmental Chemical Engineering, 11*(2), 109532. https://doi.org/10.1016/j.jece.2023.109532

Liu, C., Wei, L., Zhang, S., Xu, X., & Li, F. (2014). Effects of nanoscale silica sol foliar application on arsenic uptake, distribution and oxidative damage defense in rice (*Oryza sativa* L.) under arsenic stress. *RSC Advances, 4*(100), 57227–57234. https://doi.org/10.1039/C4RA08496A

Ma, X., Sharifan, H., Dou, F., & Sun, W. (2020). Simultaneous reduction of arsenic (As) and cadmium (Cd) accumulation in rice by zinc oxide nanoparticles. *Chemical Engineering Journal, 384*, 123802. https://doi.org/10.1016/j.cej.2019.123802

Maity, J. P., Chen, C. Y., Bhattacharya, P., Sharma, R. K., Ahmad, A., Patnaik, S., & Bundschuh, J. (2021). Advanced application of nano-technological and biological processes as well as mitigation options for arsenic removal. *Journal of Hazardous Materials, 405*, 123885. https://doi.org/10.1016/j.jhazmat.2020.123885

Menichetti, A., Mavridi-Printezi, A., Mordini, D., & Montalti, M. (2023). Effect of size, shape and surface functionalization on the antibacterial activity of silver nanoparticles. *Journal of Functional Biomaterials, 14*(5), 244. https://doi.org/10.3390/jfb14050244

Najafi, S., Razavi, S. M., Khoshkam, M., & Asadi, A. (2020). Effects of green synthesis of sulfur nanoparticles from *Cinnamomum zeylanicum* barks on physiological and biochemical factors of Lettuce (*Lactuca sativa*). *Physiology and Molecular Biology of Plants, 26*, 1055–1066. https://doi.org/10.1007/s12298-020-00793-3

Pan, D., Yi, J., Li, F., Li, X., Liu, C., Wu, W., & Tao, T. (2020). Dynamics of gene expression associated with arsenic uptake and transport in rice during the whole growth period. *BMC Plant Biology, 20*, 1–10. https://doi.org/10.1186/s12870-020-02343-1

Praveen, A., Khan, E., Ngiimei, D. S., Perwez, M., Sardar, M., & Gupta, M. (2018). Iron oxide nanoparticles as nano-adsorbents: a possible way to reduce arsenic phytotoxicity in Indian mustard plant (*Brassica juncea* L.). *Journal of Plant Growth Regulation, 37*, 612–624. https://doi.org/10.1007/s00344-017-9760-0

Rai, P. K., Song, H., & Kim, K. H. (2023). Nanoparticles modulate heavy-metal and arsenic stress in food crops: Hormesis for food security, safety, and public health. *Science of the Total Environment, 166064.* https://doi.org/10.1016/j.scitotenv.2023.166064

Rajput, V., Minkina, T., Sushkova, S., Behal, A., Maksimov, A., Blicharska, E., Ghazaryan, K., Movsesyan, H., & Barsova, N. (2020). ZnO and CuO nanoparticles: A threat to soil organisms, plants, and human health. *Environmental Geochemistry and Health, 42*, 147–158. https://doi.org/10.1007/s10653-019-00317-3

Rehman, A., Alam, M. M., Alvarado, R., Işık, C., Ahmad, F., Cismas, L. M., & Pupazan, M. C. M. (2022). Carbonization and agricultural productivity in Bhutan: Investigating the impact of crops production, fertilizer usage, and employment on CO_2 emissions. *Journal of Cleaner Production, 375*, 134178. https://doi.org/10.1016/j.jclepro.2022.134178

Sajid, M., & Płotka-Wasylka, J. (2020). Nanoparticles: Synthesis, characteristics, and applications in analytical and other sciences. *Microchemical Journal, 154*, 104623. https://doi.org/10.1016/j.microc.2020.104623

Selim, S., Abuelsoud, W., Al-Sanea, M. M., & AbdElgawad, H. (2021). Elevated CO_2 differently suppresses the arsenic oxide nanoparticles-induced stress in C_3 (*Hordeum vulgare*) and C_4 (*Zea maize*) plants via altered homeostasis in metabolites specifically proline and anthocyanin metabolism. *Plant Physiology and Biochemistry, 166*, 235–245. https://doi.org/10.1016/j.plaphy.2021.05.036

Shah, A. A., Yasin, N. A., Mudassir, M., Ramzan, M., Hussain, I., Siddiqui, M. H., Shabbir, Z., Ali, A., Ahmed, S., & Kumar, R. (2022). Iron oxide nanoparticles and selenium supplementation improve growth and photosynthesis by modulating antioxidant system and gene expression of chlorophyll synthase (CHLG) and protochlorophyllide oxidoreductase (POR) in arsenic-stressed *Cucumis melo*. *Environmental Pollution, 307*, 119413. https://doi.org/10.1016/j.envpol.2022.119413

Shamshir, F., Abbas, G., Amjad, M., Rizwan, M., Akram, M., Ahmad, S., Tahir, M., Ali, S., & Farooq, A. B. U. (2022). Physiological and biochemical characterization of Kalongi (*Nigella sativa*) against arsenic stress: Implications for human health risk assessment. *Environmental Pollution, 298*, 118829. https://doi.org/10.1016/j.envpol.2022.118829

Shirin, S., & Yadav, A. K. (2014). Physico-chemical analysis of municipal wastewater discharge in Ganga river, Haridwar district of Uttarakhand, India. *Current World Environment, 9*(2), 536–543. https://doi.org/10.12944/CWE.9.2.39

Simon, M., Joshi, H., Yadav, A. K., & Hopke, P. K. (2024). Simultaneous nitrogen and phosphorus removal from polluted water using a novel bacterial consortium (BACON-38): Promising application for wastewater remediation. *Journal of Water Process Engineering, 66*, 106044. https://doi.org/10.1016/j.jwpe.2024.106044

Singh, A., Gautam, P. K., Verma, A., Singh, V., Shivapriya, P. M., Shivalkar, S., Sahoo, A. K., & Samanta, S. K. (2020). Green synthesis of metallic nanoparticles as effective alternatives to treat antibiotics resistant bacterial infections: A review. *Biotechnology Reports, 25*, e00427. https://doi.org/10.1016/j.btre.2020.e00427

Sundararajan, N., Habeebsheriff, H. S., Dhanabalan, K., Cong, V. H., Wong, L. S., Rajamani, R., & Dhar, B. K. (2024). Mitigating global challenges: Harnessing green synthesized nanomaterials for sustainable crop production systems. *Global Challenges, 8*(1), 2300187. https://doi.org/10.1002/gch2.202300187

Tanveer, Y., Yasmin, H., Nosheen, A., Ali, S., & Ahmad, A. (2022). Ameliorative effects of plant growth promoting bacteria, zinc oxide nanoparticles and oxalic acid on *Luffa acutangula* grown on arsenic enriched soil. *Environmental Pollution, 300*, 118889. https://doi.org/10.1016/j.envpol.2022.118889

Ulhassan, Z., Bhat, J. A., Zhou, W., Senan, A. M., Alam, P., & Ahmad, P. (2022). Attenuation mechanisms of arsenic induced toxicity and its accumulation in plants by engineered nanoparticles: A review. *Environmental Pollution, 302*, 119038. https://doi.org/10.1016/j.envpol.2022.119038

Wei, C., Jiao, Q., Agathokleous, E., Liu, H., Li, G., Zhang, J., Fahad, S., & Jiang, Y. (2022). Hormetic effects of zinc on growth and antioxidant defense system of wheat plants. *Science of the Total Environment, 807*, 150992. https://doi.org/10.1016/j.scitotenv.2021.150992

Wu, K., Su, D., Liu, J., Saha, R., & Wang, J. P. (2019). Magnetic nanoparticles in nanomedicine: A review of recent advances. *Nanotechnology, 30*(50), 502003. https://doi.org/10.1088/1361-6528/ab4241

Wu, X., Hu, J., Wu, F., Zhang, X., Wang, B., Yang, Y., Shen, G., Liu, J., Tao, S., & Wang, X. (2021). Application of TiO2 nanoparticles to reduce bioaccumulation of arsenic in rice seedlings (*Oryza sativa* L.): A mechanistic study. *Journal of Hazardous Materials, 405*, 124047. https://doi.org/10.1016/j.jhazmat.2020.124047

Yan, Q., Zhang, Z., & Chen, Z. (2023). Microbial synthesized iron nanoparticles after recovering rare earth elements used for removing arsenic in mine groundwater. *Separation and Purification Technology, 327*, 124938. https://doi.org/10.1016/j.seppur.2023.124938

Yan, S., Wu, F., Zhou, S., Yang, J., Tang, X., & Ye, W. (2021). Zinc oxide nanoparticles alleviate the arsenic toxicity and decrease the accumulation of arsenic in rice (*Oryza sativa* L.). *BMC Plant Biology, 21*, 1–11. https://doi.org/10.1186/s12870-021-02929-3

Zeeshan, M., Hu, Y. X., Iqbal, A., Salam, A., Liu, Y. X., Muhammad, I., Ahmad, S., Khan, A. H., Hale, B., Wu, H. Y., & Zhou, X. B. (2021). Amelioration of AsV toxicity by concurrent application of ZnO-NPs and Se-NPs is associated with differential regulation of photosynthetic indexes, antioxidant pool and osmolytes content in soybean seedling. *Ecotoxicology and Environmental Safety, 225*, 112738. https://doi.org/10.1016/j.ecoenv.2021.112738

Zhao, X., Zhou, L., Riaz Rajoka, M. S., Yan, L., Jiang, C., Shao, D., Zhu, J., Shi, J., Huang, Q., Yang, H., & Jin, M. (2018). Fungal silver nanoparticles: Synthesis, application and challenges. *Critical Reviews in Biotechnology, 38*(6), 817–835. https://doi.org/10.1080/07388551.2017.1414141

Chapter 15
Challenges and Opportunities in Implementing Arsenic Technologies

S. C. Vetrivel, V. Sabareeshwari, K. C. Sowmiya, and V. P. Arun

Abstract Arsenic contamination in drinking water poses a significant public health risk, particularly in regions with high levels of natural arsenic. Implementing effective arsenic removal technologies is crucial to mitigate these risks. This chapter explores the multifaceted challenges and opportunities associated with the deployment of arsenic remediation technologies. The challenges encompass technical, economic, and sociocultural dimensions. Technical hurdles include variability in arsenic speciation and concentration, the presence of co-contaminants, and the need for low-cost, low-maintenance solutions suitable for rural and resource-limited settings. Economic challenges include the high initial costs of technology installation, ongoing maintenance expenses, and limited financial resources in affected communities. Sociocultural challenges highlight the need for community involvement, education, and adaptation of technologies to local contexts and preferences. Conversely, opportunities in this field are driven by advances in technology, such as the development of novel adsorbent materials and point-of-use systems, increased global awareness, and funding initiatives. Collaborative efforts between governments, nonprofits, and the private sector are critical to scaling successful implementations. By addressing these challenges and leveraging opportunities, sustainable arsenic mitigation can be achieved, improving public health outcomes and ensuring access to safe drinking water for affected populations.

S. C. Vetrivel (✉)
Department of Management Studies, Kongu Engineering College, Perundurai, India
e-mail: scvetrivel@gmail.com

V. Sabareeshwari
Department of Soil Science and Agricultural Chemistry, Amrita Vishwa Vidyapeetham, Coimbatore, India

K. C. Sowmiya
Research Department of Physics, Sri Vasavi College, Erode, India

V. P. Arun
Department of Business Administration, JKKN College of Engineering and Technology, Komarapalayam, India

A. K. Yadav and S. Shirin (eds.), *Emerging and Innovative Arsenic Removal Technologies for a Sustainable Future*, Water Science and Technology Library 131,
https://doi.org/10.1007/978-3-032-01176-3_15

Keywords Arsenic contamination · Drinking water · Arsenic removal technologies · Public health · Technical challenges · Economic challenges · Sociocultural challenges

15.1 Introduction

15.1.1 Background

Arsenic, a naturally occurring element found in the Earth's crust, has been a persistent and serious environmental contaminant. Its presence in groundwater is a major concern in many regions around the world, especially in areas with high levels of arsenic in soil and rock formations. Industrial activities, such as mining and smelting, and agricultural practices, including the use of arsenic-based pesticides, have exacerbated the problem (Agua Clara, 2016). Arsenic contamination primarily affects drinking water sources and poses significant health risks to populations that rely on such water. The severity of the problem varies by region, with South Asia, particularly Bangladesh and India, facing serious arsenic contamination issues. Addressing arsenic contamination requires an understanding of both the environmental sources and the technological solutions available to mitigate its impact.

15.1.2 Importance of Addressing Arsenic Contamination

The health impacts of arsenic contamination are profound and far-reaching. Chronic exposure to arsenic is associated with various health conditions, including cancer (especially skin, bladder, and lung cancers), cardiovascular disease, and neurological disorders. The World Health Organisation has classified arsenic as a Group 1 carcinogen. Beyond individual health, arsenic contamination has significant socioeconomic implications, affecting agricultural productivity, reducing quality of life, and putting a burden on health care systems (Aqualatina, 2017). Therefore, effective treatment of arsenic contamination is crucial not only for public health, but also for social and economic stability in affected regions. By developing and implementing effective arsenic mitigation technologies, communities can reduce health risks, improve quality of life, and foster sustainable development.

15.2 Arsenic Contamination: An Overview

15.2.1 Sources and Types of Arsenic Contamination

Arsenic contamination in the environment can originate from both natural and anthropogenic sources. Natural sources include the dissolution of arsenic-containing minerals from the Earth crust. In regions with arsenic-rich geological formations, groundwater can become contaminated as arsenic leaches from these rocks into aquifers (Bagla & Kaiser, 1996). This process is particularly prevalent in areas with high levels of sedimentary deposits or volcanic activity.

Anthropogenic sources of arsenic contamination are primarily related to industrial and agricultural activities. Mining operations, especially those that involve gold and other metals, can release arsenic into the environment. The smelting of ores and the burning of fossil fuels, such as coal, can also contribute to arsenic pollution through the emission of arsenic-containing particulates and gases (Berg et al., 2001). In agriculture, arsenic-based pesticides and fertilisers used historically have left lasting residues in soils, which can then leach into water sources. Additionally, improper disposal of industrial and agricultural waste can exacerbate arsenic contamination (Bissen & Frimmel, 2003). Arsenic contamination can take various forms, including inorganic and organic arsenic. Inorganic arsenic, mainly in the form of arsenite (As^{3+}) and arsenate (As^{5+}), is the most toxic and commonly found in contaminated groundwater. Organic arsenic, which includes compounds such as arsine (AsH_3) and arsonic acid, is less toxic but can still pose health risks. The distinction between these types of arsenic is crucial for determining the appropriate treatment and remediation strategies.

15.2.2 Health Impacts of Arsenic Exposure

Exposure to arsenic, particularly through drinking contaminated water, poses serious health risks. The severity of these health impacts is influenced by the level and duration of exposure, as well as individual susceptibility factors such as age, nutritional status, and pre-existing health conditions. Therefore, effective targeting of arsenic contamination is essential to mitigate these health risks and improve public health outcomes. Chronic exposure to arsenic has been linked to a range of serious health conditions:

- **Cancer**: Arsenic is classified as a Group 1 carcinogen by the International Agency for Research on Cancer (IARC). Long-term exposure is associated with an increased risk of several types of cancer, including skin, bladder, lung, and liver cancers (Chen et al., 1999). Skin lesions, such as hyperpigmentation and keratosis, are often among the first visible signs of arsenic-related cancer.

- **Cardiovascular Diseases**: Arsenic exposure is associated with cardiovascular conditions such as hypertension, arteriosclerosis, and heart disease. Chronic exposure can lead to atherosclerosis, increasing the risk of stroke and heart attacks.
- **Neurological Disorders**: Arsenic can negatively affect the nervous system, leading to cognitive deficits, peripheral neuropathy, and other neurological symptoms. Long-term exposure is particularly harmful to children and can affect their developmental and cognitive functions.
- **Respiratory Issues**: Inhalation of dust or fume inhalation can cause respiratory problems, including chronic cough, bronchitis, and asthma. Arsenic exposure has also been associated with increased risks of lung infections and chronic obstructive pulmonary disease (COPD).
- **Diabetes and Metabolic Disorders**: Research has shown that arsenic exposure may increase the risk of developing diabetes and metabolic syndrome (Davis & Edwards, 2014). The element can interfere with insulin signaling and glucose metabolism, contributing to the development of these conditions.
- **Reproductive and Developmental Effects**: Arsenic exposure during pregnancy can affect foetal development, leading to low birth weight, preterm delivery, and developmental delays in children. Long-term exposure can also affect reproductive health and increase the risk of miscarriage and birth defects.

15.3 Existing Arsenic Mitigation Technologies

15.3.1 Overview of Arsenic Removal Technologies

Arsenic removal technologies encompass a range of methods designed to reduce arsenic concentrations in water to safe levels. These technologies can be broadly categorised into physical, chemical, and biological methods, each with distinct mechanisms, advantages, and limitations.

- **Adsorption**: This method uses materials such as activated alumina, iron oxides, and various modified natural adsorbents to bind arsenic ions from water (Eawag, 2015). Adsorption is widely used because of its simplicity, effectiveness, and cost-efficiency. Granular ferric hydroxide (GFH) and granular activated carbon (GAC) are popular choices for household and community-level treatment systems.
- **Ion Exchange**: Ion exchange resins selectively remove arsenic ions from water by exchanging them with harmless ions. This method is highly effective in treating low to moderate arsenic concentrations and is commonly used in both household and industrial applications. However, it requires regular regeneration of the resins, which can be costly and complex.
- **Membrane Filtration**: Techniques such as reverse osmosis (RO) and nanofiltration (NF) utilize semi-permeable membranes to separate arsenic from water. These methods are highly effective at removing both organic and inorganic contaminants,

including arsenic. However, they require high operational pressure, energy, and maintenance, making them less feasible for low-income regions.

- **Coagulation-Flocculation**: This method involves adding coagulants (e.g., iron salts, aluminum salts) to water, which react with arsenic to form insoluble precipitates. These precipitates are then removed by sedimentation or filtration (Ghurye et al., 1999). This technique is effective in treating large volumes of water, but it can generate significant amounts of sludge that require proper disposal.
- **Electrochemical Methods**: Electrocoagulation and electrooxidation use electrical currents to induce the removal of arsenic from water. These methods are efficient and can be tailored for small- and large-scale applications. However, they require reliable electricity supply and careful operational control.
- **Biological Methods**: Bioremediation leverages microorganisms or plants to remove or transform arsenic into less toxic forms. This eco-friendly approach is still in the research and development phase for large-scale applications, but it shows promise for low-cost and sustainable treatment solutions.
- **Hybrid Systems**: Combining multiple methods can enhance the efficiency of arsenic removal and address the limitations of individual technologies. For example, coupling adsorption with membrane filtration can provide robust treatment for varying water qualities.

15.3.2 Traditional and Indigenous Methods

Traditional and indigenous methods for arsenic removal are rooted in local knowledge and practices, offering sustainable and culturally acceptable solutions, particularly in rural settings and limited resources. While traditional and indigenous methods may not always achieve the same removal efficiencies as modern technologies, they offer practical and accessible solutions that can be integrated into comprehensive water management strategies (Guo et al., 2009). Combining these methods with modern scientific knowledge can enhance their effectiveness and ensure sustainable arsenic mitigation in affected communities. These methods, although often less technologically advanced, play a crucial role in community-level water treatment.

- **Sand Filtration**: A common traditional method involves using sand filters, sometimes combined with layers of charcoal and gravel, to filter out arsenic and other contaminants. The bio-sand filter, which uses biological processes in a sand column, has been particularly effective in reducing arsenic levels and improving water quality.
- **Iron-Coated Sand**: Some communities have adapted sand filtration by using iron-coated sand, which enhances arsenic adsorption (Gutierrez, 2015). The iron oxides in the sand particles react with arsenic to form stable compounds that can be filtered out. This method takes advantage of locally available materials and is cost-effective.
- **Use of Natural Adsorbents**: Various natural materials, such as clays, laterite soil, and agricultural by-products (e.g., rice husk ash, coconut shell charcoal),

have been traditionally used to remove arsenic from water. These materials are often modified or activated to improve their adsorption capacities and are favoured because of their low cost and availability.

- **Sari Filtration**: In some regions, particularly in South Asia, simple cloth filtration using old saris has been found to reduce arsenic levels. The cloth acts as a prefilter, removing larger particles and some arsenic-laden compounds before the water undergoes further treatment.
- **Traditional herbal treatments**: Some indigenous practices involve the use of specific plants and herbs known for their medicinal and detoxifying properties (Habuda-Stanic et al., 2007). For example, *Moringa oleifera* seeds (drumstick tree) have been used for their coagulating properties to reduce the levels of turbidity and arsenic in water.
- **Community-Based Dug Wells**: Traditional community wells, if properly maintained and protected from surface contamination, can provide safer alternatives to shallow-tube wells, which are often more susceptible to arsenic contamination. These dug wells are typically deeper and can access groundwater with lower arsenic levels.

15.3.3 Modern Technological Solutions

Modern technological solutions for arsenic removal leverage advanced scientific principles and engineering designs to ensure effective and reliable treatment. Adsorption technologies, such as granular ferric hydroxide (GFH) and activated alumina, are widely used because of their high affinity for arsenic and ease of use in both household and large-scale systems. Membrane filtration methods, including reverse osmosis (RO) and nanofiltration (NF), offer a high removal efficiency for arsenic and other contaminants, although they require regular maintenance and can be expensive (Jain & Singh, 2012). Electrochemical methods such as electrocoagulation and electrooxidation use electrical currents to generate coagulants or oxidizing agents that effectively remove arsenic from water. Hybrid systems, which combine different technologies such as adsorption with membrane filtration, provide multiple barriers against contaminants and enhance overall treatment efficiency. Innovations such as titanium dioxide nanoparticles and photocatalytic oxidation represent the cutting edge of arsenic removal technologies, offering promising solutions for both current and future challenges (Table 15.1).

Table 15.1 Comparative analysis of various arsenic removal technologies

Technology	Mechanism	Efficiency	Cost	Maintenance	Scalability	Advantages	Limitations
Granular ferric hydroxide (GFH)	Adsorption	High	Moderate	Low to moderate	Small to medium	High adsorption capacity, regenerable	Requires disposal of spent media
Activated alumina	Adsorption	High for As(V)	Moderate	Low to moderate	Small to medium	Effective for arsenate, widely available	Less effective for arsenite, needs frequent regeneration
Reverse osmosis (RO)	Membrane filtration	Very high	High	High	Small to large	Removes multiple contaminants, reliable	High operational cost, energy-intensive
Nanofiltration (NF)	Membrane filtration	High	High	Moderate to high	Small to large	Effective for As(V), lower pressure than RO	Still requires significant maintenance
Ultrafiltration (UF)	Membrane filtration	Moderate	Moderate	Moderate	Small to large	Removes larger particles, low-cost	Less effective for dissolved arsenic
Electrocoagulation	Electrochemical	High	Moderate	Moderate	Small to large	Effective for both As(III) and As(V)	Requires power supply, operational control

(continued)

Table 15.1 (continued)

Technology	Mechanism	Efficiency	Cost	Maintenance	Scalability	Advantages	Limitations
Electrooxidation	Electrochemical	High	Moderate to High	Moderate to high	Small to large	Converts As(III) to As(V), then removable	Requires advanced infrastructure
Titanium dioxide nanoparticles	Adsorption (Nanotechnology)	High	High	Low to moderate	Small to medium	High surface area, highly reactive	Emerging technology, high initial cost
Combined adsorption and membrane filtration	Hybrid (Adsorption + Filtration)	Very high	High	Moderate to high	Small to large	Multiple barriers, enhanced performance	Complex system, higher overall cost
Photocatalytic oxidation	Photocatalysis (Nanotechnology)	High	High	Low to moderate	Small to medium	Innovative, effective for various contaminants	Emerging technology, requires UV light source

15.4 Technological Challenges in Arsenic Mitigation

15.4.1 Technical Limitations of Current Technologies

Despite advances in arsenic removal technologies, several technical limitations hinder their widespread implementation and effectiveness. One major limitation is the efficiency of arsenic removal across different types of water source (Jones et al., 2008). Technologies such as adsorption and membrane filtration are highly effective for specific types of arsenic, such as arsenate (As(V)), but less so for arsenite (As(III)), which requires preoxidation. This variability in effectiveness necessitates multistep treatment processes, increasing complexity and cost. Another significant technical challenge is the disposal of arsenic-laden waste. Adsorption technologies, for example, generate spent media that require safe disposal to prevent secondary contamination. Similarly, methods such as coagulation-flocculation produce sludge that must be appropriately managed. Inadequate disposal methods can lead to environmental hazards, negating the benefits of arsenic removal. The durability and maintenance of treatment systems also pose technical challenges. Membrane filtration systems, such as reverse osmosis (RO), are prone to fouling and scaling, which requires regular cleaning and replacement of membranes. This not only increases operational costs, but also demands technical expertise, which may not be readily available in low-resource settings (Katsoyiannis et al., 2015). Additionally, the energy requirements for processes such as RO and electrochemical treatments can be prohibitive, especially in areas with an unreliable power supply. Furthermore, the effectiveness of biological methods, such as bioremediation, is often limited by environmental factors such as temperature, pH, and the presence of competing ions. These factors can significantly impact the performance and reliability of these methods, making them less viable for large-scale, consistent applications.

15.4.2 Cost and Affordability Issues

Cost and affordability are significant barriers to the adoption of arsenic mitigation technologies, particularly in low-income and resource-limited regions, where the burden of arsenic contamination is often the highest. Many advanced technologies, such as reverse osmosis (RO) and nanofiltration (NF), involve high initial capital investment, operational costs, and maintenance expenses (Katsoyiannis et al., 2008). The need for specialised equipment and materials, such as high-quality membranes and adsorbents, further drives up the cost. Operational costs include the expenses for electricity, consumables, and skilled labour for maintenance and monitoring. For instance, membrane-based systems require regular replacement of membranes, which can be costly. Similarly, the regeneration or replacement of adsorbent materials such as granular ferric hydroxide (GFH) and activated alumina adds to the recurring expenses. These costs can be prohibitive for small communities and households, making it difficult to sustain the long-term operation of such systems. In many developing regions, financial constraints limit the availability of funds for both the installation and maintenance of arsenic removal technologies. Public funding and subsidies can be inadequate, and private investment is often deterred by the low financial returns and high risks associated with these projects. The lack of affordable financing options and financial support mechanisms further exacerbates the situation (Le et al., 2007). Furthermore, the cost of safe disposal of arsenic-laden waste products, such as sludge from coagulation-flocculation processes or spent adsorption media, adds another layer of financial burden. Improper disposal practices can lead to environmental contamination, creating additional public health risks and remediation costs. The challenge of cost and affordability is compounded by socioeconomic factors, such as low household incomes and limited access to financial resources. In rural and underserved areas, communities can prioritize immediate needs, such as food and basic healthcare, over investing in water treatment technologies. This economic reality underscores the need for cost-effective and sustainable solutions that can be tailored to the financial capacities of affected populations. Addressing cost and affordability issues requires a multifaceted approach, including the development of low-cost technologies, financial subsidies, community-based funding mechanisms, and public–private partnerships. Innovations that reduce operating and maintenance costs, improve system efficiency, and enhance the durability of arsenic removal technologies are critical to making these solutions more accessible and sustainable for communities in need.

15.4.3 Maintenance and Sustainability Challenges

Adapting arsenic mitigation technologies to local conditions is crucial for their effectiveness, sustainability, and acceptance by the communities they serve (Malik et al., 2009). Local conditions include a variety of factors such as the specific characteristics of arsenic contamination, socioeconomic status, cultural practices, availability of resources, and environmental conditions. One of the primary considerations is the specific type of arsenic contamination present in the water supply. In many regions, both arsenite (As(III)) and arsenate (As(V)) are present, requiring treatment methods that can effectively target both forms. Technologies must be tailored to local water chemistry, including pH, the presence of competing ions, and other contaminants that could interfere with arsenic removal processes. Socioeconomic factors play an important role in technology adaptation. In low-income areas, cost-effective and low-maintenance solutions are essential. Technologies that require minimal financial investment, are easy to operate, and do not require frequent maintenance are more likely to be adopted and sustained. For instance, traditional and indigenous methods using locally available materials, such as iron-coated sand or biosand filters, can be highly effective and affordable options (Meng et al., 2000). Cultural practices and community participation are also critical to successful adaptation. The involvement of local communities in the selection, implementation, and maintenance of arsenic removal technologies ensures that the solutions are culturally acceptable and meet the specific needs of the people. Education and awareness programs can help communities understand the risks of arsenic contamination and the benefits of using the technologies provided, fostering a sense of ownership and responsibility. The availability of local resources, including materials, technical expertise, and infrastructure, influences the choice of arsenic removal technologies. In areas with limited access to electricity, energy-intensive methods such as reverse osmosis (RO) might not be feasible, whereas gravity-based systems or solar-powered options might be more suitable. The use of locally sourced materials for adsorbents or the construction of treatment units with local labor can reduce costs and enhance sustainability. Environmental conditions, such as climate and geography, must also be considered (Rodriguez-Lado et al., 2013; Shirin & Yadav, 2021). For example, in regions with seasonal variations in water availability, technologies must be resilient to changes in water quality and quantity. Rainwater harvesting combined with arsenic treatment systems can provide a reliable water source throughout the year. In areas prone to natural disasters, robust and portable treatment units can ensure the continuity of safe water supply (Fig. 15.1).

Fig. 15.1 Maintenance and sustainability challenges

15.5 Socioeconomic and Cultural Barriers

15.5.1 Community Awareness and Acceptance

Community awareness and acceptance are critical for the successful implementation of arsenic mitigation technologies. The lack of awareness of the health risks associated with arsenic contamination often leads to resistance or indifference to the adoption of new technologies. Educational campaigns and community engagement initiatives are essential to inform the population about the dangers of arsenic exposure and the benefits of using arsenic removal technologies (Rowland et al., 2011). These campaigns should be culturally sensitive and use local languages and communication channels to reach a broad audience. Additionally, involving community leaders and stakeholders in the planning and implementation process can foster trust and acceptance, making it easier to introduce and sustain new technologies.

15.5.2 Socio-economic Factors Affecting Implementation

Socioeconomic factors have a significant impact on the adoption and sustainability of arsenic mitigation technologies. In many affected regions, poverty and limited financial resources limit the ability of households and communities to invest in and maintain arsenic removal systems. High initial costs and ongoing maintenance expenses can be prohibitive, especially in low-income areas (Sancha, 2006). Furthermore, access to funding and financial support is often limited, reducing the feasibility of long-term solutions. Addressing these issues requires the development of low-cost and efficient technologies and the provision of financial assistance or subsidies to make these solutions accessible (Kumar et al., 2023; Singh et al., 2015). Microfinance options and community financing mechanisms can also play a vital role in overcoming economic barriers.

15.5.3 Cultural Beliefs and Practices

Cultural beliefs and practices can influence the acceptance and effectiveness of arsenic mitigation efforts. In some communities, traditional beliefs and practices around water use and purification can conflict with modern technological solutions. For instance, certain communities may have long-standing practices of using specific water sources that are culturally significant, even if those sources are contaminated (Tresintsi et al., 2013a). To address this, it is important to integrate traditional knowledge with modern science, creating hybrid solutions that respect and incorporate local customs. Engaging community leaders and incorporating culturally appropriate methods can help bridge the gap between traditional practices and modern technology, ensuring that solutions are accepted and sustained.

15.5.4 Gender-Specific Challenges

Gender-specific challenges play a crucial role in the implementation of arsenic mitigation technologies. In many rural and low-income communities, women are primarily responsible for water collection and management (Tresintsi et al., 2013b). However, they often have limited decision-making power and access to resources, hindering their ability to adopt and maintain arsenic removal technologies. Gender biases and inequalities can further exacerbate these challenges, making it difficult for women to participate in training and educational programmes related to water purification. Addressing these challenges requires a gender-sensitive approach that empowers women through targeted education, capacity building, and involvement in decision-making processes. Ensuring that women have equal access to resources and opportunities is essential for the successful adoption and sustainability of arsenic mitigation technologies.

15.6 Policy and Regulatory Framework

15.6.1 International Guidelines and Standards

International guidelines and standards play a crucial role in shaping national policies and practices related to arsenic mitigation. Organisations such as the World Health Organisation (WHO) and the United Nations (UN) set benchmark guidelines for arsenic levels in drinking water, which are widely adopted by countries around the world. The WHO, for example, recommends a maximum arsenic concentration of 10 microgrammes per litre (μg/L) in drinking water (Tubic et al., 2010). These guidelines are based on comprehensive research and aim to protect public health by minimizing exposure to arsenic. International standards also provide a framework for developing and evaluating arsenic removal technologies. Standards set by the International Organization for Standardization (ISO) and other bodies ensure that technologies are effective and reliable. Compliance with these standards helps to build trust and ensure the safety of arsenic mitigation solutions.

15.6.2 National Policies and Regulations

National policies and regulations are essential for effective management and mitigation of arsenic contamination. These policies typically encompass standards for permissible arsenic levels, guidelines for water quality monitoring, and frameworks for the implementation of mitigation strategies. Countries with significant arsenic contamination issues often have specific regulations tailored to address local conditions and challenges (United Nations, 2016). For example, in countries like Bangladesh and India, where arsenic contamination is a serious problem, governments have established national action plans that include extensive testing, identification of affected areas, and the deployment of arsenic removal technologies. These plans often involve multisectoral collaboration between government agencies, research institutions, and non-governmental organizations (NGOs) to ensure comprehensive coverage and effective implementation. National policies also play a critical role in the allocation of funds and resources. Governments may provide subsidies or financial assistance for the development and deployment of arsenic mitigation technologies, ensuring that even low-income communities can access safe drinking water.

15.6.3 Role of Government and Non-governmental Organizations

Both government and non-governmental organizations (NGOs) are pivotal in the fight against arsenic contamination. Governments are responsible for setting and enforcing regulations, providing funding and resources, and coordinating national efforts to address the issue. They conduct large-scale water quality tests, identify contaminated areas, and oversee the deployment of mitigation technologies. Government agencies often collaborate with local authorities to ensure that policies are implemented effectively at the grassroots level. NGOs, on the other hand, often play a complementary role by filling gaps in government efforts, particularly in remote or underserved areas (Wilkie & Hering, 1998). They are instrumental in raising awareness about arsenic contamination, educating communities about health risks, and training local populations on the use and maintenance of arsenic removal technologies. NGOs also advocate for policy changes and increased funding, leveraging their grassroots connections to bring attention to the needs of affected communities. Successful arsenic mitigation efforts typically involve strong partnerships between governments and NGOs, ensuring that policies are not only implemented, but also adapted to the local context and needs.

15.7 Opportunities for Improvement and Innovation

15.7.1 Technological Innovations and Emerging Solutions

Technological innovation is crucial to advance arsenic mitigation strategies. Recent developments have introduced promising solutions to address the limitations of traditional arsenic removal methods. For instance, advanced materials such as nanoscale zerovalent iron and graphene-based adsorbents have shown exceptional efficacy in removing arsenic from contaminated water sources. These materials offer higher removal capacities and faster kinetics in comparison to conventional adsorbents. Furthermore, emerging electrochemical technologies, such as electrodialysis and electrocoagulation, provide energy-efficient alternatives for arsenic treatment (Zouboulis & Katsoyiannis, 2002). These methods not only improve arsenic removal but also reduce the operational costs associated with large-scale water treatment facilities. Furthermore, integration of artificial intelligence (AI) and machine learning in monitoring and control systems can improve the accuracy of arsenic detection and optimise the performance of treatment technologies. The continuous advancement in sensor technology also holds promise, enabling real-time monitoring and early detection of arsenic contamination, which is vital for timely intervention.

15.7.2 Community Engagement and Education

Effective arsenic mitigation extends beyond technological solutions to include community engagement and education. Raising awareness of the health risks associated with arsenic exposure and the available mitigation technologies is essential for community participation. Educational programmes should focus on informing local populations about safe water practices and the importance of regular testing for arsenic levels. Community-based participatory approaches, where local stakeholders are actively involved in decision making processes, can enhance the adoption of arsenic removal technologies and ensure that solutions are culturally appropriate and accepted (Yadav et al., 2011). Workshops, seminars, and school programs can play an important role in educating individuals about the importance of arsenic contamination and the benefits of preventive measures. Empowering communities with knowledge and tools allows them to take proactive steps in managing and mitigating arsenic exposure, thus fostering a collaborative approach to addressing this critical issue.

15.7.3 Public–Private Partnerships

Public–private partnerships (PPPs) represent a powerful mechanism for advancing arsenic mitigation efforts. Collaborations between government agencies, private companies, and nongovernmental organizations can lead to the development and deployment of innovative technologies and solutions. Private sector involvement can bring technical expertise, financial resources, and operational efficiency, while public institutions can provide regulatory support, infrastructure, and policy frameworks. Successful PPPs can facilitate the large-scale implementation of arsenic removal technologies by using combined resources and expertise. For example, joint ventures between technology developers and water utilities can accelerate the deployment of cutting-edge filtration systems in affected regions. In addition, partnerships can drive research and development initiatives, leading to new innovations in arsenic treatment and monitoring. By fostering collaboration between various stakeholders, PPPs can address the multifaceted challenges of arsenic contamination and improve the overall effectiveness of mitigation strategies.

15.7.4 Funding and Investment Opportunities

The secure acquisition of adequate funding and investment is critical to the successful implementation and scaling of arsenic mitigation technologies. Public funding sources, including government grants and international aid, play a vital role in supporting research and development, as well as in deploying new technologies in affected areas. In addition, private investments from corporations and philanthropic organizations can provide the capital necessary to advance innovative solutions and expand their reach. Establishing funding mechanisms such as impact investment funds and green bonds can attract capital specifically earmarked for environmental and health-related projects. Additionally, fostering partnerships with financial institutions can facilitate access to low-interest loans and other financial instruments designed to support large-scale infrastructure projects. By creating a robust funding ecosystem and encouraging investment in arsenic mitigation efforts, stakeholders can ensure the development and widespread adoption of effective technologies, ultimately leading to improved public health outcomes and environmental protection.

15.8 Case Studies and Best Practices

15.8.1 Successful Implementations

The successful implementation of arsenic mitigation technologies often depends on a combination of effective technology, community participation, and supportive policy frameworks. Examining successful case studies provides valuable insight and best practices that can be adapted to other regions facing similar challenges.

15.8.1.1 Bangladesh's Arsenic Mitigation Programme

In Bangladesh, the problem of arsenic contamination of groundwater has been addressed through a comprehensive national strategy that includes both technological and community-based solutions. The government, in collaboration with international organizations such as the World Bank and UNICEF, implemented a large-scale arsenic mitigation program. This program included the installation of low-cost community-based arsenic removal units, such as the modified version of the activated alumina filter and the use of iron-based filtration systems. Additionally, the programme emphasised community education on arsenic risks and safe water practices. Public health campaigns, coupled with training for local technicians, helped ensure the proper use and maintenance of these systems. The success of this programme can be attributed to the integration of technological innovation with strong community involvement and sustained government support, resulting in improved water quality for millions of people.

15.8.1.2 The Arsenic Removal Project in West Bengal, India

West Bengal, India, faced severe arsenic contamination in its groundwater sources, prompting a targeted intervention. The state government, in collaboration with NGOs and private sector partners, initiated the Arsenic Removal Project, which utilised a combination of technologies including the use of iron-activated alumina filters and reverse osmosis systems. A key aspect of this project was its focus on decentralization: installing small-scale treatment units at the village level. This approach not only made technology more accessible but also empowered local communities to take ownership of the maintenance and operation of the systems. Furthermore, the project included a strong component of community engagement, involving local leaders in educational campaigns on arsenic risks and water safety. Integration of technology with local knowledge and active participation of the community led to significant improvements in water quality and public health.

15.8.1.3 The Use of Solar-Powered Arsenic Removal Systems in Nepal

In Nepal, remote and rural areas faced challenges with arsenic contamination and limited access to electricity. To address this, the Solar Arsenic Removal Project was implemented, which incorporated solar-powered filtration systems. These systems used a combination of adsorption and filtration technologies, specifically designed to operate independently of the electrical grid. Solar-powered units were installed in community centers, where they provided clean and arsenic-free water to local populations (Vetrivel et al., 2024). The project was successful due to its innovative use of renewable energy, which provided a sustainable solution in areas with limited infrastructure. Furthermore, the project included extensive training for local operators and ongoing support to ensure long-term functionality of the systems. This case highlights how the combination of technology with renewable energy solutions can address both arsenic contamination and infrastructure challenges in remote areas.

15.8.1.4 The "Arsenic Safe" Initiative in Vietnam

Vietnam's "Arsenic Safe" initiative is another example of successful arsenic mitigation. This project involved the development and deployment of low-cost, household-scale filtration systems using locally available materials. The initiative was driven by a collaboration between the government, academic institutions and local NGOs. The project focused on creating affordable and easy-to-use solutions that could be rapidly deployed in affected communities. Educational programmes were also integral, as they helped raise awareness about the dangers of arsenic and trained users on how to properly maintain and use the filters. The success of the "Arsenic Safe" initiative can be attributed to its emphasis on affordability, local material use, and community involvement, demonstrating that even low-cost solutions can have a significant impact when effectively implemented.

15.9 Future Directions and Recommendations

15.9.1 Research and Development Priorities

Future research and development (R&D) should focus on several key areas to advance arsenic mitigation technologies. First, the development of more efficient and cost-effective arsenic removal materials is crucial. Innovations in nanomaterials, advanced composites, and hybrid systems can offer improved performance in terms of removal capacity and speed. Research should also explore the integration of arsenic removal technologies with other water treatment processes, such as desalination and purification, to enhance the overall efficiency of the system. Additionally, the development of inexpensive and user-friendly arsenic detection methods in drinking water can improve monitoring and early warning systems. It is also essential to investigate the long-term impacts of various treatment technologies on water quality and human health. Collaborative efforts between researchers, engineers, and public health experts can help identify and address gaps in current technologies and lead to the development of novel solutions that are both effective and sustainable.

15.9.2 Enhancing Technology Transfer and Adoption

To enhance technology transfer and adoption, it is important to create a supportive environment that facilitates the dissemination of innovations from research laboratories to real-world applications. This involves establishing clear pathways for scaling up successful pilot projects and facilitating partnerships between technology developers, governments, and local stakeholders. Engaging with local communities early in the process can ensure that technologies are adapted to meet specific needs and cultural contexts. Additionally, providing technical training and support to local operators and maintenance personnel is crucial for ensuring the long-term success of new technologies. Creating platforms for knowledge sharing and collaboration between countries and regions can also accelerate the adoption of effective solutions. When these factors, the transition from research to practical implementation can be more seamless, leading to widespread adoption and impact.

15.10 Conclusion

The implementation of arsenic mitigation technologies presents a complex array of challenges and opportunities. Although advances in technology offer promising solutions to remove arsenic from contaminated water, these innovations must be paired with effective community engagement, robust policy frameworks, and sustainable operational practices to ensure their success. Challenges such as technical limitations, high costs, and maintenance issues must be addressed through ongoing research and development, coupled with strategies to improve technology transfer and adoption. Opportunities lie in leveraging public–private partnerships, increasing funding and investment, and fostering a collaborative approach to technology implementation. By addressing these challenges and capitalising on available opportunities, stakeholders can improve the effectiveness and reach of arsenic mitigation efforts, ultimately protecting public health and ensuring access to safe drinking water for affected communities.

Acknowledgements The authors sincerely thank our prestigious institute and other people who are involved in this study for their direct and indirect contributions.

References

Agua Clara. (2016). Atima plant. http://monitor.wash4all.org/ (July 6, 2024).

Aqualatina. (2017). Qualità dell'acqua per Comune. http://www.acqualatina.it/qualita-dellacqua/qualita-dellacqua-per-comune/

Bagla, P., & Kaiser, J. (1996). Epidemiology—India's spreading health crisis draws global arsenic experts. *Science, 274*(5285), 174–175.

Berg, M., Tran, H. C., Nguyen, T. C., Pham, H. V., Schertenleib, R., & Giger, W. (2001). Arsenic contamination of groundwater and drinking water in Vietnam: A human health threat. *Environmental Science & Technology, 35*(13), 2621–2626.

Bissen, M., & Frimmel, F. H. (2003). Arsenic—A review. Part II: Oxidation of arsenic and its removal in water treatment. *Acta Hydrochimica et Hydrobiologica, 31*(2), 97–107.

Chen, H. W., Frey, M. M., Clifford, D., McNeill, L. S., & Edwards, M. (1999). Arsenic treatment considerations. *Journal of the American Water Works Association, 91*(3), 74–85.

Davis, C. C., & Edwards, M. (2014). Coagulation with hydrolyzing metal salts: Mechanisms and water quality impacts. *Critical Reviews in Environmental Science and Technology, 44*(4), 303–347.

Eawag (Swiss Federal Institute of Aquatic Science and Technology). (2015). *Geogenic contamination handbook—Addressing arsenic and fluoride in drinking water.*

Ghurye, G. L., Clifford, D. A., & Tripp, A. R. (1999). Combined arsenic and nitrate removal by ion exchange. *Journal of the American Water Works Association, 91*(10), 85–96.

Guo, X., Wu, Z., & He, M. (2009). Removal of antimony(V) and antimony(III) from drinking water by coagulation-flocculation-sedimentation (CFS). *Water Research, 43*(17), 4327–4335.

Gutierrez, S. C. (2015). *Arsenic in drinking water: An overview of U.S. regulation and removal technologies.* U.S. Environmental Protection Agency.

Habuda-Stanic, M., Kules, M., Kalajdzic, B., & Romic, Z. (2007). Quality of groundwater in eastern Croatia. The problem of arsenic pollution. *Desalination, 210*(1–3), 157–162.

Jain, C. K., & Singh, R. D. (2012). Technological options for the removal of arsenic with special reference to South East Asia. *Journal of Environmental Management, 107*, 1–18.

Jones, H., et al. (2008). Case reports: Arsenic pollution in Thailand, Bangladesh, and Hungary. In D. M. Whitacre (Ed.), *Reviews of environmental contamination* (pp. 163–187). Springer.

Katsoyiannis, I. A., Mitrakas, M., & Zouboulis, A. I. (2015). Arsenic occurrence in Europe: Emphasis in Greece and description of the applied full-scale treatment plants. *Desalination and Water Treatment, 54*(8), 2100–2107.

Katsoyiannis, I. A., Zikoudi, A., & Hug, S. J. (2008). Arsenic removal from groundwaters containing iron, ammonium, manganese and phosphate: A case study from a treatment unit in northern Greece. *Desalination, 224*(1–3), 330–339.

Kumar, A., Chava, R., Gupta, S., Shirin, S., Jamal, A., & Yadav, A. K. (2023). Environmental and health effects of fluoride contamination and treatment of wastewater using various technologies. In A. K. Yadav, S. Saba, & V. P. Singh (Eds.), *Advanced treatment technologies for fluoride removal in water* (pp. 323–341). Springer Nature Switzerland. https://doi.org/10.1007/978-3-031-38845-3_18

Le, V. C., et al. (2007). Arsenic removal from groundwater by a combination of chlorination and consecutive sand filtration. In *Core university program between Japan society for the promotion of science (JSPS) and Vietnamese academy of science and technology (VAST)* (pp. 343–347).

Malik, A. H., Khan, Z. M., Mahmood, Q., Nasreen, S., & Bhatti, Z. A. (2009). Perspectives of low cost arsenic remediation of drinking water in Pakistan and other countries. *Journal of Hazardous Materials, 168*(1), 1–12.

Meng, X. G., Bang, S., & Korfiatis, G. P. (2000). Effects of silicate, sulfate, and carbonate on arsenic removal by ferric chloride. *Water Research, 34*(4), 1255–1261.

Rodriguez-Lado, L., et al. (2013). Groundwater arsenic contamination throughout China. *Science, 341*(6148), 866–868.

Rowland, H. A. L., et al. (2011). Geochemistry and arsenic behaviour in groundwater resources of the Pannonian Basin (Hungary and Romania). *Applied Geochemistry, 26*(1), 1–17.

Sancha, A. M. (2006). Review of coagulation technology for removal of arsenic: Case of Chile. *Journal of Health, Population and Nutrition, 24*(3), 267–272.

Shirin, S., & Yadav, A. K. (2021). Assessment and suitability for the dumping of flyash in an abandoned mine. *Environmental Challenges, 4*, Article 100111. https://doi.org/10.1016/j.envc.2021.100111

Singh, R., Singh, S., Parihar, P., Singh, V. P., & Prasad, S. M. (2015). Arsenic contamination, consequences and remediation techniques: A review. *Ecotoxicology and Environmental Safety, 112*, 247–270.

Tresintsi, S., Simeonidis, K., Zouboulis, A., & Mitrakas, M. (2013a). Tetravalent manganese feroxyhyte: A novel nanoadsorbent equally selective for As(III) and As(V) removal from drinking water. *Environmental Science & Technology, 47*(17), 9699–9705.

Tresintsi, S., Simeonidis, K., Zouboulis, A., & Mitrakas, M. (2013b). Comparative study of As(V) removal by ferric coagulation and oxy-hydroxides adsorption: Laboratory and full-scale case studies. *Desalination and Water Treatment, 51*(13–15), 2872–2880.

Tubic, A., Dalmacija, B., Agbaba, J., Ivancev-Tumbas, I., Klasnja, M., & Dalmacija, M. (2010). Tracking disinfection by-products and arsenic removal during various drinking water treatment trains. *Water Science and Technology, 61*(12), 3169–3177.

United Nations. (2016). Sustainable development goal 6. https://sustainabledevelopment.un.org/sdg6 (July 6, 2016).

Vetrivel, S. C., Sowmiya, K. C., Sabareeshwari, V., & Arun, V. P. (2024). Human resources optimization for public space security: A GANs approach. In S. Ponnusamy, J. Antari, P. Bhaladhare, A. Potgantwar, & S. Kalyanaraman (Eds.), *Enhancing security in public spaces through generative adversarial networks (GANs)* (pp. 274–295). IGI Global. https://doi.org/10.4018/979-8-3693-3597-0.ch019

Wilkie, J. A., & Hering, J. G. (1998). Rapid oxidation of geothermal arsenic(III) in streamwaters of the eastern Sierra Nevada. *Environmental Science & Technology, 32*(5), 657–662.

Yadav, I. C., Dhuldhaj, U. P., Mohan, D., & Singh, S. (2011). Current status of groundwater arsenic and its impacts on health and mitigation measures in the Terai basin of Nepal: An overview. *Environmental Reviews, 19*, 55–67.

Zouboulis, A., & Katsoyiannis, I. (2002). Removal of arsenates from contaminated water by coagulation-direct filtration. *Separation Science and Technology, 37*(12), 2859–2873.

Chapter 16
Current Research and Future Directions in Affordable Arsenic Removal Technologies

Praveen Kumar, Ashish Kapoor, and Muthukumar Raghunathan

Abstract Arsenic pollution in water for human consumption is a major public health hazard worldwide, especially in developing countries where access to potent water treatment facilities is poor. This chapter reviews ongoing research lines and future developments in finding affordable technologies for arsenic removal, covering socioeconomic and health impacts and limitations of such treatments present today. Focus is given to low-cost and cost-effective approaches that can be applied in low resource settings. New developments from the field of materials science have resulted in several options for arsenic remediation. Novel materials- new adsorbents, nonconventional iron species, i.e. biochar, iron oxide coated sands, nanomaterials gave arsenic removal up-gradation, efficiencies. Bioremediation techniques like phytoremediation using arsenic-accumulator plants and microbes hold good promise for environment-friendly options. The use of membrane-based technologies for the removal of arsenic is also discussed. The integration of these technologies into existing water infrastructure is explored and written with modular and scalable designs that can be adapted to the various needs of communities. The importance of interdisciplinary research in bringing laboratory-scale innovation into field applicability is also highlighted. Research directions for the future are identified, and areas of process intensification for more effective treatment of arsenic pollution are outlined. Extensive field studies and pilot project are required to test the laboratory research outcome and optimize the transfer of technology to practical application. In addition, the incorporation of socioeconomic elements into the design and application of arsenic removal technologies is suggested, so that they might be sustainable and acceptable to communities in affected areas.

P. Kumar · M. Raghunathan (✉)
Department of Biotechnology, School of Bioengineering, SRM Institute of Science and Technology, Kattankulathur, Chennai, Tamil Nadu, India
e-mail: muthukur@srmist.edu.in

A. Kapoor
Department of Chemical Engineering, Harcourt Butler Technical University, Kanpur, Uttar Pradesh, India

A. K. Yadav and S. Shirin (eds.), *Emerging and Innovative Arsenic Removal Technologies for a Sustainable Future*, Water Science and Technology Library 131,
https://doi.org/10.1007/978-3-032-01176-3_16

Keywords Arsenic · Contamination · Drinking water · Low-cost removal technologies

16.1 Introduction

Arsenic is one of the most abundant elements on Earth and is toxic to terrestrial and marine life at certain concentrations. It has a silvery gray appearance and is a hard, brittle crystalline solid (Yin et al., 2017). Arsenic has a boiling point of 614 °C and a melting point of 817 °C. Its specific gravity is 5.73 g/cm^3, and its atomic weight is 74.9 g/mol. Water contamination, food grown in soils polluted with arsenic or irrigated with contaminated water, and other forms of environmental arsenic pollution are major problems caused by human activities and pose serious health risks to millions of people (Zakhar et al., 2018). Arsenate contamination of drinking water in sedimentary aquifers has become a global issue due to the far-reaching effects of arsenic (As) poisoning. The World Health Organization (WHO) and the United States Environmental Protection Agency (USEPA) have established a safe limit of arsenate in drinking water at 10 μg/l, as arsenate is harmful to human organ systems. Health studies carried out in the Netherlands have shown that the standard value of 10 μg/l present in drinking water is unsafe for human consumption and can cause poisoning. Furthermore, the Dutch government aims to reduce arsenic levels to below 1 μg/l. It is cost-effective to reduce arsenate levels in drinking water to below 1 μg/l for health reasons. It is reasonable to try to reduce the contaminant level to less than 1 μg/l due to the benefits for people with health problems (Ahmad & Bhattacharya, 2019). To protect potentially vulnerable populations, current inorganic toxicity guidelines need to be revised based on new data from in vitro and field studies. Given the unknown consequences of lower concentrations and the possibility of other exposure sources, it is prudent to limit the arsenic (As) levels in drinking water to the greatest extent feasible. Humans and other animals may not be completely protected against concentrations in drinking water that reach 10 μg/L. Human arsenic exposures have been associated with its accumulation in keratin-rich regions, leading to clinical manifestations such as hyperpigmentation, keratosis, and several types of cancer (Kapaj et al., 2006). Serious health and socioeconomic implications can result from the presence of ions such as arsenic, fluoride, and uranium. Two occurrences of Klinefelter's disease were found as a result of the effect that arsenic-enriched water had on human chromosomes, according to the research (Singh et al., 2013). Arsenic poisoning of surface and groundwater can be attributed to both natural and human-made sources; however, the biggest threat to human populations across the world comes from the former. Arsenic contamination in natural environmental settings is greatly influenced by the hydrological, geochemical, and geological features of aquifers. To reduce the negative impact, it is necessary to understand these aquifer characteristics in different environmental contexts to create cost-effective arsenic removal technologies (Singh et al., 2021).

16.2 Exposure to Arsenic (As) and Health Risks

Worldwide, arsenic's natural or geogenic occurrence poses a threat to biotic health in a variety of ways. Because it does not participate in any beneficial metabolic processes, inorganic arsenic in water can cause cancer, skin disorders, vascular diseases, neurological disorders, and other illnesses. People who use groundwater as their main source are at risk due to the high levels of arsenic in it (Podgorski & Berg, 2020). Because it contaminates soils, it also threatens agricultural activities. Groundwater is the main source of agriculture and drinking in irrigated semi-arid and desert areas, where hazardous exposure has affected millions of people globally. Water contamination is an important source of inorganic arsenic in the diet; rice is a typical food item that contains this poisonous element. Prolonged exposure to water with high concentrations of arsenic concentrations (> 100 mg/L) has been associated with non-melanoma skin, lung, and bladder cancers. Well water contaminated with arsenic at 10 mg/L poses a threat to drinking water of 20–45 million people. This is particularly evident in Bangladesh, where arsenic is responsible for the deaths of one adult in every eighteen (Zeng et al., 2020). There is no industry that can solve this worldwide problem. A cost–benefit analysis led by WHO and other groups to recommend a federal maximum contamination level (MCL) for arsenic of 10 mg/L. However, because there are not enough fresh water sources in some nations, the allowed maximum is 50 mg/L. The large disparity in MCLs is concerning. More recent studies on human health have shown that even at 10 mg/L, it may not be safe, particularly for developing organisms. Harming human cells and genetic material, arsenic, a carcinogen, has the potential to cause serious diseases. The harmful effects of inorganic arsenicals are due to their ability to bind to sulfhydryl groups on proteins and to replace phosphate groups with arsenate. Some of the health issues associated with them include cancer, heart disease, lung disease, immune system disorders, genotoxicity, arsenics, skin infections, neurological disorders, liver disease, gastrointestinal abnormalities, neurological disruptions, and endocrine disruptions (Liu et al., 2018). Inorganic arsenic can cause cancer, skin diseases, and gastrointestinal problems when it is methylated. Research shows that arsenic, even at low doses (< 50 mg/L), can affect neurodevelopment in early life and cause birth abnormalities, according to empirical investigations. An increased chance of developing diabetes in the future may be due to exposure to arsenic during pregnancy, which alters gene expression pathways related to the disease. A genome-wide DNA methylation study that included 396 participants in the Health consequences of Arsenic (HEALS) (Demanelis et al., 2019) sheds light on the epigenetic consequences of arsenic poisoning.

16.3 Physico-chemical and Emerging for Arsenic Removal Technologies

16.3.1 *Adsorption*

When molecules in a gas or liquid attach themselves to a solid or liquid surface by binding, this process is called adsorption. Physical and chemical adsorptions are differentiated by the nature of the contact. Conditions such as surface area, porosity, temperature, and pressure all play a role in adsorption. Two hallmarks of physical adsorption are energy release and reversible binding. In most cases, van der Waals forces are responsible for the adhesion of many adsorbate layers to the surface of the adsorbent, as shown in Fig. 16.1.

Because of its simple procedure, efficacy, and low cost, this approach has been widely used to remove arsenic. Several parameters influence arsenic adsorption, such as arsenic concentration, oxidation state, temperature, medium pH, competing ions, and contact time (Dabrowski, 2001). In most situations, arsenate shows a stronger affinity for particular adsorbents than arsenate. Arsenic adsorption onto different adsorbents may be better understood by the use of adsorption isotherm models such as Freundlich and Langmuir. When it comes to the processes of metal and metalloid mobilization and sorption, adsorption isotherms give the most accurate picture.

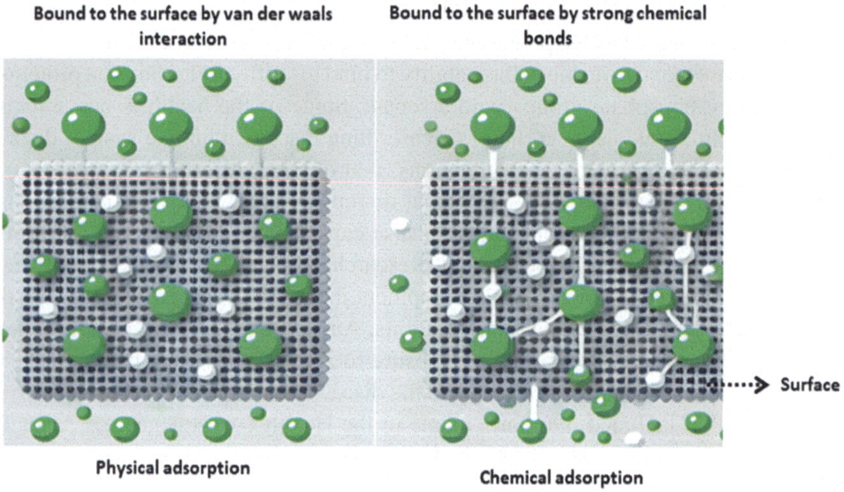

Fig. 16.1 A diagrammatic representation of adsorptive phenomenon for arsenic decontamination

16.3.2 Arsenic Removal Using Nanomaterials

Iron filings can be utilized to remove arsenic 100 times or more because of their strong absorption capabilities. Effective arsenic scavengers include biologically produced iron and manganese oxides, and adsorption processes can be facilitated by dissolved iron concentrations of 100–300 mg/L (Ramaswami et al., 2001). This procedure is ideal for large-scale applications that want to provide arsenic-free water, since it is efficient, simple, and uses recyclable iron filings. This method works flawlessly without oxygenation, even when dissolved sulfate ions are present. Among their suggestions were inexpensive yet effective adsorbents, synthetic and commercial activated carbons, and a number of other possibilities. Getting arsenic removal levels up to WHO guidelines has been the subject of much research and development. Relatedly, Hajji et al. investigated arsenic and chromium concurrent sorption utilizing iron nanominerals. The fact that arsenide could partially oxidize to arsenate without the need of an external oxidizing agent suggests that it might be a viable option for industrial-scale purification on a large scale. New adsorbents for effective arsenic removal from polluted water have been developed, thanks to advances in improving binding affinity and nanostructural changes. Technological advances in water filtration have been greatly facilitated by nanomaterials (Hajji et al., 2019). New nanomaterials with remarkable arsenic binding effectiveness were created in a recent study. Across a pH range of 2 to 10, these nanoparticles removed arsenic with 100% efficacy. Around 300 mg of arsenic per gram of nanoadsorbent were successfully removed from the solution.

16.3.3 Nanoparticle-Based Adsorption and Removal

The creation of sensors that can detect arsenic pollution is one area where nanotechnology has completely transformed water treatment methods. Because of their tiny size and huge surface area, nanomaterials have piqued the curiosity of researchers looking into water purification. Nanotechnology may one day improve existing arsenic removal methods and provide practical, low-cost strategies for the long-term control of contaminated surface and groundwater sources. As evidence of the growing interest and use of nanotechnology in arsenic remediation, a 2018 search for "nanoparticles in arsenic removal" produced 1510 results. It is worth noting that a number of these methods make use of nanoparticles derived from iron oxide, which are well known for their ability to remove arsenic and other prevalent inorganic and organic contaminants. The elimination of arsenate and arsenate can be achieved effectively by means of Fe_3O_4 nanoparticles. Approximately 46.06 mg of arenite and 16.56 mg of arsenate can be captured by just one gram of these nanomaterials. Arsenic(III) polluted aquifers can be effectively treated using iron-based nanoadsorbents in reducing environments due to their high affinity for the reduced form of arsenic form (Prathna et al., 2018).

Research on nanoadsorbents such as MnO_2 and CuO_2 has been carried out in parallel with research on iron-based nanomaterials. Studies have shown that recently developed gels containing cerium and titanium nanoparticles can effectively adsorb and remove arsenic. The adsorbent achieved 99% removal in just 10 min and reduced the total arsenic content to less than 7 μg/l. Scientists discovered that these nanoparticles functioned efficiently even in the presence of a large number of competing ions and at different pH levels. Significantly, the nanoadsorbent produced exhibited a significant attraction to arsenic, retaining its bound state even after 24 h. The ability of the nanomaterial to quickly remove arsenic is based on the oxidation of As(III) and its extensive interconnected pore network (Mishra & Rai, 2019). However, this approach has several limitations. For example, changes in the physical and chemical properties of treated water will affect its actual effect. The sorption effectiveness of iron nanoparticles deposited onto filter medium may also be diminished as a result of agglomeration.

16.3.4 As Adsorption Using Nano-Zerovalent Iron

Several factors are to be considered while selecting adsorbents for As removal, as illustrated in (Fig. 16.2). Nanozerovalent iron (nZVI) is one of the promising materials for adsorptive remediation applications. When nZVI to remove arsenic, the effectiveness depends on a number of variables, such as the adsorption process, the physicochemical qualities, the pH level, the availability of oxygen, and the adsorbent-to-adsorbate ratio. Arsenic collection in aquifers has been achieved by using permeable reactive barriers (PRBs) containing zero-valent iron particles. They are also used in household filters for arsenic removal, especially in underdeveloped areas. Their large surface area allows for efficient removal and retention of arsenic from natural environments (Zhu et al., 2009). As a result of their nanometer size, nanomaterials (nZVI) have the potential to contribute to an increase in iron content in drinking water and cause iron loss.

Therefore, it is important to adsorb them on a suitable carrier matrix to improve performance. In the combination of biochar and nZVI, arsenate is mostly eliminated through electrostatic interactions. Arsenate adsorbed on the biochar-nZVI composite can be easily extracted from the medium because of the magnetic properties of the adsorbent. The use of biochar to promote arsenic adsorption onto zero-valent iron has also been the subject of recent research. The pH value of the solution is an important factor for arsenic binding (Wu et al., 2017). Recent studies have shown that the arsenic adsorption capacity of nanozerovalent iron (nZVI) is significantly lower than that of zerovalent iron (ZVI) due to changes in pH. In discussion of the improved arsenic binding effectiveness of nZVI in relation to electrostatic interactions, they pointed out that corrosion plays a major role. Regardless of pH level, nZVI is an excellent arsenic decontaminant. The ability of nZVI to bind to arsenic is significantly diminished when its particles aggregate, despite the nanosize and magnetic

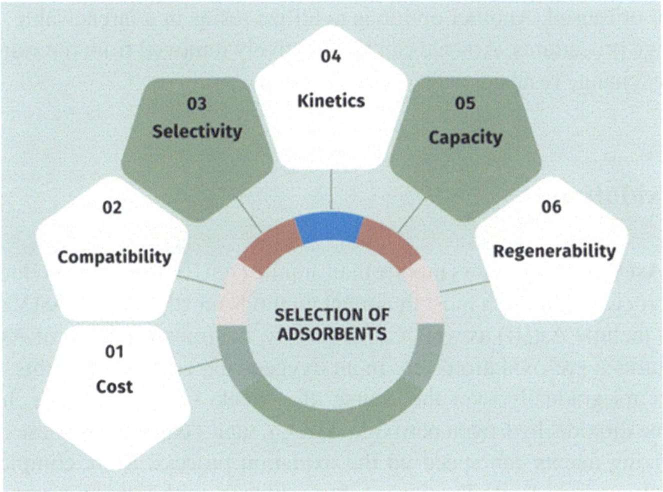

Fig. 16.2 Factors influencing the choice of adsorbents for arsenic removal

characteristics. However, these problems can be lessened if nZVI is firmly attached to a nontoxic backing material.

16.4 Ion-Exchange

The physicochemical method of ion exchange is used to remove contaminants and arsenic. Electrostatically bound ions to the surface of the solid phase can be swapped with solution-bound ions with a comparable charge. Water hardness is effectively reduced and contaminants including arsenate, selenium, nitrate, and chromate are removed from wastewater and contaminated water using this procedure. One step in the process is to treat resins with dirty water in order to regenerate them and improve the exchanged ions. To ensure that the resins remain in optimal condition and prevent any degradation, organic substances are removed before entering the ion exchange chamber by using solids, impurities, and filtering treatments. A high concentration of total dissolved solids (salts) can reduce the efficacy of arsenic removal in ion-exchange methods (Al-Jubouri & Holmes, 2020; Jadhav et al., 2015; Shirin & Yadav, 2014); however, ion exchange preferentially removes arsenate from water because of its negative charge, while arsenate, which mostly exists as uncharged H_3AsO_3, is not successfully removed. For the removal of arsenic, the EPA suggests the use of certain ion exchange resins, especially those based on chloride. In this approach, chloride-ion-loaded ion exchange resins are packed at exchange sites through which arsenic-containing water passes. Ions that compete with one another affect how well ions exchange. To restore depleted resins, aqueous NaCl was used. Extra treatment is necessary before regeneration water and waste resin with high arsenic levels may be disposed of or reused. Another option is to utilize resins in nonrenewable, one-time ion exchange procedures. Arsenic can be effectively removed from the water supply using ion-exchange resins.

16.5 Oxidation

Arsenate (As(V)), which is less mobile than arenite (As(III)), may be produced via an oxidative process. Although most treatment methods mostly remove As(V), groundwater does include As(III) as well. Consequently, the transformation of As(III) into As(V) requires a pre-oxidation step. In an oxygen-rich environment, this oxidation process occurs gradually over the course of a week. Ozone, chlorine, hypochlorite, chlorine dioxide, hydrogen peroxide (H_2O_2), sand-coated manganese oxide and other oxidizing agents can speed up the oxidation process. More complex oxidation methods include Fenton's reagent, hypochlorite, and permanganate. In many experimental settings, fast oxidation of As(III) to As(V) is best accomplished with oxidants such as chlorine and permanganate. Iron and manganese are often removed from polluted water by oxidation. These elements can be effectively scavenged by oxidizing them into insoluble oxides. Adsorption onto hydroxides and oxides of iron and manganese removes the oxidized arsenic from polluted water. The oxidation of arsenate(III) to less dangerous arsenate(V) is aided by iron and manganese oxides. In these settings, the amount of soluble iron determines the effectiveness of arsenic removal by oxidation. Research has shown that arsenic may be removed at rates ranging from 80 to 95% when the iron to arsenic weight ratio is up to 20:1. The effect of manganese oxides on the elimination of arsenic is very substantial. They are excellent at treating polluted aquifers on the spot because they promote arsenic oxidation and increase its subsequent scavenging (Jain & Singh, 2012). Manganese oxides undergo a second-order process kinetic transformation from arenite to arsenate. To oxidize and remove arenite at neutral pH levels ranging from 2 to 60 μg/L, Zhang et al. created a metal oxide complex using Ti(IV) and Mn(IV). Arsenate (107 mg/g metal oxide) and arsenate (87.2 mg/g binary metal oxide) were both removed more effectively using this method (Zhang et al., 2017). The oxidation and elimination of arsenic can be effectively catalyzed by burnside, a naturally occurring Mn oxide. As an alternative to current methods of reducing arsenic exposure, laboratory synthesis of these naturally occurring mineral forms may be possible, but in economically depressed areas of the Indian subcontinent, the method could not be practical for in situ arsenic removal due to its high cost. Mild oxidation with chlorine and other oxidants can increase operating expenses and produce harmful by-products. Arsenic oxidation under sunlight has recently been achieved by precipitation techniques of the synthesis of Cu-doped materials. These materials demonstrated a remarkable affinity for binding arsenic in real-world groundwater circumstances. Upon treatment, the arenite underwent full oxidation in under 2 h (Vaiano et al., 2018). Arsenic-contaminated water can be effectively treated to concentrations below 1 μg/L using advanced oxidation-coagulation filtration (AOCF), according to both pilot and batch tests. The researchers used sands with different grain sizes, different concentrations of oxidizing agents, including ferric chloride and potassium permanganate, and different pH values in their experiments. Reduced arsenic levels was significantly more effective when sand coated with metal oxides. However, the AOCF method proved ineffective in removing typical water contaminants such as ammonium, iron, and manganese. To convert arenite into a more adsorbent arsenate variant, it must be treated in a series with MnO_4^- and Fe^{3+}. Groundwater purification methods employing an aeration-rapid sand filtration unit, along with oxidizing agents (1.2 mg/L permanganate and 1.8 mg/L ferric ions), successfully reduced arsenic concentrations to below 1 μg/L (Ahmad et al., 2020).

16.6 Membrane Technology

It is widely known that membrane technology reduces arsenic levels in groundwater by 96%. It both inhibits hazardous germs and eliminates arsenic from contaminated water. The membrane simplifies waste management maintenance by eliminating chemicals and low operational demands. The challenges in converting arsenicate to arsenicite, a procedure that potentially compromising the durability of the membrane, were mentioned earlier. Recent studies have explored several membrane technologies, such as reverse osmosis, microfiltration (MF), nanofiltration (NF), and ultrafiltration (UF), in an effort to successfully remove arsenic from water sources. Nanofiltration and reverse osmosis are two methods that are being considered as potential effective techniques to remove arsenic from contaminated water. The force required to pass water through a hydrophilic membrane and remove contaminants such as arsenic is produced by applying pressure to the feed side in both NF and RO procedures. Membrane distillation (MD), which employs controlled heat and air pressure, is a prominent approach to treating arsenic-containing water. This technique is effective in cleaning groundwater of pollutants, including heavy metals and arsenic, while simultaneously improving water quality (Pal & Manna, 2010). Additionally, Membrane distillation (MD) produces fresh water using geothermal heat, making it ideal for affordable heating solutions.

16.7 Electrokinetic Technique

A new and promising approach to soil remediation is the electrokinetic technique, which effectively eliminates free pollutants. This method utilizes electrical fields to facilitate processes like electromigration, electrophoresis, electrolysis of water, and electroosmotic flow, specifically for the removal of contaminant. Research has also explored the use of electrokinetic methods for the extraction of arsenic. However, dissolved arsenic, a substance that can increase the mobility of harmful heavy metals, poses challenges for treatment. Despite this, the electrokinetic method is shown to be more environmentally sustainable, efficient, and cost effective when integrated with other techniques (Xu et al., 2019). A permeable reactive barrier (PRB) and an electrokinetic approach were utilized to extract arsenic from the soil. The combination of bioremediation and the electrokinetic method improved the effectiveness of soil treatment for chromium contamination. In efforts to remediate soil affected by arsenic and chromium, a chelating agent is applied alongside Electrokinetic Technology and the Permeable Reactive Barrier (EK-PRB). More research is needed to assess the impact of fulvic and humic acids on improving electrokinetic processes for the removal of oxoanionic arsenate from contaminated soil (Yao et al., 2020). Much of the current research has focused on integrating PRB and EK methods to improve the effectiveness of arsenic remediation while keeping costs in check.

16.8 Electrocoagulation

Electrocoagulation (EC) is a cutting-edge interconnected process that requires minimal space for treatment, is fully automated, and boasts impressive removal rates. However, the process faces challenges such as electrode passivation, the need for an effective EC reactor design, and high-power usage in large-scale settings. Electrochemical chelation (EC) is still a viable option for purifying arsenate and arenite, with removal rates between 93 and 99.9%. Sacrificial anode electrodes in electrocoagulation can be made of iron or aluminium; however, iron is known to be more effective than aluminum (Mohora et al., 2018). When used as electrical energy, EC can cause heavy metals to dissolve, destabilize colloidal suspensions, and improve impurity flotation and flocculation. By dissolving electrolytes at the sacrificial anode, cationic monomer species are formed when the cathode releases hydroxide ions due to its electrical charge, as shown in (Fig. 16.3) by generating Fe^{3+} products, the Fe electrode in EC eliminates arsenic. Synthetic solutions made from drinking water have different end results, according to researchers. This is because different species in the water have different effects on the arsenic treatment process.

Most CE research has been conducted in controlled laboratory settings, with treated water undergoing brief 8-h trials (> 100 L), rather than in the field for longer periods of time. This method has made evaluating the effectiveness of continuous EC reactors more difficult and has left large gaps in our understanding of arsenic removal. Electrocoagulation using iron, aluminum, or titanium may successfully remove arsenic from water, with different metals that have different efficiency (Nidheesh & Singh, 2017). Arsenic is more effectively adsorbed when aluminum is used as a substitute because it inhibits the surface crystallization of iron oxides. The electrode composed of zinc and copper was utilized in a detailed experiment to extract arsenic. To improve the efficiency of arsenic removal from wastewater, a response surface methodology was used. Recent research shows that electrocoagulation can effectively eliminate arsenic from water sources. Metal-Air Fuel Cell Electrocoagulation (MAFCEC) offers an alternative method aimed at addressing various issues, especially the significant energy needs associated with traditional

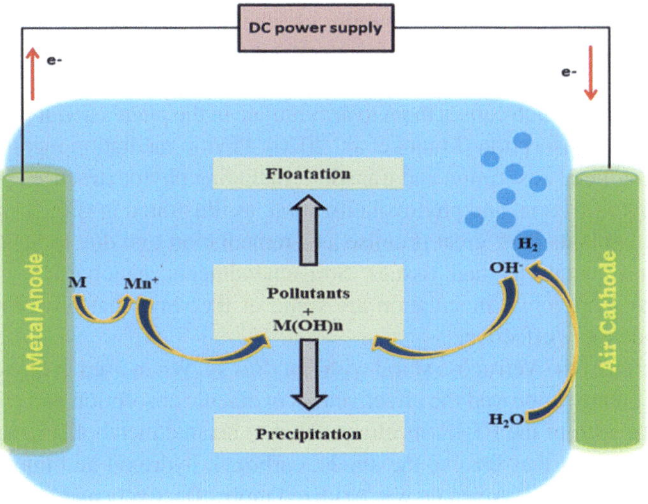

Fig. 16.3 Electrocoagulation mechanism for arsenic removal

EC techniques. This approach not only produces electricity but also cleans arsenic-contaminated water. When the electrochemical process with an air–fuel cell, arsenic can be removed efficiently without the need for additional energy. Challenges related to traditional EC methods include power management, sludge disposal, and the exploration of other arsenic removal techniques. The use of renewable energy sources can reduce the energy consumption necessary for the operations of the EC, making them more cost effective. This strategy also lowers the costs associated with the purification of the water for vegetation. Furthermore, EC reliably meets the WHO Standard for Safe Drinking Water (10 μg/L) without compromising human health. More research is needed to fully explore the effectiveness of the CE technique in treating arsenic-polluted water supplies (Kobya et al., 2020; Simon et al., 2024).

16.9 Biological Approaches

16.9.1 Phytoremediation

One well-known technique for eliminating toxins with plant-based methods is phytoremediation. This approach promotes environmental sustainability, requires minimal nutrients, and is relatively simple to implement. In phytoremediation, specific plants that can absorb and neutralize arsenic from contaminated soils are used. Plants have a large system, exhibit high tolerance levels, grow rapidly, adapt well, and are capable of extracting toxins from polluted soils. Hyperaccumulators facilitate the movement of metals to the root surface through various processes, including osmoregulation, redistribution, exclusion, translocation, and sequestration, all of which take place in the surrounding soil solutions. Although some plant species use transporter proteins to transfer metals, others take them in directly by water absorption, which causes them to accumulate in the plant's aerial parts through a process called adsorption (Manoj et al., 2020). Phytoremediation methods rely on processes for metal absorption and transport, including phytoextraction, phytostabilization, phytofiltration, and phytovolatilization, as illustrated in (Fig. 16.4). Hyperaccumulating plants have great promise as a remediation tool due to how well they handle heavy metals in their tissues. Soil amendments, nanoparticle treatments, chelators, and microbial inoculation are some of the ways that phytoremediation may be made more effective.

Inoculation of the Verrucose spiral with *Bacillus* sp. When used in phytoremediation, XZM greatly improved the effectiveness of arsenic absorption and elimination. The changes seen in the FT-IR spectra pointed to arsenic metabolism and detoxification mechanisms that involve the amide, carboxyl, hydroxyl and thiol groups in V. denseerrulata leaves. Arsenate was oxidized more effectively and taken up by the *Cupriavidus basilensis* strain when *P. vittata* colonized. A dramatic 171% increase in arsenic build-up in *P. vittata* and a greater count of aioA-like genes in the rhizosphere followed after inoculation with the *Cupriavidus basilensis* strain? Advanced single-molecule real-time sequencing (Irshad et al., 2020) was used to build a comprehensive transcriptome and etonoplast proteome database to pinpoint the genetic elements

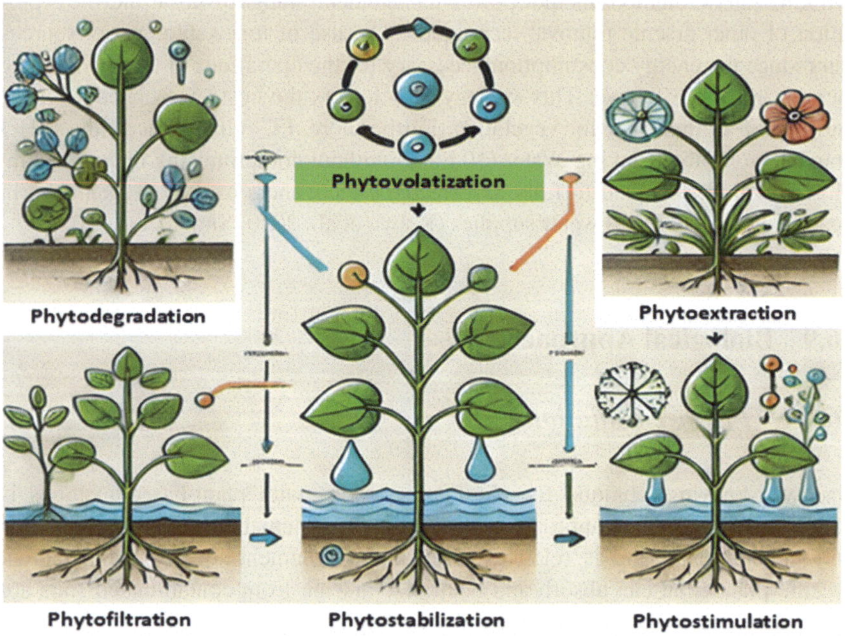

Fig. 16.4 Illustrative diagram of phytoremediation strategies for environmental restoration

involved in arsenic detoxification in *Pteris vittata*. Phytoremediation, a technique designed to restore agricultural soils polluted with metals such as arsenic, is one of the uses for this study. Improving the efficiency of heavy metal phytoremediation depends on a better understanding of the processes by which Rhizobacteria promote plant development. Moreover, exploring the molecular dynamics behind phytoremediation offers significant opportunities to create transgenic hyperaccumulators that exhibit greater tolerance to various pollutants. Additionally, nanoparticles are crucial in this arena. Through nano phytoremediation (NP), the combination of nanomaterials with plants has been shown to effectively decrease a variety of pollutants, including pesticides, heavy metals, petroleum, and polycyclic aromatic hydrocarbons.

Nanoparticles have the potential to significantly improve phytoremediation systems, according to some research; however, this is a lengthy process. Combining nanomaterials with plants and bacteria is a relatively new approach to bioremediation, but it shows great potential to improve on more conventional approaches. However, there is still a dearth of studies on Nano phytoremediation's ability to remove arsenic and the processes that control its synergistic effects. In order to treat contaminated regions in an environmentally sustainable manner, Nano phytoremediation (NP) combines nanotechnology with phytoremediation. To shorten cleaning periods and reduce the need for additional treatment, novel nanomaterials in nanotechnology can efficiently remove and repair harmful contaminants from water and soil polluted with heavy metals.

16.10 Conclusion

Many people, especially those living in rural areas, may not realize that arsenic-contaminated water is poisonous because it has no discernible smell or flavor. The health might be jeopardized due to this lack of understanding. Regrettably, metropolitan regions have similar difficulties in obtaining safe arsenic water due to inadequate finance, a lack of political will, and a dearth of technical support. Affordable, sustainable and socially acceptable methods should be developed and implemented at the community or family level to reduce arsenic exposure in drinking water. There are societal factors, such as trash creation and management, that these technologies need to consider. This is an ideal situation for eco-friendly technologies that are easy to implement, employ inexpensive tools, and are based on materials that are readily accessible in the area. Technologies that may be used in small-scale domestic applications or in remote rural or peri-urban locations should be prioritized for advancement. Red mud, charcoal, and chitosan-based adsorbents are examples of adsorption-based technologies that show potential, but are seldom implemented in countries. While there are some cutting-edge methods for arsenic cleanup, they have only been evaluated in small-scale settings and have not yet been used on a larger scale. Hence, to find out whether they can be implemented in the real world, it is suggested to conduct experiments on a smaller scale and keep an eye on them over time.

Acknowledgements The authors acknowledge their respective institutes for providing the necessary support for the contribution.

References

Ahmad, A., & Bhattacharya, P. (2019). Environmental arsenic in a changing world. *Groundwater for Sustainable Development, 8*, 169–171. https://doi.org/10.1016/j.gsd.2018.11.001

Ahmad, A., Heijnen, L., de Waal, L., Battaglia-Brunet, F., Oorthuizen, W., Pieterse, B., & Van Der Wal, A. (2020). Mobility and redox transformation of arsenic during treatment of artificially recharged groundwater for drinking water production. *Water Research, 178*, Article 115826. https://doi.org/10.1016/j.watres.2020.115826

Al-Jubouri, S. M., & Holmes, S. M. (2020). Immobilization of cobalt ions using hierarchically porous 4A zeolite-based carbon composites: Ion-exchange and solidification. *Journal of Water Process Engineering, 33*, Article 101059. https://doi.org/10.1016/j.jwpe.2019.101059

Dabrowski, A. (2001). Adsorption—From theory to practice. *Advances in Colloid and Interface Science, 93*(1–3), 135–224. https://doi.org/10.1016/S0001-8686(00)00082-8

Demanelis, K., Argos, M., Tong, L., Shinkle, J., Sabarinathan, M., Rakibuz-Zaman, M., & Pierce, B. L. (2019). Association of arsenic exposure with whole blood DNA methylation: An epigenome-wide study of Bangladeshi adults. *Environmental Health Perspectives, 127*(5), Article 057011. https://doi.org/10.1289/EHP3849

Hajji, S., Montes-Hernandez, G., Sarret, G., Tordo, A., Morin, G., Ona-Nguema, G., & Mzoughi, N. (2019). Arsenite and chromate sequestration onto ferrihydrite, siderite and goethite nanostructured minerals: Isotherms from flow-through reactor experiments and XAS measurements. *Journal of Hazardous Materials, 362*, 358–367. https://doi.org/10.1016/j.jhazmat.2018.09.031

Irshad, S., Xie, Z., Wang, J., Nawaz, A., Luo, Y., Wang, Y., & Mehmood, S. (2020). Indigenous strain Bacillus XZM assisted phytoremediation and detoxification of arsenic in Vallisneria denseserrulata. *Journal of Hazardous Materials, 381*, Article 120903. https://doi.org/10.1016/j.jhazmat.2019.120903

Jadhav, S. V., Bringas, E., Yadav, G. D., Rathod, V. K., Ortiz, I., & Marathe, K. V. (2015). Arsenic and fluoride contaminated groundwaters: A review of current technologies for contaminants removal. *Journal of Environmental Management, 162*, 306–325. https://doi.org/10.1016/j.jenvman.2015.07.020

Jain, C. K., & Singh, R. D. (2012). Technological options for the removal of arsenic with special reference to South East Asia. *Journal of Environmental Management, 107*, 1–18. https://doi.org/10.1016/j.jenvman.2012.04.016

Kapaj, S., Peterson, H., Liber, K., & Bhattacharya, P. (2006). Human health effects from chronic arsenic poisoning–a review. *Journal of Environmental Science and Health Part A, 41*(10), 2399–2428. https://doi.org/10.1080/10934520600873571

Kobya, M., Soltani, R. D. C., Omwene, P. I., & Khataee, A. (2020). A review on decontamination of arsenic-contained water by electrocoagulation: Reactor configurations and operating cost along with removal mechanisms. *Environmental Technology & Innovation, 17*, Article 100519. https://doi.org/10.1016/j.eti.2019.100519

Liu, Q., Leslie, E. M., Moe, B., Zhang, H., Douglas, D. N., Kneteman, N. M., & Le, X. C. (2018). Metabolism of a phenylarsenical in human hepatic cells and identification of a new arsenic metabolite. *Environmental Science & Technology, 52*(3), 1386–1392. https://doi.org/10.1021/acs.est.7b05081

Manoj, S. R., Karthik, C., Kadirvelu, K., Arulselvi, P. I., Shanmugasundaram, T., Bruno, B., & Rajkumar, M. (2020). Understanding the molecular mechanisms for the enhanced phytoremediation of heavy metals through plant growth promoting rhizobacteria: A review. *Journal of Environmental Management, 254*, Article 109779. https://doi.org/10.1016/j.jenvman.2019.109779

Mishra, P. K., & Rai, P. K. (2019). Ultrafast removal of arsenic using solid solution of aero-gel based Ce1-XTi$_x$O$_{2\text{-}Y}$ oxide nanoparticles. *Chemosphere, 217*, 483–495. https://doi.org/10.1016/j.chemosphere.2018.11.003

Mohora, E., Rončević, S., Agbaba, J., Zrnić, K., Tubić, A., & Dalmacija, B. (2018). Arsenic removal from groundwater by horizontal-flow continuous electrocoagulation (EC) as a standalone process. *Journal of Environmental Chemical Engineering, 6*(1), 512–519. https://doi.org/10.1016/j.jece.2017.12.042

Nidheesh, P. V., & Singh, T. A. (2017). Arsenic removal by electrocoagulation process: Recent trends and removal mechanism. *Chemosphere, 181*, 418–432. https://doi.org/10.1016/j.chemosphere.2017.04.082

Pal, P., & Manna, A. K. (2010). Removal of arsenic from contaminated groundwater by solar-driven membrane distillation using three different commercial membranes. *Water Research, 44*(19), 5750–5760. https://doi.org/10.1016/j.watres.2010.05.031

Podgorski, J., & Berg, M. (2020). Global threat of arsenic in groundwater. *Science, 368*(6493), 845–850. https://doi.org/10.1126/science.aba1510

Prathna, T. C., Sitompul, D. N., Sharma, S. K., & Kennedy, M. (2018). Synthesis, characterization and performance of iron oxide/alumina-based nanoadsorbents for simultaneous arsenic and fluoride removal. *Desalination and Water Treatment, 104*, 121–134. https://doi.org/10.5004/dwt.2018.21960

Ramaswami, A., Tawachsupa, S., & Isleyen, M. (2001). Batch-mixed iron treatment of high arsenic waters. *Water Research, 35*(18), 4474–4479. https://doi.org/10.1016/S0043-1354(01)00168-3

Shirin, S., & Yadav, A. K. (2014). Physico-chemical analysis of municipal wastewater discharge in Ganga river, Haridwar district of Uttarakhand, India. *Current World Environment, 9*(2), 536–543. https://doi.org/10.12944/CWE.9.2.39

Simon, M., Joshi, H., Yadav, A. K., & Hopke, P. K. (2024). Simultaneous nitrogen and phosphorus removal from polluted water using a novel bacterial consortium (BACON-38): Promising application for wastewater remediation. *Journal of Water Process Engineering, 66*, Article 106044. https://doi.org/10.1016/j.jwpe.2024.106044

Singh, A. L., Singh, V. K., & Srivastava, A. (2013). Effect of arsenic contaminated drinking water on human chromosome: A case study. *Indian Journal of Clinical Biochemistry, 28*, 422–425. https://doi.org/10.1007/s12291-013-0330-3

Singh, P., Borthakur, A., Singh, R., Bhadouria, R., Singh, V. K., & Devi, P. (2021). A critical review on the research trends and emerging technologies for arsenic decontamination from water. *Groundwater for Sustainable Development, 14*, Article 100607. https://doi.org/10.1016/j.gsd.2021.100607

Vaiano, V., Iervolino, G., & Rizzo, L. (2018). Cu-doped ZnO as efficient photocatalyst for the oxidation of arsenite to arsenate under visible light. *Applied Catalysis b: Environmental, 238*, 471–479. https://doi.org/10.1016/j.apcatb.2018.07.026

Wu, C., Tu, J., Liu, W., Zhang, J., Chu, S., Lu, G., & Dang, Z. (2017). The double influence mechanism of pH on arsenic removal by nano zero valent iron: Electrostatic interactions and the corrosion of Fe^0. *Environmental Science: Nano, 4*(7), 1544–1552. https://doi.org/10.1039/C7EN00240H

Xu, Y., Li, J., Xia, W., Sun, Y., Qian, G., & Zhang, J. (2019). Enhanced remediation of arsenic and chromium co-contaminated soil by eletrokinetic-permeable reactive barriers with different reagents. *Environmental Science and Pollution Research, 26*, 3392–3403. https://doi.org/10.1007/s11356-018-3842-9

Yao, W., Cai, Z., Sun, S., Romantschuk, M., Sinkkonen, A., Sun, Y., & Wang, Q. (2020). Electrokinetic-enhanced remediation of actual arsenic-contaminated soils with approaching cathode and Fe 0 permeable reactive barrier. *Journal of Soils and Sediments, 20*, 1526–1533. https://doi.org/10.1007/s11368-019-02459-4

Yin, H., Kong, M., Gu, X., & Chen, H. (2017). Removal of arsenic from water by porous charred granulated attapulgite-supported hydrated iron oxide in bath and column modes. *Journal of Cleaner Production, 166*, 88–97. https://doi.org/10.1016/j.jclepro.2017.08.026

Zakhar, R., Derco, J., & Cacho, F. (2018). An overview of main arsenic removal technologies. *Acta Chimica Slovaca, 11*(2), 107–113. https://doi.org/10.2478/acs-2018-0016

Zeng, H., Yu, Y., Wang, F., Zhang, J., & Li, D. (2020). Arsenic (V) removal by granular adsorbents made from water treatment residuals materials and chitosan. *Colloids and Surfaces a: Physicochemical and Engineering Aspects, 585*, Article 124036. https://doi.org/10.1016/j.colsurfa.2019.124036

Zhang, L., Qin, X., Tang, J., Liu, W., & Yang, H. (2017). Review of arsenic geochemical characteristics and its significance on arsenic pollution studies in karst groundwater, Southwest China. *Applied Geochemistry, 77*, 80–88. https://doi.org/10.1016/j.apgeochem.2016.05.014

Zhu, H., Jia, Y., Wu, X., & Wang, H. (2009). Removal of arsenic from water by supported nano zero-valent iron on activated carbon. *Journal of Hazardous Materials, 172*(2–3), 1591–1596. https://doi.org/10.1016/j.jhazmat.2009.08.031

The manufacturer's authorised representative in the EU is Springer Nature Customer Service Centre GmbH, Europaplatz 3, 69115 Heidelberg, Germany. If you have any concerns regarding our products, please contact ProductSafety@springernature.com

Printed and bound by CPI Group (UK) Ltd, Croydon, CR0 4YY
01/10/2026
02225973-0007